HIGH-TEMPERATURE CHEMISTRY OF SILICATES AND OTHER OXIDE SYSTEMS

VYSOKOTEMPERATURNAYA KHIMIYA SILIKATNYKH I DRUGIKH OKISNYKH SISTEM

ВЫСОКОТЕМПЕРАТУРНАЯ ХИМИЯ СИЛИКАТНЫХ И ДРУГИХ ОКИСНЫХ СИСТЕМ

HIGH-TEMPERATURE CHEMISTRY OF SILICATES
AND OTHER OXIDE SYSTEMS

Nikita Aleksandrovich Toropov

and

Valentin Pavlovich Barzakovskii
Leningrad Institute of Silicate Chemistry
Academy of Sciences of the USSR

Translated from Russian by
C. Nigel Turton and Tatiana I. Turton

CONSULTANTS BUREAU
NEW YORK
1966

Library of Congress Catalog Card Number 65-25264

The Russian text, originally published for the Leningrad Institute of
Silicate Chemistry by the press of the Academy of Sciences of the USSR
in 1963, has been extensively corrected and updated by the authors
for this edition.

Никита Александрович Торопов и
Валентин Павлович Барзаковский

ВЫСОКОТЕМПЕРАТУРНАЯ ХИМИЯ СИЛИКАТНЫХ
И ДРУГИХ ОКИСНЫХ СИСТЕМ

ISBN 978-1-4684-7211-0   ISBN 978-1-4684-7209-7 (eBook)

DOI 10.1007/978-1-4684-7209-7

PREFACE TO THE AMERICAN EDITION

The ever-increasing importance of chemical reactions at high and superhigh temperatures in crystalline, amorphous, and semicrystalline solids, as well as the reactions of these solids with gases, prompted the authors of this book to examine critically the literature available in this field and to present a general review of the subject.

In this monograph we discuss those chemical and physicochemical points which we consider to be most important for solving a series of problems in the preparation and use of new inorganic materials.

We hope that this book will be of interest to the many specialists working on inorganic materials.

N. A. Toropov

PREFACE

Modern technology demands ever more materials with high mechanical strength, heat and chemical resistance, fire resistance, special electrical properties, particular behavior toward active radiations, etc. The search for such materials requires the study of various chemical compounds, metallic alloys, and other fused inorganic systems, especially oxide systems. Materials based on oxides begin to assume increasing importance in many fields of the new technology. In this connection the investigation of oxides and systems consisting of two and more oxides is expanding greatly.

Of particular importance are investigations of oxide systems at high temperatures when the evaporation of oxides and their dissociation begin to play an important part. Under these conditions it is impossible to ignore the gas phase in the study of oxides and the system "condensed phase−gas" has to be investigated. The importance of the gas phase (primarily oxygen) is particularly great in systems with oxides of elements of variable valence such as iron, manganese, etc. Chapter IV is devoted to such systems. It now begins to appear that for their complete description, all high-temperature oxide systems should not be investigated as condensed systems $Me_n'O_n$ (cond) + $Me_p''O_q$ (cond), but as the systems "$Me'−Me''−$oxygen." The upper temperature limit of such systems is naturally raised and the system may ultimately change completely into a gaseous state.

One of the most important problems facing investigators of these systems is the creation of particularly strong materials. At the present time work on the creation of such materials is developing in at least two directions. Systems are being sought in which with the aid of special methods crystallization from the glassy state proceeds with the formation of extremely fine crystals. The material obtained in this way, which is usually called "sitall" in Soviet technical literature, has very high mechanical and thermal properties. The second route to the production of particularly strong materials is the preliminary synthesis of thin fibrous crystals by means of which (for example, by using some binder) it is possible to obtain a material with, according to the statement of the specialists, the highest (of all that is now available in technology).

Chapters I, II, and III give a theoretical account of the problems which are of importance in the production of glass ceramic materials (sitalls). In addition to crystallization, layer formation in the liquid phases is very important here. Chapter XI is devoted to fibrous crystals of highly refractory oxides ("whiskers").

At the present time scientists in many countries are making great efforts to find methods of determining the thermodynamic characteristics of reactions in oxide systems. A new method in this direction proposed in 1957 is the method of studying electromotive forces of galvanic cells from solid electrolytes. Work is continuing on the study of oxidation-reduction reactions involving gaseous substances. Chapter V is devoted to these problems.

Chapters VII−X review work on the evaporation of both oxides themselves and more volatile products of lower valence obtained, in particular, by heating oxides under reducing conditions. Silicon monoxide SiO which is obtained, for example, by the reaction $SiO_2 + Si = 2SiO$ is the most characteristic volatile oxide of silicon. Chapter VII is devoted predominantly to the properties of this compound. In the rest of these chapters we examine recent work on the evaporation of oxides of Group II− VI oxides. The investigation of the evaporation of oxides is one of the most rapidly developing sections of high-temperature chemistry. The penetration of technology into the field of very high temperatures requires a knowledge of not only the conditions for conversion of a solid or liquid oxide into the gaseous state, but also a knowledge of the structure, properties and stability of gaseous oxides.

It seems to us that this book deals with the most urgent problems in the high-temperature chemistry of oxide systems. We did not undertake to give a full and systematic account of this rapidly developing scientific field and selected only the regions of it in which there is the most intense research work, as reflected by the continuous appearance of a large number of articles. In our review we tried to cover the very latest literature.

Apart from some translated collections (for example, Research at High Temperature, 1962), in the Russian scientific literature there are no special works on progress in the chemistry of oxides at high temperatures. It is hardly necessary to justify critical scientific reviews, especially in scientific fields which, like high-temperature chemistry, are developing vigorously at the present time.

We admit that our book ignores some important problems in the chemistry of oxides. Thus, we have not examined research on the systems Me—MeO and compounds of oxides with other substances such as carbides, nitrides, etc. Very little space has been given to practical problems.

The first four chapters were written by N. A. Toropov and the rest by V. P. Barzakovskii.

BIOGRAPHICAL NOTE

Nikita Alexandrovich Toropov, one of the leading Soviet scientists in physical chemistry and silicate technology, was born in 1908. His major work has dealt with the mineralogy of silicates and physicochemical investigations of silicate systems. Director of the Leningrad Institute of Silicate Chemistry, he is also head of the physicochemical laboratory at the Institute. In 1952, Toropov received the State Prize for his work on ferrite materials. He was elected Corresponding Member of the Academy of Sciences of the USSR in 1962. N. A. Toropov is the author and editor of more than 300 scientific works, among them is Structural Transformations in Glasses at High Temperatures, Volume 5, in the Structure of Glass series (co-edited by E. A. Porai-Koshits), published in English translation by Consultants Bureau in 1965.

Valentin Pavlovich Barzakovskii, born in 1906, is the author of more than 100 works and is the editor of several collections on the electrochemical production of light metals and the physicochemical study of fused salts. Associated with the Academy of Sciences of the USSR since 1933, he has worked at the Laboratory of Silicate Chemistry, headed by Academician I. V. Grebenshchikov. Since 1948, Barzakovskii has been at the Institute of Silicate Chemistry.

CONTENTS

PUBLISHER'S NOTE

The following Soviet journals cited in this book are available in cover-to-cover translation:

Russian Title	English Title	Publisher
Atomnaya énergiya	Soviet Journal of Atomic Energy	Consultants Bureau
Doklady Akademii Nauk SSSR	Doklady Chemical Technology	Consultants Bureau
	Doklady Chemistry	Consultants Bureau
Fizika tverdogo tela	Soviet Physics—Solid State	American Institute of Physics
Izvestiya Akademii Nauk SSSR: Otdelenie khimicheskikh nauk	Bulletin of the Academy of Sciences of the USSR: Division of Chemical Science	Consultants Bureau
Izvestiya Akademii Nauk SSSR: Seriya fizicheskaya	Bulletin of the Academy of Science of the USSR: Physical Series	Columbia Technical Translations
Kristallografiya	Soviet Physics— Crystallography	American Institute of Physics
Optika i spektroskopiya	Optics and Spectroscopy	American Institute of Physics
Steklo i keramika	Glass and Ceramics	Consultants Bureau
Uspekhi fizicheskikh nauk	Soviet Physics—Uspekhi	American Institute of Physics
Zhurnal fizicheskoi khimii	Russian Journal of Physical Chemistry	The Chemical Society (London)
Zhurnal neorganicheskoi khimii	Russian Journal of Inorganic Chemistry	The Chemical Society (London)
Zhurnal obshchei khimii	Journal of General Chemistry of the USSR	Consultants Bureau
Zhurnal prikladnoi khimii	Journal of Applied Chemistry of the USSR	Consultants Bureau
Zhurnal strukturnoi khimii	Journal of Structural Chemistry	Consultants Bureau

CHAPTER I

LIQUATION OR THE FORMATION OF IMMISCIBLE LIQUIDS
IN SILICATE SYSTEMS

In this field, first of all, there has been considerable development of work on systems in which there is the formation of two or more immiscible liquid phases or regions of liquation. It is characteristic that in most of the silicate systems studied up to the present time there is the formation of liquation regions, which are of particular importance in connection with the development of the theory of formation of new glass-crystalline materials, which are otherwise known as sitalls, pyroceramics, or devitroceramics.

These materials are characterized by a finely granular structure, and consist of fine crystals, obtained by catalyzed or controlled crystallization, and residual interlayers of glass, which cement the crystalline concretion, reinforcing the structure of an object made from sitall.

The basic chemical system for preparing a sitall is usually a crystallized glass from regions corresponding on phase diagrams to concentration sections where liquation phenomena are observed or sections adjacent to them. The catalysts for controlled crystallization, which make it possible to obtain a vast number of crystal nuclei in the mass of the starting glass, are usually finely dispersed metals, namely gold, silver, and platinum, or oxides of chromium, titanium, cerium, vanadium, nickel, and zirconium, which are introduced in tenths or hundredths of a percent, and also some sulfides of heavy or transition metals or some fluorides. In some cases glasses with catalysts introduced into them are treated with active radiation, namely ultraviolet, gamma, or x rays. In other cases the activating irradiation is not essential.

The glasses with the catalysts are then annealed under definite temperature conditions. As a result of the thermal treatment there grow fine crystals, whose nature depends mainly on the chemical composition of the glass used. These crystals give to sitalls unusually high mechanical and dielectric strength, chemical stability, and increased resistance to sharp changes in temperature.

Sitalls also have high softening points, which reach 1200-1300°C. Materials with such a combination of physical properties, which also have a low specific gravity and a high abrasion resistance, are extremely valuable for producing various constructions, building components, domestic articles, etc.

While the formation of nuclei or crystallization centers is determined mainly by structural conditions and to some extent by the correspondence between the cells of the crystal lattice of the catalyst and the crystal phase arising in the glass, during the subsequent course of crystallization the decisive part is played by phase relations, which are described by the phase diagram and determine the character of the processes occurring in the system.

Despite the frequent or even typical formation of metastable states in silicate systems, the course of crystallization in them is determined in general by the diagram of stable equilibria. For the realization of the finely crystalline structure typical of sitalls, it is of great importance to select glass compositions immediately adjacent to regions of liquation on the corresponding phase diagrams.

The possibility of physicochemical liquation or layer formation of natural silicate melts is accepted theoretically and even estimated as the main factor providing one of the reasons for the variety of igneous rocks forming the upper zones of the earth's crust by many great petrographers of the second half of the 19th and first half of the 20th centuries (F. U. Levinson-Lessing, I. Vogt, et al.). However, the first experimental demonstrations of the formation of immiscible liquids in silicate systems were provided by Greig (1927) only in 1927.

Table 1. Liquation in Silicate Systems with Alkaline Earth Oxides

System	Composition of first liquid		Composition of second liquid		Layer formation temperature, °C	Critical point, °C
	SiO₂	RO	SiO₂	RO		
MgO—SiO₂ . .	99.2	0.8	69.0	31.0	1695	2200
CaO—SiO₂ . .	99.4	0.6	71.8	28.2	1698	2100
SrO—SiO₂ . .	97.6	2.4	70.0	30.0	1693	1920

With the separation of silica crystals as a result of cooling two immiscible liquids, the following characteristics are observed: of the two liquid phases, that which contains least silica generally does not form a clear or pure glass on quenching; the same is also observed in liquids formed on cooling three component silicate glasses on condition that their compositions are close to the boundaries of the region of immiscibility and they do not contain too much silica. On cooling such melts it is possible to obtain slightly opalescent glass, milky glass, or opaque, porcelain-like material. In microscope samples it is obvious that this phenomenon is caused by the separation of fine points with a refractive index much lower than that of the surrounding glass.

On slow cooling there are formed relatively coarse crystals, which are readily determined as cristobalite under a microscope. It is therefore natural to assume that the points formed with rapid cooling are also cristobalite. There is also another possibility, namely that there is supercooling of the liquid and its separation into two instead of the formation of cristobalite. Thus it is possible to realize the metastable equilibrium, namely the continuation of the region of immiscibility below the liquidus surface. The liquid thus formed must contain much silica and have a refractive index lower than that of the glass surrounding it. However, observations of coarser inclusions of this type showed that they did not have a spherical form.

In the case of coarser formations it was possible to see that they did not consist of groups of separate particles, but appeared as elementary formations composed of small rods radiating from a common center. The cross sections of these rods were very small at their middle, but they thickened toward the ends and assumed the form of small dumbbells. In cases where it was possible to count the number of rods radiating from the common center, this was found to equal six. In other cases there were more of them and the aggregates sometimes assumed a more complex character.

The relative orientation of the rods could not be determined accurately, but it apparently corresponded to crystallographic axes passing through the apices of an octahedron. Thus, according to these observations the phase which separated on cooling consisted of cristobalite and not a second liquid in a metastable state. X-ray analysis also showed the presence of cristobalite. However, a microscopic check was necessary here as many other phases are structurally similar to cristobalite and an assumption based on x-ray diffraction data alone is clearly inadequate. As a result of accurate experiments by many investigators it was established that liquation phenomena are present in systems formed by silica with magnesium oxide, strontium oxide, etc. Table 1 gives numerical data characterizing liquation phenomena in these systems.

In a series of papers by Kracek (1930) it was shown that in systems formed by silica with alkaline oxides Na_2O, K_2O, Rb_2O, and Cs_2O, and also with Li_2O, no formation of two immiscible liquids is observed. However, it was observed that the course of the liquidus curve of cristobalite in these systems changes regularly, as is shown in Fig. 1. While in systems with cesium and rubidium this curve is almost linear, in systems with potassium, and sodium, and particularly with lithium, there are appreciable deviations from linearity. The liquidus curve takes on a sigmoid form, which is particularly clearly expressed for the system $Li_2O—SiO_2$.

Experiments confirming the existence of liquation phenomena in silicate systems aroused great interest and in the literature there soon appeared a series of theoretical papers in which attempts were made to provide a scientific basis for the appearance of these phenomena, and to assess them quantitatively. In addition to silicate systems, these phenomena were found in a series of borate, phosphate, and other systems formed by oxides or fluorides. It is characteristic that in systems with beryllium fluoride BeF_2, which is known to be a crystal

Table 2. Ratio of Valences of Cations to Their Radii

Cation	R, Å	z	z/R	Type of liquidus curve of cristobalite
Cs^+	1.65	1	0.61	
Rb^+	1.49	1	0.67	Almost linear
K^+	1.33	1	0.75	
Na^+	0.98	1	1.02	
Li^+	0.78	1	1.02	Sigmoid
Ba^{2+}	1.43	2	1.40	
Sr^{2+}	1.27	2	1.57	
Ca^{2+}	1.06	2	1.89	Immiscibility
Mg^{2+}	0.78	2	2.56	

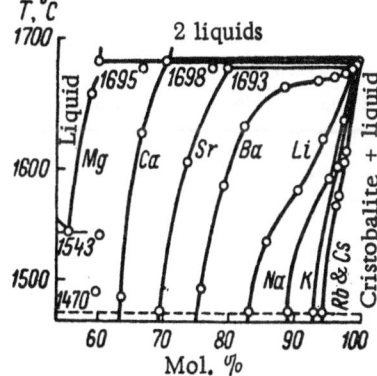

Fig. 1. Liquidus curves of cristobalite in systems with alkaline and alkaline earth oxides, mol.%.

chemical model for silica, the liquidus curve for BeF_2 in a series of cases has a very flat form, indicating the presence of a microheterogeneous breakdown in the liquid phase. As a result of more accurate experiments, this was subsequently demonstrated by Vogel (1958).

The first attempt at any interpretation of liquation phenomena in silicate systems in terms of crystal chemical concepts on the structure of glass was made by Warren and Pinkus (1940). According to these authors, complete miscibility in molten and glassy silicate systems is promoted by the tendency of cations of the glass formers Si, B, and P to bind all the nearest oxygen ions in the melt by strong chemical bonds of a predominantly ionic character. On introduction into the melt of other cations, which are called modifiers, there is a redistribution of the oxygen ions, which enter the melt in oxides of two sorts, namely glass formers and modifiers.

If the modifier ion forms a bond with oxygen of high energy, then it forms independent and chemically individual cation—oxygen regions with a low content of the glass-former cation. Second regions are formed by the glass-former cation with the corresponding oxygen ions of the melt, and this leads to immiscibility in the melt. If the energy of the bond of the modifier cation with oxygen is low, this differentiation does not occur and the melt retains its homogeneity and consists of silicon—oxygen anions and individual modifier cations.

The interatomic bond in glasses is characterized as a bond with a predominantly ionic character. The energy of the interionic electrostatic bond in glasses is described by the expression

$$E = z_1 z_2 e^2 / R_{1,2},$$

where $R_{1,2}$ represents the interionic distances (distances between centers of ions) and $z_1 z_2$ is the valence of the ions. The magnitude of the bond energy increases with an increase in the valence and falls in proportion to an increase in the size of the ions.

The Si—O bond is regarded as extremely stable as the valences here equal 4 and 2, while the distance of 1.62 Å is very small. On the other hand, the Na—O bond is relatively weak. Therefore, in melts formed by silica with sodium oxide, the components mix in all proportions. Thus, the homogeneity of these melts results from the fact that the sodium ion is a large monovalent cation and in the melt the tendency to form complexes of silicon—oxygen ions predominates over the tendency to form sodium—oxygen bonds. When a divalent cation such as Ca^{2+} is introduced into the melt, the modifier cation—oxygen bond is much stronger and produces a tendency for the formation of groupings in which a considerable part of the oxygens are coordinated about calcium ions.

In an investigation of the course of the liquidus curves in some binary systems formed by silica with alkaline earth and some other oxides (FeO and ZnO), it was established that two immiscible liquids were formed in a series of cases, and this was reflected on the corresponding phase diagrams. The formation of immiscible liquids was found in the systems formed by silica with calcium, magnesium, strontium, ferrous, and zinc oxides. In all cases one of the liquids was very rich in silica and the temperature of complete melting in all systems was close to 1700°C. No liquation was observed in the system $BaO-SiO_2$, but the form of the liquidus curve had a characteristic sigmoid form, which indicated the possibility of the presence of layer formation in this system at higher temperatures, as was confirmed by Argyle and Hummel (1963). No liquation was observed in systems formed by silica with beryllium, plumbous, and stannic oxides. Qualitative experiments without the determination of the concentrations of the components in the layer-forming liquids showed the presence of liquation in the systems SiO_2-MnO, SiO_2-CoO, and SiO_2-NiO.

The most detailed experiments were carried out with the system $CaO-SiO_2$. It was established that of the two immiscible liquids formed at high temperatures (above 1698°C), one, which is close in composition to pure SiO_2, is very viscous, while the other, which contains about 30 wt.% CaO, is quite fluid. This second phase is extremely difficult to quench at such a rate that crystal formation does not occur. A microscopic examination of samples from this region that had been cooled rapidly showed that they consisted of two glasses. One was a pure homogeneous glass with a low refractive index (phase richer in silica) and the other phase was dispersed through this glass in the form of spheres of various sizes. The smaller spheres were readily quenched and consisted of a glass with a refractive index higher than that of the glass rich in silica. However, all these spheres, even the smallest, contained "small points" with a refractive index lower than that of the surrounding glass. These points were very fine crystals of cristobalite, formed during cooling.

If the composition of the starting glass was such that a large amount of the phase with a low silica content was formed, then a certain amount of this collected around spheres of the second phase. Very fine spheres of this glass with extremely fine grains of cristobalite thickly distributed through them appear bluish in reflected light and brownish in transmitted light under a microscope. The coarser grains are generally opaque. Macroscopically these samples appear as opaque, white, porcelain-like glass. The boundaries between the two phases described are quite sharp, and there are no transitions between them.

The investigation of ternary systems showed that on addition of such components as alumina or alkaline oxides, the compositions of the two immiscible liquids and consequently their physical properties gradually get closer and finally become identical, so that on addition of a definite amount of the additive, the glass becomes quite homogeneous.

As was pointed out above, no liquation phenomena are observed in the system $BaO-SiO_2$, but the liquidus curve has a very characteristic form which is typical of cases where the components lose their complete miscibility at temperatures almost adjacent to the horizontal liquidus line as in the $Ga_2O_3-SiO_2$ system (see Toropov and Lin' Tszu-syan, 1960; Glasser, 1959). Immiscibility of the liquid phases in such systems may be found at temperatures above or below the liquidus temperature.

If immiscibility arises at a temperature below the liquidus, this is naturally a metastable state, which is possible only in cases where the solid is not able to crystallize on cooling, i.e., when some supercooling occurs. However, cristobalite crystallizes from such systems at such a rate that it is very difficult to achieve this metastable immiscibility experimentally.

In the first investigations of regions of immiscibility in three-component silicate systems it was established that if both of the silicate systems forming the three-component system have liquation regions, then these regions merge and the overall liquation region extends through the whole field of the system though, as was first shown by Roozeboom-Bakhuis (1903), in these cases there is the theoretical possibility of a break in the continuity of the region of immiscibility. If liquation is observed only in one of the binary silicate systems bounding the ternary system examined, then in the ternary field the corresponding region will consist of only a narrow band adjacent to one side of the triangle. In accordance with the phase rule, in a three-component system there must be temperatures at which cristobalite and the two liquids will be at equilibrium with each other. Then, to each such temperature there will correspond two conjugate liquid phases of definite chemical composition. A rough experimental check of the direction of the tie lines or conodes is carried out by measuring the refractive indices of the silica (homogeneous) glasses.

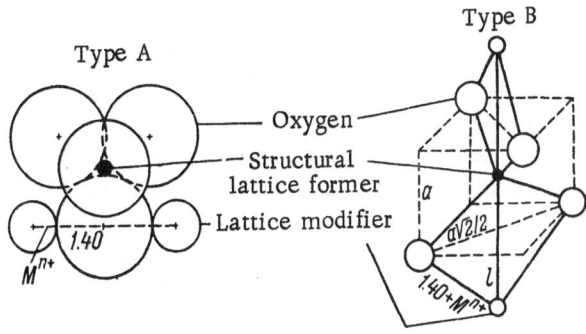

Fig. 2. Coordination types A and B in silicate glass.

Table 3. Electrostatic Field of a Cation
According to Dietzel

Cation	z	Ionic radius r_c (for coordination number 6), A	Electrostatic field of cation z/a^2, where $a = r_a + r_c$
K+	1	1.33	0.13
Na+	1	0.98	0.19
Li+	1	0.78	0.23
Ba2+	2	1.43	0.24
Sr2+	2	1.27	0.28
Ca2+	2	1.06	0.33
Mn2+ . . .	2	0.91	0.40
Fe2+ . . .	2	0.83	0.43
Mg2+ . . .	2	0.78	0.45
Fe3+ . . .	3	0.67	0.76
Al3+ . . .	3	0.57	0.84
Ti3+	3	0.64	1.04
B3+	3	0.20	1.34
Si4+ . . .	4	0.39	1.57
P5+	5	0.34	2.01

The ratio of the valence of an ion to its radius determines the strength of the bond of oxygen with an alkali or alkaline earth ion and the higher the ratio z/R, the more completely the modifier cation will be surrounded by unsaturated oxygens and the more marked will become the tendency for breakdown of the melt into two liquid phases (Table 2).

Attempts to calculate the limiting compositions of the region of immiscibility in silicate systems, based on the above analogies in the atomic structure of crystalline and glassy silicates (ionic radii and electrostatic bond) gave quite satisfactory agreement with experimental results for these values.

Then, by using as the characteristic of the electrostatic field of the cation the expression z/r^2 (where z is the valence of the cation and r its ionic radius in A), or z/a^2 (where $a = r_a + r_c$), Dietzel (1942) showed that this value is better for characterizing the tendency for the cation to form two immiscible liquids in silicate systems (Table 3).

According to Dietzel, favorable conditions for the formation of immiscible glasses in a system are created when the electrostatic forces of the cation fields differ little and consequently there is a tendency for the division of the oxygen ions between the competing cations. A relation between the radius of the modifier cation and the tendency for the formation of immiscible liquids was also pointed out by Esin (1949).

In a series of papers by Levin and Block (1957), the tendency of modifier cations to form immiscible liquids was examined in relation to the strength of the electrostatic bond of the cation and anion S = ze / n, where n is the coordination number of the cation with respect to oxygen in the glass. For modifier cations in the glass, these authors assume that there are two types of coordination in compositions corresponding to the composition of the limiting liquid with the lowest silica content (see Table 1 for composition of second liquid). These two types of coordination are denoted by the letters A and B. Coordination of type A is characterized by the fact that two modifier cations are bound to the same oxygen in the silicon—oxygen tetrahedron of the glass. Coordination of type B is characterized by the fact that two modifier cations are bound to different oxygen ions of the silicon—oxygen tetrahedron so that each modifier ion binds two of the four oxygens of the tetrahedron (Fig. 2).

The appearance of these types of coordination is regarded as the consequence of existing concepts on the relative positions of the ions in the structure in accordance with the theory of the atomic structure of glass. In pure glass formers such as SiO_2, B_2O_3, GeO_2, P_2O_5, etc., anions are coordinated about the cations, forming polyhedra, and the theory of these was developed in the work of Pauling and Belov. These polyhedra are attached solely through the apices and not through the edges or faces. In a silicate glass, as in crystalline silicates, silicon is always tetrahedrally surrounded by four oxygens. When an oxide modifier (Li_2O, BaO, CaO, MgO, etc.) is added to a pure glass former, the oxygen of the oxide modifier may be coordinated about the Si^{4+} only if the "transverse" Si—O—Si bonds existing previously in the pure glass former SiO_2 are broken. In this way arise ". saturated" oxygens, whose excess charges must be compensated by the modifier cations introduced into the gl.

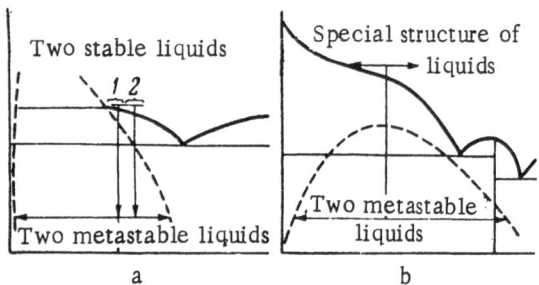

Fig. 3. Regions of possible subsolidus metastable
liquation.

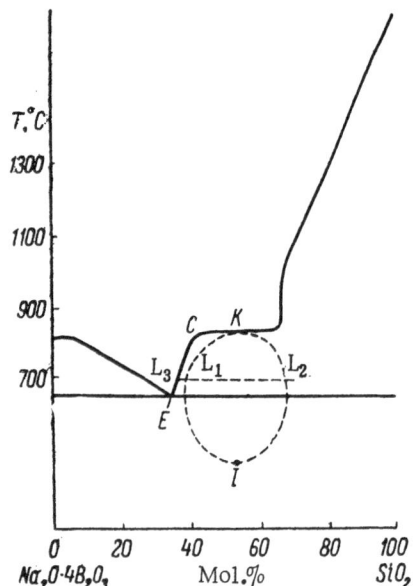

Fig. 4. Approximate region of metastable liqua-
tion in the system $Na_2O \cdot 4B_2O_3 - SiO_2$. L_1, L_2, and
L_3-liquids of different compositions.

As we saw above, it is assumed that there exist two types of coordination, A and B. All divalent cations with ionic radii greater than that of Ca^{2+} assume coordination of type B (weaker bond of the cation with oxygen), while all cations with an ionic radius less than that of Ca^{2+} assume coordination of type A. Divalent calcium itself may assume both types of coordination. The results of calculating the compositions of the limiting liquids in liquation silicate systems from the types of the modifier ion coordination adopted and the corresponding ionic radii agree well with data obtained by using other calculation methods and with experimental results. This confirms the validity of the theoretical assumptions made.

Further development of the theory of liquation phenomena in silicate systems has been made in the work of Iwase and Fukusima (1936), Roy (1960), and most recently, Galakhov (1962).

Roy examined the possibility of the appearance of metastable formations in typical glass-forming systems in the production of two immiscible liquid phases and indicated the great value of such phenomena in the technology of glass-crystalline materials.

In silicate systems where two immiscible liquids exist there often arises the possibility of a metastable state of one of them. Such observations of Greig were mentioned at the beginning of this chapter. In an investigation of the system $CaO - TiO_2 - SiO_2$ (De Vries, Roy, and Osborn, 1955) there was observed the formation of enamel-like glasses in samples with compositions adjacent to the region of equilibrium coexistence of two liquid phases.

It is interesting that the Japanese investigators Iwase and Fukusima had previously predicted and then demonstrated experimentally metastable liquidus lines in the same silicate system. It is believed that the structure of the liquid in a region directly adjacent to a section of the field of equilibrium of two macro-layer forming liquids is to some extent special, and that this structure becomes normal as the composition moves away from the region of compositions with two immiscible liquids. It may be surmised that the structure consists of particles 10 to 100 A in size of glass of one type rich in SiO_2 mixed with or forming inclusions in a matrix of another glass rich in the modifier cation.

The further we depart from the region of stable coexistence of two immiscible liquids, the more homogeneous becomes the glass. In connection with the subsolidus crystallization of glasses, in Fig. 3 it is possible to see the possible conversions which will occur in glasses with a relatively fine heterogeneous structure on cooling. The liquid may be cooled to a relatively clear glass, which will be characterized by a structure with larger sections of less viscous glass.

Figure 3a shows the position of the region of metastable liquation when there is also liquation in the system under stable conditions, i.e., above the temperature of the liquidus curves. Glass 1 will crystallize spontaneously during cooling and the sample will have a porcelain-like structure. Glass 2 has a composition which is more remote from the region of stable liquation and therefore here it is necessary to anneal the samples, especially after they have been cooled for the metastable separation of two vitreous phases. Figure 3b shows the position of the region of metastable liquation in a system where no separation of two liquid phases is observed above the temperature of the liquidus curves.

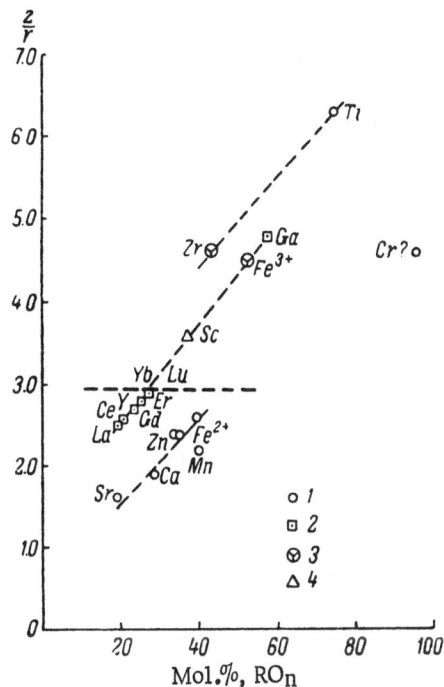

Fig. 5. Relation between ionic potential z/r and concentration of RO_n. 1) Literature data; 2) data of Warchaw, Glasser, and Roy; 3) predicted data; 4) data of Glasser.

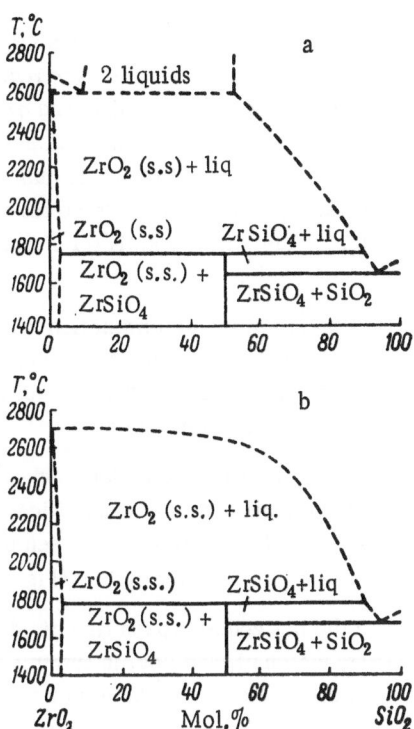

Fig. 6. Phase diagram of the ZrO_2-SiO_2 system. a) According to Roy et al.; b) according to Geller and Lang (calculated data). (s.s. = solid solution.)

In recently published work of Galakhov (1962), which was carried out in the Institute of Silicate Chemistry, a detailed analysis was made of the relation between the capacity of some sodium borosilicate glasses for selective extraction and tendencies for the formation of submicroliquation regions in the system $Na_2O-B_2O_3-$ $-SiO_2$, which is reflected in the morphology of individual regions of its phase diagram.

Compositions giving porous glasses lie in a region of the diagram characterized by a very flat liquidus surface, which is known to be a clear sign of the tendency of a system to form immiscible liquid phases. The structure of porous glasses obtained as a result of acid treatment of a starting glass from this region depends on its preliminary annealing conditions. Thus, according to the data of Porai-Koshits, Zhdanov, and Andreev (1960), and also Dobychin and Kiseleva (1957), a porous glass obtained from a material quenched from no lower than 750°C consists of a silica framework penetrated by small pores up to 50 A across. Glasses annealed at 600-750°C and extracted have pores up to 1000 A and these in their turn are filled with a fine silica framework, forming finer secondary pores. Annealing below 600°C gives a monodisperse structure with relatively coarse pores.

According to Galakhov, this behavior of sodium borosilicate glasses, which are analogous to the known "Vicor" glasses, is caused by the presence of a metastable subsolidus region of microliquation of these glasses in the part of the diagram examined. This region is shown schematically in Fig. 4, which represents the partial section $Na_2O \cdot 4B_2O_3-SiO_2$ of the above three-component diagram. The region of layer formation within the limits of the section is shown by a broken binodal curve with the critical points K and J. This curve describes an unstable state of the system and consequently the metastable coexistence of two liquids, which is intermediate and precedes special crystallization (see p. 8).

As a result of prolonged annealing, as the equilibrium state is approached, the two coexisting metastable liquids will disappear and in the temperature range of approximately 800 to 675°C there will be formed a liquid whose composition lies on the liquidus between the points C and E (Fig. 4) and SiO_2 crystals; below the temperature of the eutectic point E (675°), as equilibrium is reached, crystals of sodium tetraborate $Na_2O \cdot 4B_2O_3$ and SiO_2 separate.

Depending on which of the three regions, designated in accordance with the diagram in Fig. 4: a) above the submicroliquation region, b) within the region of metastable submicroliquation, but above the solidus temperature, i.e., above the eutectic point, and c) within the region of metastable submicroliquation, but below the solidus, according to Galakhov, the composition examined lies, the formation of the crystallization structure consisting of the readily extracted sodium borate component $Na_2O \cdot 4B_2O_3$ and SiO_2 will proceed differently. There will arise precisely those structures whose presence was established by electron microscopy (Bondarev, 1961), low-angle x-ray diffraction (Porai-Koshits, Zhdanov, and Andreev, 1960), and adsorption analysis (Zhdanov, 1955a, 1955b).

In this connection, the determination of metastable regions of submicroliquation assumes particular importance for the development of the theoretical foundations of the production of glass-crystalline materials, regardless of whether they lie above the liquidus temperature or are truly metastable, i.e., they are found at temperatures below the liquidus curves.

From an analysis of literature data and some of their own additional experiments to determine the limits of the regions of immiscibility in binary silicate systems, Glasser, Warchaw, and Roy (1960) came to the conclusion that there is a linear relation between the ionic potential z/r and the concentration of the liquid enriched in the second cation of the system (Fig. 5).

In Fig. 5 the horizontal broken line divides the regions with $z/r > 3$ and $z/r < 3$. According to the concepts of Glasser, Warchaw, and Roy, in systems from the first region the crystalline phase in equilibrium with the two immiscible liquids should be either a silicate or oxide of the second cation, but not cristobalite. The latter is the crystalline phase in equilibrium with liquids in systems where $z/r < 3$ for the second cation.

For cations with different charges the lines of the functional relation of ionic potential to limiting concentration of the liquation region are different, but according to Glasser, Warchaw, and Roy they are generally straight lines. However, the results of some of our recent experimental work on binary silicate systems, in which the presence of liquation regions was established, do not correspond to the graphs of Glasser, Warchaw, and Roy.

Thus, in particular, in determining the position of Zr on the diagram, the American authors did not use the data of Toropov and Galakhov, though they came to the same conclusion as in our work, that the diagram of Geller and Lang (1956) is not in accord with the well-known liquation phenomena in the same system. The liquation phenomena were described even earlier, though only qualitatively, by Barlett (1931) and were mapped out for the first time by us on the phase diagram of the ZrO_2-SiO_2 system (Fig. 7). The position of Zr on the diagram of Glasser, Warchaw, and Roy (Fig. 6a), which was determined indirectly, does not correspond to the composition that we determined experimentally (Fig. 7). This raises doubt on the value of the slope of the Zr——Ti line in Fig. 5. It is therefore necessary to study systems formed by SiO_2 with oxides of other tetravalent elements such as ThO_2, HfO_2, and UO_2 for a conclusive solution of this problem.

The second discrepancy concerns the position of the point of Sc in Fig. 5. According to the investigation of Toropov and Vasil'eva (1961), the phase diagram of the $Sc_2O_3-SiO_2$ system has the form shown in Fig. 8. The limiting concentration of the region of two immiscible phases corresponds to the composition 68.5 mol.% SiO_2 and 31.5 mol.% Sc_2O_3, which clearly does not correspond to the rough data of the American authors.

According to Glasser, Warchaw, and Roy, the crystalline phase in equilibrium with the two liquid phases in the same system should be scandium oxide as the point of Sc lies above the broken horizontal line in Fig. 5.

However, in our work it was established that: 1) scandium oxide forms a series of silicates, whose field of stability lies between the region of two immiscible liquids and the field of crystalline scandium oxide, and 2) the region of two immiscible liquids is in direct contact with the field of crystallization of cristobalite. These circumstances make it necessary to assume that there are more complex relations between the energy characteristics of the ionic fields in silicate melts and the limiting compositions of immiscible liquids than those given in the work of Glasser, Warchaw, and Roy.

Liquation phenomena in multicomponent systems are of great importance for the development of the physicochemical theory of the production of glass—crystalline materials. In connection with the use here of

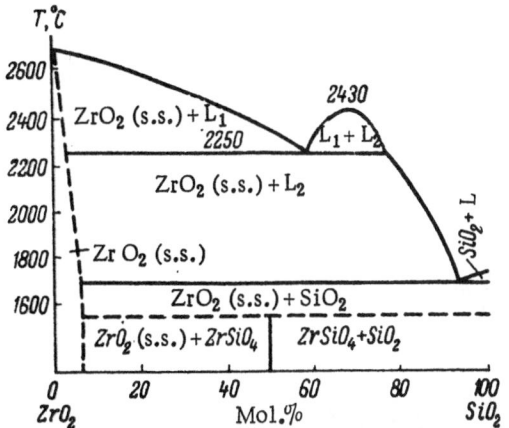

Fig. 7. Calculated liquation regions on the phase diagram of the system ZrO_2-SiO_2 (according to the data of Toropov and Galakhov, 1958). (s.s. = solid solution.)

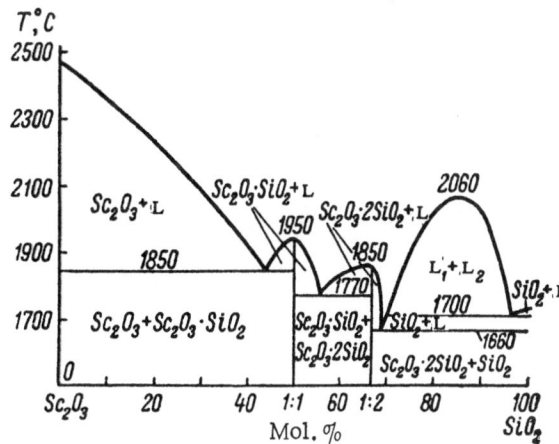

Fig. 8. Phase diagram of the system $Sc_2O_3-SiO_2$.

Table 4. Viscosity of the Glass
$CaO \cdot 0.312Al_2O_3 \cdot 2SiO_2$, in Poises

Temperature, °C	With addition of 5% CaF$_2$	Without addition of CaF$_2$
1440	470	331
1420	480	339
1400	500	346
1300	660	457
1200	1420	725
1180	2000	815

blast-furnace slags and the application of fluorides as active catalysts for crystallization, we should re-examine the results of the work of Toropov and Bondar' (1959b) on the effect of calcium fluoride on phase equilibria in the system $CaO-Al_2O_3-SiO_2$.

The part of the system we studied covers the compositions of both acid and basic blast-furnace slags; this region is adjacent to the $CaO-SiO_2$ side of the ternary system $CaO-Al_2O_3-SiO_2$ and extends from 20 to 65% SiO_2 and 80% Al_2O_3. The investigation was carried out by adding a constant amount of CaF_2 (5 and 10%) to ternary compositions containing CaO, Al_2O_3, and SiO_2. In this way we studied two sections through the four-component system $CaO-Al_2O_3-SiO_2-CaF_2$ with 5 and 10% CaF_2, lying parallel to the base triangle of $CaO-Al_2O_3-SiO_2$. Chemically pure materials were used for preparing the samples.

Fluorine-containing substances are known to have a high volatility and capacity to interact with atmospheric oxygen and therefore it was necessary to use anhydrous materials and hermetically sealed crucibles. In such a vessel mixtures of given composition were fused in a vacuum furnace and then cooled slowly. Control analyses showed that the fluorine losses did not exceed 0.05 wt.%.

Figure 9 shows that the plane ABC (5% CaF_2) in the part investigated of the four-component system $CaO-Al_2O_3-SiO_2-CaF_2$ passes through regions of primary crystallization of the following phases: calcium metasilicate $CaSiO_3$, anorthite $CaAl_2Si_2O_8$, rankinite $Ca_3Si_2O_7$, and also mullite $Al_6Si_2O_{13}$, gehlenite $Ca_2Al_2SiO_7$, and calcium orthosilicate Ca_2SiO_4.

As the investigation showed, calcium fluoride is a very effective mineralizer. Small additions of it (up to 1.5 wt.%) promote considerable growth of the crystals of the phases separating and accelerate polymorphic conversions of dicalcium silicate. No appreciable increase in the crystals is observed with a further increase in the CaF_2 content up to 5%. It was established that the introduction of 5% of CaF_2 into the charge reduces the crystallization temperature of melts and this was connected with a decrease in their viscosity. The viscosity of individual melts was considerably reduced on addition of 5% CaF_2, sometimes by a factor of more

than 2. Thus, for example, a glass with the composition $CaO \cdot 0.312Al_2O_3 \cdot 2SiO_2$ had the viscosities given in Table 4.

Additions of calcium fluoride considerably reduce the liquidus temperatures in the system (of the order of 50-70° for additions of 5% CaF_2). Figure 10 shows isotherms for the section with 5% CaF_2 and the distribution of the primary crystallization fields of the separate phases of the given section of the system.

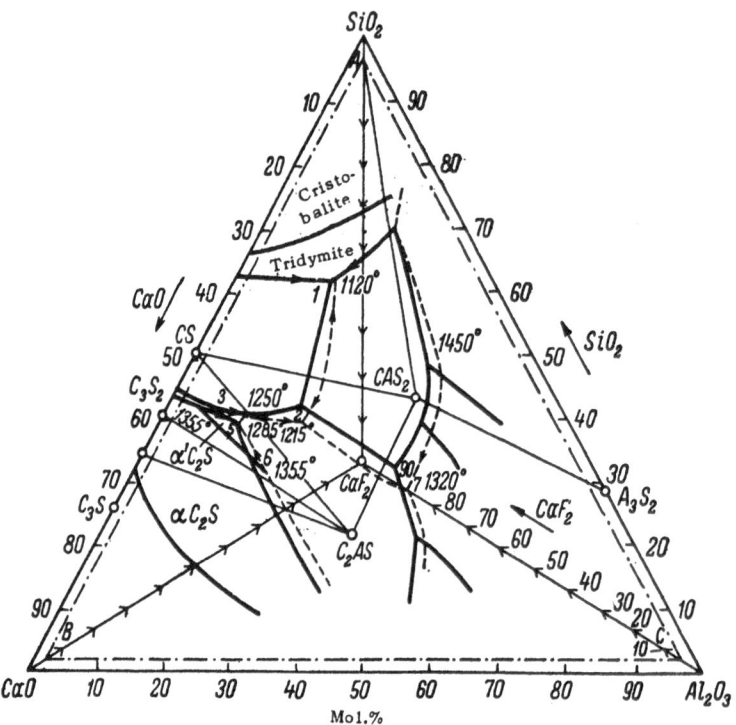

Fig. 9. Projection of the system $CaO-Al_2O_3-SiO_2-CaF_2$ (5% CaF_2).

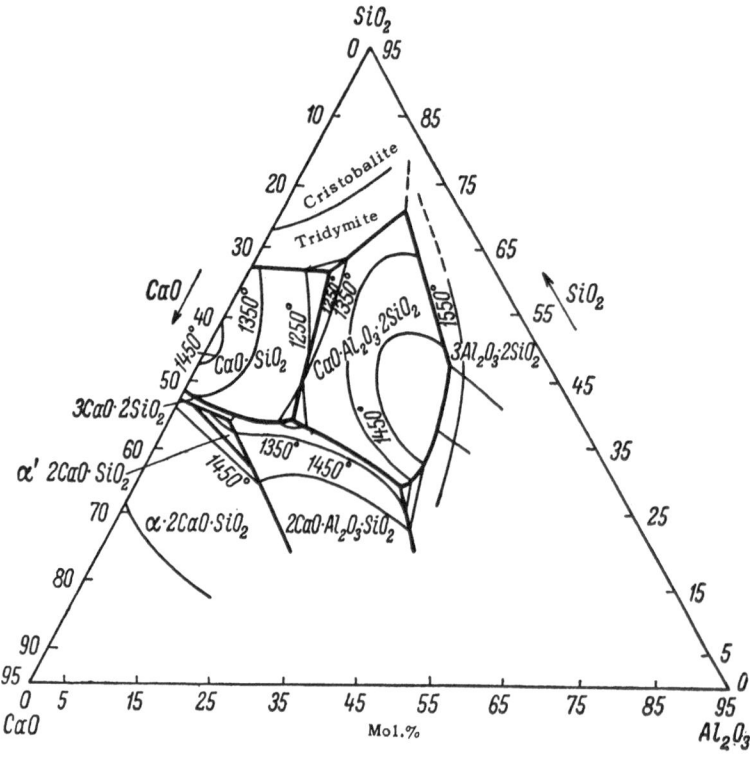

Fig. 10. Isotherms in the system $CaO-Al_2O_3-SiO_2-CaF_2$ (5% CaF_2).

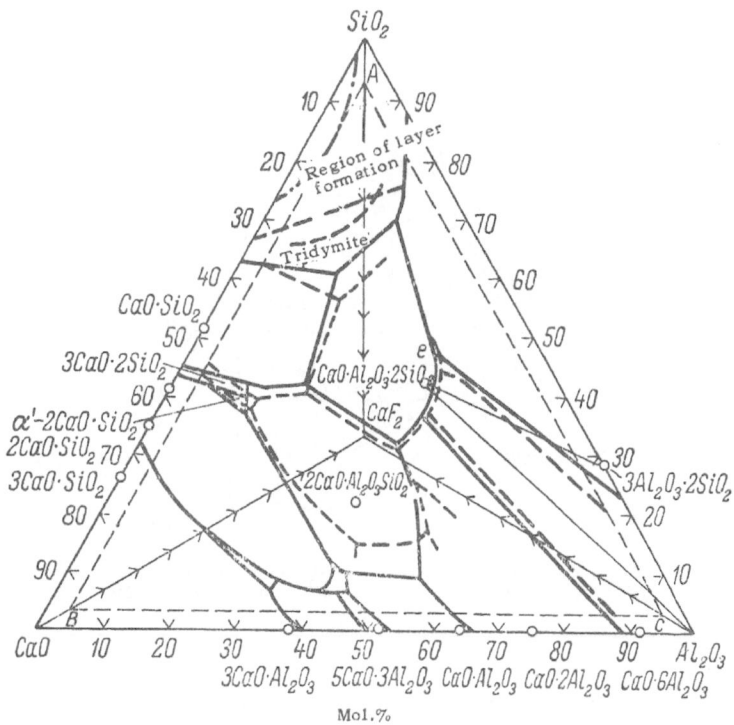

Fig. 11. Quaternary system $CaO-Al_2O_3-SiO_2-CaF_2$ (10% CaF_2).

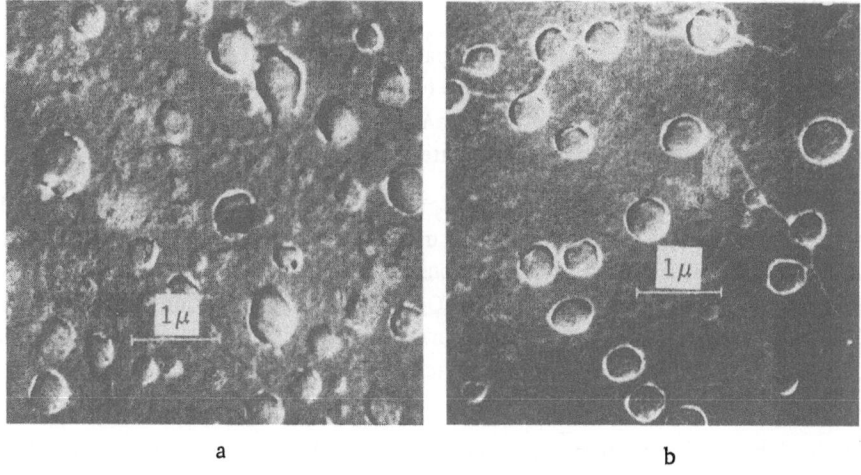

a   b

Fig. 12. Electron microscope photographs of two glasses. a) 2% CaO, 8% Al_2O_3, 80% SiO_2, 10% CaF_2; b) 20% CaO, 5% Al_2O_3, 65% SiO_2, 10% CaF_2. Magnification 18,000.

As a result of the investigation, it was established that the addition of 5% CaF_2 does not affect the character of the fusion of the individual compounds and no transitions from cases of congruent to incongruent melting or vice versa were observed. A shift was observed in the boundaries of the fields of primary crystallization of individual phases: thus, the field of anorthite (broken line in Fig. 9) was shifted by about 2 wt.% toward the $Al_2O_3-SiO_2$ side of the triangle. There was also a shift in the boundaries of the fields of calcium metasilicate and anorthite to regions of compositions with a lower SiO_2 content.

In accordance with these changes in the positions of the boundary curves there was also a change in the position of the separate five-phase invariant points. Figure 9 shows the triangulation of the part of the system

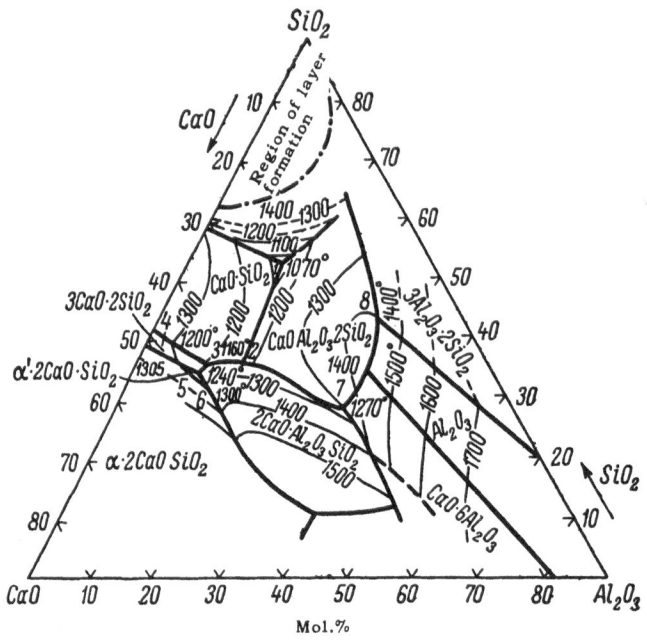

Fig. 13. Isotherms in the system $CaO-Al_2O_3-SiO_2-CaF_2$ (10%
CaF_2).

investigated, which makes it possible to determine the sequence of the crystallization of various melts of the system and to predict the final products of equilibrium crystallization in it. The results of experiments on samples containing 10% CaF_2 are given in Fig. 11.

The plane of the section ABC (10% CaF_2) in the part of the system $CaO-Al_2O_3-SiO_2-CaF_2$ investigated cuts through the primary crystallization volumes of the following phases: SiO_2, calcium metasilicate $CaSiO_3$, rankinite $Ca_3Si_2O_7$, gehlenite $Ca_2Al_2SiO_7$, anorthite $CaAl_2Si_2O_8$, mullite $Al_6Si_2O_{13}$, and also partly corundum, dicalcium silicate Ca_2SiO_4, and calcium hexa-aluminate $CaAl_{12}O_{19}$.

The region of liquation in the system $CaO-Al_2O_3-SiO_2$, according to the data of Greig (1927), occupies a narrow section adjacent to the $CaO-SiO_2$ side of the triangle of $CaO-Al_2O_3-SiO_2$. On addition of 10% CaF_2 to the three-component compositions, this section is considerably extended. In Fig. 11 it is bounded by the binodal curve which begins at a composition with 5% Al_2O_3 on the $Al_2O_3-SiO_2$ side of the triangle and ends at the point of 27.6% CaO on the $CaO-SiO_2$ side. In an investigation of samples from this region under a polarizing microscope, drops of the main glass with a high refractive index were found distributed in a glass rich in SiO_2 with a low refractive index.

The fine drops of the main glass remained round and quite clear on quenching, while the coarse drops contained grains of cristobalite, formed during cooling. By observing the samples with an electron microscope the boundaries of the region of layer formation were determined more accurately, as here it was possible to observe heterogeneities which could not be detected under an optical microscope. In the electron microscope investigation we were able to extend the boundaries of the region of coexistence of two glasses by approximately 2% in comparison with determinations obtained with normal optics.

Figure 12 gives photographs obtained with an electron microscope using replicas. On introduction of calcium fluoride into the charge, the fields of primary crystallization of the individual compounds changed as in the first case. A fall in the crystallization temperature of the melts of approximately 100-120°C was observed.

Figure 13 shows liquidus isotherms of the section of the system containing 10% CaF_2 and the distribution of the fields of stability of the individual phases. The introduction of 10% CaF_2 did not affect the character of the melting of the individual chemical compounds and only led to a change in the disposition of the boundaries of the fields of stability of some phases. Thus, the field of gehlenite (broken line in Fig. 11) was considerably reduced in size and its boundaries were somewhat shifted toward the $Al_2O_3-SiO_2$ side of the triangle. There was also some decrease in the area of the fields of primary crystallization of calcium metasilicate, rankinite, and anorthite. The position of the boundary curve between the fields of mullite and corundum again confirms the conclusion of Toropov and Galakhov (1958) that mullite melts congruently.

Further experiments on the effect of higher concentrations of calcium fluoride (20 wt.% of CaF_2) on crystallization in the system $CaO-Al_2O_3-SiO_2$ were carried out by the author together with Lin' Tszu-syan. For these compositions, the use of fusion in hermetically sealed crucibles in a vacuum furnace with subsequent slow cooling in the furnace was found to be unsuitable because of dendritic metastable crystallization of CaF_2. Therefore, the crucibles were heated in a normal furnace and the samples then cooled rapidly in dishes of water.

We investigated compositions from the region adjacent to the $CaO-SiO_2$ side, extending from 40 to 60 wt.% SiO_2 and up to 40 wt.% Al_2O_3. These compositions covered mainly the region of basic blast-furnace slags. In the region of the system studied we found four fields of primary crystallization, namely those of calcium metasilicate, calcium fluoride, cuspidine, and anorthite. No field of rankinite was found at all in compositions with 20 wt.% CaF_2.

BIBLIOGRAPHY

Argyle, T. F., and F. A. Hummel. Phys. Chem. Glasses 4(3): 103 (1963).

Barlett, H.B., J. Am. Ceram. Soc. 14(1): 11 (1931).

Bondarev, K. T. Steklo Byull. Gos. Inst. Stekla 1: 10 (1961).

Dietzel, A. Z. Elektrochem. 48: 9-23 (1942).

Dobychin, D.P., and N. N. Kiseleva. Dokl. Akad. Nauk SSSR 113: 372 (1957).

Esin, O. A., Proceedings of the 2nd All-Union Conference on Theoretical and Applied Electrochemistry, Izd. Akad. Nauk UkrSSR, Kiev (1949), p. 215.

Galakhov, F. Ya. Bull. Soc. Franç. Ceram. 38: 11 (1958).

Galakhov, F. Ya. Izv. Akad. Nauk SSSR, Otd. Khim. Nauk 5: 743 (1962).

Geller, R.F., and S. M. Lang. Phase Diagrams for Ceramists, Edited by the American Ceramic Society, Inc. (1956), p. 67.

Glasser, F. P. J. Phys. Chem. 63(12): 2085 (1959).

Glasser, F. P. I. Warchaw, and R. Roy. Phys. Chem. Glasses 1(2): 39 (1960).

Greig, J. W. Am. J. Sci. 13(73): 1-44 (1927); 13(74): 133-154 (1927).

Iwase, K., and M. Fukusima Sci. Rept. Tohoku Univ. First Ser. Handa Anniversary (1936), pp. 454-464.

Kracek, F. C. J. Am. Chem. Soc. 52(4): 1436-1442 (1930).

Levin, E.M., and S. Block. J. Am. Ceram. Soc. 40(5): 95-106 (1957).

Levin, E.M., and S. Block. J. Am. Ceram. Soc. 40(4): 113 (1958).

Ol'shanskii, Ya. I. Proceedings of the 5th Conference on Exper. Techn. Mineral. Petrog., Izd. Akad. Nauk SSSR, Moscow (1958), p. 114.

Porai-Koshits, E. A., S. P. Zhdanov, and N. S. Andreev. In collection: The Glassy State, Proceedings of the 3rd Conference on the Structure of Glass, Izd. Akad. Nauk SSSR, Moscow-Leningrad (1960), pp. 517-522.

Porai-Koshits, E. A., S. P. Zhdanov, and D. I. Levin. Izv. Akad. Nauk SSSR, Otd. Khim. Nauk (1955), p. 395.

Roozeboom-Bakhuis, H. W. Die Heterogenen Gleichgewichte, Vol. III (1903).

Roy, R. J. Am. Ceram. Soc. 43: 670-671 (1960).

Toropov, N. A., and I. A. Bondar'. Proceedings of the Conference on Exper. Techn. and Methods of High-Temperature Research, Akad. Nauk SSSR, Izd. Akad. Nauk SSSR, Moscow (1959), pp. 205-212.

Toropov, N.A., and I. A. Bondar'. Izv. Akad. Nauk SSSR, Otd. Khim. Nauk 9: 1520-1525 (1959).

Toropov, N.A., and V. A. Vasil'eva. Kristallografiya 6(6): 968-972 (1961).

Toropov, N. A., and F. Ya. Galakhov. Izv. Akad. Nauk SSSR, Otd. Khim. Nauk 1: 8-11 (1958).

Toropov, N. A., and Lin' Tszu-syan. Zh. Neorgan. Khim. 5(11): 2462 (1960).

Vogel, W. Symposium sur la Fusion du Verre, Union Scientifique Continentale du Verre, Charleroi, Belgium (1958), pp. 741-770.

De Vries, R. C., R. Roy, and E. E. Osborn. J. Am. Ceram. Soc. 38(5): 158-171 (1955).

Warren, B. E., and A. G. Pinkus. J. Am. Ceram. Soc. 23(10): 301 (1940).

CHAPTER II

THE THREE-COMPONENT SYSTEM LITHIUM OXIDE–ALUMINA–SILICA

This system has now assumed exceptionally great importance as it contains a series of compositions with negative or zero coefficients of thermal expansion, so that, on the basis of them it is possible to make ceramic or glass—crystalline materials of exceptionally high thermal stability. The formation of complex metastable products during the crystallization of supercooled glasses is also very characteristic of this system. This requires a series of additional investigations.

This system includes the three natural minerals petalite $Li_2O \cdot Al_2O \cdot 8SiO_2$, spodumene $Li_2O \cdot Al_2O_3 \cdot 4SiO_2$, and eucryptite $Li_2O \cdot Al_2O_3 \cdot 2SiO_2$, which is an analog of the sodium aluminosilicate nepheline. The first systematic physicochemical investigation of the phase diagram was made by Hatch (1943), though it was rough. It was mainly concerned with the $Li_2O \cdot Al_2O_3$—SiO_2 section, within which lie the compositions of all these three minerals. Further refinement was made by Roy, Roy, and Osborn (1950), but up to now the study of this system has been very rough and many of its details require further refinement.

Syntheses of individual lithium aluminosilicates found as natural minerals were carried out much earlier by many other authors. Thus, Hautefeuille (1880) synthesized lithium aluminosilicates by fusing the components in the presence of lithium vanadate or tungstate as fluxes. Ten years later, Hautefeuille and Perry (1890) reproduced natural eucryptite by fusing metakaolinite with lithium vanadate. Another rhombic form of eucryptite was obtained by Weyberg (1905) by fusing lithium sulfate with kaolinite. A series of authors, namely Stein (1907), Ginsberg (1912), Endell and Rikke (1912), Ballo and Dittler (1912), and Jaeger and Simek (1914) also synthesized lithium aluminosilicates, which differed, however, from the above natural minerals in their physical properties.

The conversion of these minerals has been studied mainly on the example of spodumene. Thus, Brun (1902) and Tammann (1903) observed that the density of spodumene falls appreciably during brief heating at temperatures up to 1000°C. Ballo and Dittler, Jaeger and Simek, and Meissner (1920) came to the conclusion that the conversion of natural α-spodumene into the high-temperature β-form has a monotropic character.

In the work of Hatch, lithium aluminosilicate glasses were prepared by three methods: 1) by the addition of a constant amount of aluminum oxide and various amounts of silica to previously prepared lithium aluminosilicate glass or crystallized material; 2) by mixing two lithium aluminosilicate glasses; 3) by adding various amounts of silica to crystalline lithium aluminate.

The advantage of the first method is the absence of volatilization of lithium oxide, which makes it possible to fix a definite amount of lithium oxide at quite low temperatures. The method of mixing two glasses is simple and rapid. The third method is most convenient for preparing samples with a low silica content. The samples were fused five times for attaining complete homogeneity.

In the section $Li_2O \cdot Al_2O_3$—SiO_2 studied from 90% SiO_2 to spodumene (64.6% SiO_2), the system has a binary character (Fig. 14) and consists of fields of silica and solid solutions of the β-spodumene type with a eutectic between them, lying at 84.5% and 1356°C. Silica crystallizes in the form of tridymite. The β-spodumene itself has a sharply expressed temperature maximum at 1423°C.

Within the range of 64.6 to 47.7% SiO_2 (composition of eucryptite), above the solidus line the system also has a binary character and is divided into fields of solid solutions of β-spodumene and β-eucryptite. However, in samples from these fields an unknown fibrous product was found. Here, β-eucryptite itself is an

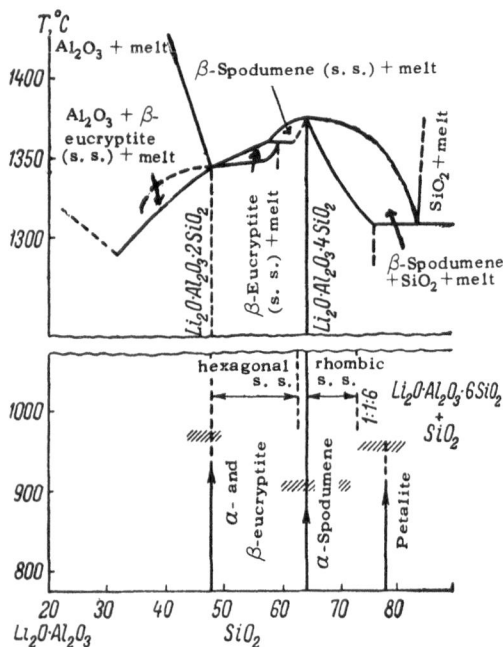

Fig. 14. Phase diagram of the system
$Li_2O \cdot Al_2O_3 - SiO_2$. (s.s. = solid solution.)

unstable phase and decomposes to γ-alumina and a eucryptite-like product below 1397°C. With a lower silica content the system completely loses its binary character and γ-alumina and eucryptite solid solutions are found as crystallization products.

Determinations of the solidus and liquidus temperatures in the regions of solid solutions were greatly hampered by the tendency of the melts to superheat and the similarity of the refractive indices of the glasses and the crystals.

With prolonged heating of crystals of solid solutions of both the β-spodumene and β-eucryptite types there was separation of fibrous crystals somewhat similar to mullite. Hatch believed that this may indicate the metastability of these solid solutions.

The tendency for the solid solutions to decompose partly, which is intensified with an increase in the alumina concentration, may be explained by weakening of the cation–oxygen chemical bond with a decrease in the silica content and when the concentration reaches 47.7 wt.%, complete decomposition of these solid solutions begins. Investigations of the optical properties of crystals of the two types of solid solution revealed definite differences between these phases and crystals of solid solutions of β-spodumene form tetragonal bipyramids, which are well formed when crystallized from melts of sufficient fluidity.

Crystals of solid solutions of β-spodumene are optically monoaxial and positive; the values of the refractive indices vary from 1.516 to 1.518 for n_0 and from 1.517 to 1.523 for n_e; the birefringence is very low, namely 0.001-0.005.

Crystals of solid solutions of β-eucryptite form granular aggregates similar to spodumene solid solutions; the crystals are optically monoaxial and negative; the refractive indices differ little from those of spodumene; the refractive indices of the glasses are higher than those of the crystals of corresponding composition.

A new synthesis method and the crystal structure of β-eucryptite were described by Winkler (1948). This author studied a series of synthetic minerals of the nepheline group and used lithium fluoride as the normal flux-mineralizer. Thus, in the synthesis of nepheline from melts containing more than 86 mol.% of lithium fluoride, Winkler obtained hexagonal bipyramids of eucryptite on heating the material for 3 h at 920°C. The size of the crystals did not exceed 1-5 mm and the product obtained was shown to be identical to eucryptite by comparing x-ray diffraction patterns of these crystals with those obtained from a $Li_2CO_3 : Al_2O_3 : SiO_2$ melt of given proportions with 10% LiF added as mineralizer. Together with LiF lines, the x-ray diffraction patterns showed only lines of nepheline-like $LiAlSiO_4$.

A mixture consisting of 73 wt.% $LiAlSiO_4$, 13.7% cryolite, and 13.3% LiF was used for obtaining well-formed crystals of β-eucryptite. The melt was heated for 30 min at 1200°C in an open crucible and at 1110°C the eucryptite crystallized in 1-5 h as large clear crystals with mirror faces. Crystallization from other melts yielded less clear crystals. Good results were obtained when glass was absent from the final melt. Here, the accompanying crystal phases were lithium fluoride and lithium cryolite.

Crystals of eucryptite obtained in this way form hexagonal bipyramids of the first order $(1\bar{1}22)$ and, as was established by x rays, prisms $(1\bar{1}20)$ are also found. No similarity to natural eucryptite crystals (monoclinic) was found.

The crystal structure was determined using Laue diffraction patterns taken along the a and c axes of the synthetic eucryptite crystals, Weissenberg patterns, and rotation patterns. It was established that there is rotation of the plane of polarization. The rotatory power equaled 5 deg/mm.

Table 5. Crystallographic Characteristics of Quartz and β-Eucryptite

Characteristics	β-Eucryptite	High-temperature quartz
Crystallographic axes, A:		
c	2 · 5.625	5.446
a	5.270	4.986
c/a	2 · 1.067	1.092
n_e	1.5195	1.5404
n_0	1.524	1.5328
Density, g/cm^3	2.352 (15°)	2.518 (600°C)
Optical rotatory power, deg/mm	10.5	24.3

The parameters of the elementary cell were as follows: a = 5.27 A and c = 5.625 ± 0.005 A. The number of molecules of $LiAlSiO_4$ in the elementary cell is three, and this requires doubling of the parameter along the c axis. In this case there is a similarity to the structure of high-temperature quartz, for which the axial ratio (with z = 3), c/a = 1.092, while for eucryptite, c/a = 1.067. Therefore, the structure of β-eucryptite may be regarded as a superstructure of high-temperature quartz in which the parameter along the c axis is approximately doubled. The structure in general is identical to that of high-temperature quartz and the $[SiO_4]^{4-}$ tetrahedra form right-hand and left-hand helical chains, extending along the c axis.

The structure of β-eucryptite may be represented by replacing half the silicon atoms in quartz by aluminum atoms and filling the spiral spaces in this structure by compensating lithium atoms in accordance with the scheme $Si^{4+} \rightarrow Al^{3+}Li^{1+}$.

Thus, half of the $[SiO_4]$ tetrahedra are replaced by $[AlO_4]$ tetrahedra. Each oxygen simultaneously belongs to a tetrahedron with silicon and one with aluminum. One of the charges of the oxygen O^{2-} is used in the bond with the charge of the adjacent silicon atom, while the adjacent aluminum atom compensates only $\frac{3}{4}$ of the second charge of the oxygen. Thus, electrostatic neutrality may be achieved by surrounding the oxygen by four lithium atoms and in this case Pauling's electrostatic compensation rule holds.

If we compare the structures of all the nepheline family, it is found that none of the other representatives have the structure of high-temperature quartz. This is explained by the fact that the potassium and sodium atoms are too large and will not fit into the spaces of the quartz structure. Therefore, these minerals require a more open structure for the silica base within which it is possible to fit these larger cations. In actual fact, such forms are, for example, modifications of tridymite, whose specific volume is 15.5% greater than that of quartz. As a rough approximation, nepheline and tridymite have similar structures. The physical constants of β-eucryptite and high-temperature quartz are given in Table 5.

As the data presented show, the introduction of lithium atoms into SiO_2 by the scheme

$$2SiO_2 \rightarrow LiAlSiO_4$$

leads to some expansion of the lattice with the change in structures quartz → β-eucryptite. Before the investigation of synthetic eucryptite, the structure of quartz was regarded as unique and quartz itself, which is one of the commonest minerals in nature, was long regarded as a rare example of a compound with a constant chemical composition, which contains in rare cases only traces of chemical impurities which give its crystals various colors. Thus, traces of manganese oxides give amethyst its violet color, nickel oxide gives chrysoprase its green, etc.

For geological conditions all this is true to a considerable extent, but experimental investigations of recent years, whose beginning is the solution of this structure of β-eucryptite, showed that in synthetic silicate systems there are increasing numbers of structures of the quartz type or quartz-like structures in compounds, which differ substantially in chemical composition from pure silicon dioxide. Subsequently, the work of Roy and Osborn on hydrothermal syntheses in the system $Li_2O—Al_2O_3—SiO_2$ gave preparations containing quartz with very mixed diffraction lines on the x-ray patterns and the authors also explained this by the introduction of lithium atoms into the structure.

Table 6. Parameters of Hexagonal Solid Solutions of the β-Eucryptite Type

Composition of sample	a, A	c, A	c/a	Volume of elementary cell, A^3
High-temperature quartz	4.989	10.89	2.184	233.3
Synthetic product Li$_2$O · Al$_2$O$_3$ · 9SiO$_2$	5.058	10.88	2.150	241.0
Petalite, natural, roasted.	5.130	10.89	2.123	248.3
The same, fused	5.183	10.92	2.120	251.2
Natural spodumene	5.225	10.92	2.090	258.3
Synthetic product Li$_2$O · Al$_2$O$_3$ · 3.5SiO$_2$ . . .	5.220	11.14	2.134	260.30

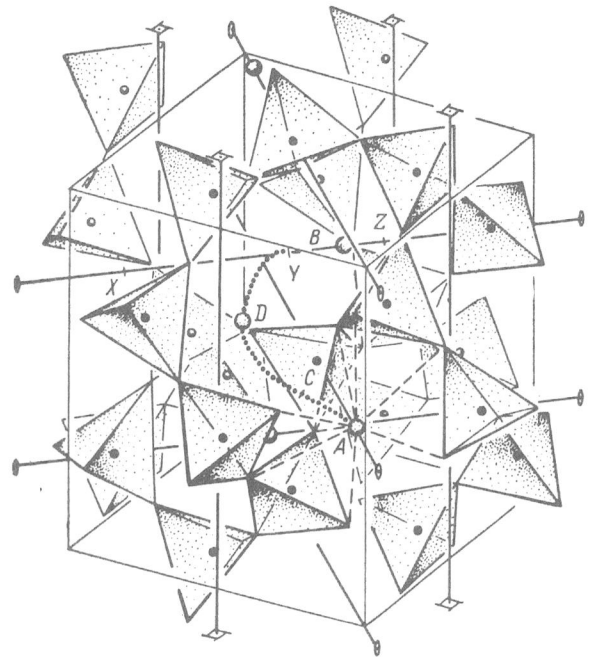

Fig. 15. Structure of keatite.

In an investigation of the system Li$_2$O—Al$_2$O$_3$—SiO$_2$, Heinglein (1956) came to the conclusion that there exists in it a continuous series of solid solutions of SiO$_2$ and LiAlSiO$_4$ with structures of high-temperature quartz. According to Skinner and Ewans (1960), some members of this series are stable under certain experimental conditions, while others are only metastable.

In 1948, Hummel (1951) investigated the thermal expansion of the natural lithium minerals spodumene and petalite, roasted at 1250°C, and established that the values are very low. In some cases a negative thermal expansion coefficient was found. It is interesting to note that such a phenomenon has long been known for high-temperature quartz. This physical phenomenon confirms the similarity in the structures of strongly roasted lithium aluminosilicates and quartz.

An x-ray investigation demonstrated to Hummel the similarity in the structures of β-eucryptite and β-spodumene. The compositions Li$_2$O · Al$_2$O$_3$ · 6SiO$_2$ and Li$_2$O · Al$_2$O$_3$ · 8SiO$_2$ gave the diffraction picture of the structure of β-spodumene with an appreciable shift in the strongest line from d = 3.47 to d = 3.44. The composition Li$_2$O · Al$_2$O$_3$ · 10SiO$_2$ showed cristobalite with the line d = 4.05 and quartz with d = 4.26, 3.34, 1.82, and 1.39. It is interesting that the quartz from this sample showed a transition between 400 and 500°C, which is considerably lower than the usual one at 573°C.

Three types of structure with large spaces, guaranteeing a low thermal expansion, may be listed here for different representatives of the silicate class, namely beryl, high-temperature quartz, and β-eucryptite.

While it was stated by Winkler that the quartz-like structure of β-eucryptite is characteristic of only the strictly stoichiometric lithium analog of nepheline, Heinglein showed that this type of structure is also possible for a wide range of compositions from pure quartz to eucryptite. Heinglein obtained a whole set of x-ray patterns, calculations from which gave a series of regular transitions in the hexagonal type of lattice (Table 6).

The next development in the problem of β-spodumene and β-eucryptite solid solutions resulted from the discovery by Keat (1954) of a new polymorphic tetragonal modification of silica, which was later called keatite.

Keatite was obtained by recrystallization of silica over quite a wide range of temperatures (380-585°C) and pressures (500-18,000 psi). * The addition of a mineralizer (Li$_2$CO$_3$, Na$_2$CO$_3$, LiOH, NaOH, KOH, Na$_2$WO$_4$)

*500-18,000 psi ≈ 35.3-1273 kg/cm^2. — Ed.

Table 7. X-Ray Data on Crystals of O-Series of β-Eucryptite Solid Solutions

hkl	Solid solutions with high SiO_2 content		Solid solutions with low SiO_2 content		hkl	Solid solutions with high SiO_2 content		Solid solutions with low SiO_2 content	
	d, Å	I/I_0	d, Å	I'		d, Å	I/I_0	d, Å	I/I_0
100	4.32	40	4.55	22	112	1.842	60	1.914	38
101	3.38	100	3.53	100	202	1.696	3	1.736	3
110	2.498	10	2.621	10	211	1.568	25	1.643	24
102	2.303	5	2.384	3	212	1.404	20	1.463	11
200	2.165	18	2.274	9	203	1.393	25	1.441	16
201	2.013	12	2.105	8					

Table 8. Dimensions of Elementary Cell of β-Spodumene and β-Spodumene Solid Solution

SiO_2 content of solid solution, wt.%		Crystallization temperature, °C	Crystallization time, h	$a \pm 0.0008$, Å	$c \pm 0.0008$, Å	$d_{303} \pm 0.003$, Å
62.04	0.3580	1330	3	7.5584	9.1720	1.9442
62.46	0.3539	1380	3	7.5548	9.1680	1.9430
62.89	0.3497	1350	1.75	7.5500	9.1656	1.9419
63.32	0.3455	1380	3	7.5452	9.1620	1.9411
63.73	0.3415	1350	1.75	7.5426	9.1608	1.9408
64.16	0.3373	1380	3	7.5396	9.1540	1.9392
64.58 (composition of spodumene)	0.3333	1350	1.75	7.5332	9.1540	1.9376
68.12	0.2990	1350	1.75	7.5224	9.188	1.9345
71.66	0.2649	1350	1.75	7.5120	9.0964	1.9313
75.21	0.2310	1350	1.75	7.5020	9.0568	1.9267
78.75	0.1973	1350	1.75	7.4932	9.0348	1.9236
82.29	0.1639	1350	1.75	7.4832	9.0004	1.9179
Petalite spodumene	0.2000	1350	1.75	7.4976	9.0372	1.9240
Natural spodumene	0.3333	1350	1.75	7.5368	9.1696	1.9403

Note. The compositions are given in wt.% of SiO_2 for the series $Li_x Al_x Si_{1-x} O_2$.

had to be quite definite. Thus, for example, the concentration of NaOH had to lie within the range of 0.0012-0.0015 mole/liter to obtain good crystals of keatite. The density of keatite (2.50 g/cm³) is intermediate between that of cristobalite (2.32 g/cm³) and that of quartz (2.66 g/cm³).

From rotation x-ray diffraction patterns of Shropshire, Keat, and Vaughan (1959), it was established that keatite belongs to the tetragonal syngony (space group $P4_1 2_1$ or $P4_3 2_1$) with the parameters of the elementary cell a = 7.456 A and c = 8.604 A. The structure of keatite consists of fourfold helical chains formed by $[SiO_4]$ tetrahedra. This structure of keatite crystals indicates their considerable similarity to high-temperature spodumene, which was studied in more detail by Skinner and Ewans in 1960. The structure of β-spodumene may be regarded as derived from the structure of keatite by replacement in the latter of some of the silicon atoms by aluminum atoms and the introduction of the equivalent number of lithium atoms to maintain the electrostatic balance of the lattice in the same way that the structure of β-eucryptite is derived from the structure of quartz.

As is shown in Fig. 15, the structure of keatite consists of elementary cells, each of which contains twelve $[SiO_4]$ tetrahedra. Eight of them form a helical chain about a fourfold screw axis, lying at the centers of the side planes of the elementary cell, while the other four tetrahedra lie on horizontal diagonal binary axes of inversion, intersecting the helical chains.

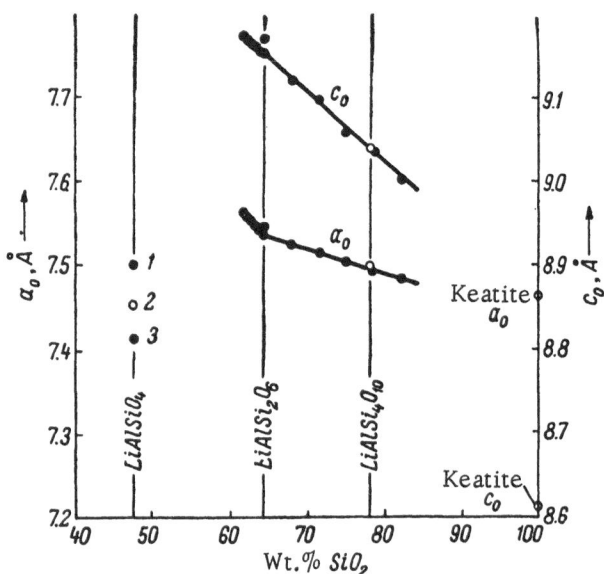

Fig. 16. Parameters of the elementary cell of a spodumene
solid solution. 1) Glass; 2) petalite; 3) spodumene.

Even in 1957, before the detailed description of the structure of keatite, Roy put forward the hypothesis
that keatite is the last member of the β-spodumene series of solid solutions, which are characterized by tetragonal symmetry. Roy named these solid solutions the K-series of lithium aluminosilicate solid solutions; he
named the hexagonal quartz-like solid solutions the O-series of solid solutions. They were obtained, as stated
above, by Heinglein by heating natural α-spodumene and α-petalite. Moreover, these phases arise in technical pyroceramics as primary or secondary metastable products and may be obtained from the mixtures Li_2O:
Al_2O_3: $8-10SiO_2$, formed from gels and calcined for 30 min to 12 h between 750-900°C. According to a communication of Rindone and Roy (1959), glasses of the Li_2O-SiO_2 system may also contain O-solid solutions when
crystallized at very low temperatures.

There is an interesting report of Roy that crystals of O-solid solutions were also found in industrial foam
glass and in the form of pure SiO_2 they may be obtained at room temperature by neutron irradiation of quartz.
The irradiation must stop at the moment when the size of the elementary cell becomes identical to that of
high-temperature quartz and naturally this is not readily controlled.

Table 7 gives x-ray data for metastable β-eucryptite solid solutions (O-series according to Roy).

A characteristic difference between the β-eucryptite and β-spodumene solid solutions is the absence of
lines at 3.85 and 3.17 A for the eucryptite series.

Skinner and Ewans (1960) obtained fine crystals of β-spodumene and solid solutions of it, which made it
possible to use them for single-crystal x-ray studies. The single crystals obtained had a tetragonal elementary
cell with $a = 7.50$ A and $c = 9.03$ A ($c/a = 1.20:1$), and the space group $P4_32_2(D_4^8)$ or its enantiomorphic form
$P4_22_1(D_4^4)$. These data are very close to those obtained by Shropshire, Keat, and Vaughan.

By precise measurements, the size of the elementary cell for β-spodumene of stoichiometric composition
was determined as $a = 7.5332 \pm 0.0008$ A and $c = 9.1540 + 0.008$ A at 25°C. The calculated density of 2.38 g
per cm^3 agrees well with the value of 2.35 g/cm^3, found experimentally by Hummel. The crystallization of
glasses from the $Li_2O \cdot Al_2O_3-SiO_2$ series was carried out for 1.75-3 h at 1350°C. The glasses crystallized rapidly after a few minutes, but as the primary product there always separated hexagonal crystals of solid solutions of
β-eucryptite (or O-silica, according to Roy). On further annealing, these solid solutions were rapidly converted
into solid solutions of β-spodumene, so that after 15 min it was possible to obtain a homogeneous β-spodumene
solid solution. It is evident that what Roy named optically negative β-spodumene is actually a β-eucryptite
solid solution formed initially, and which is metastable.

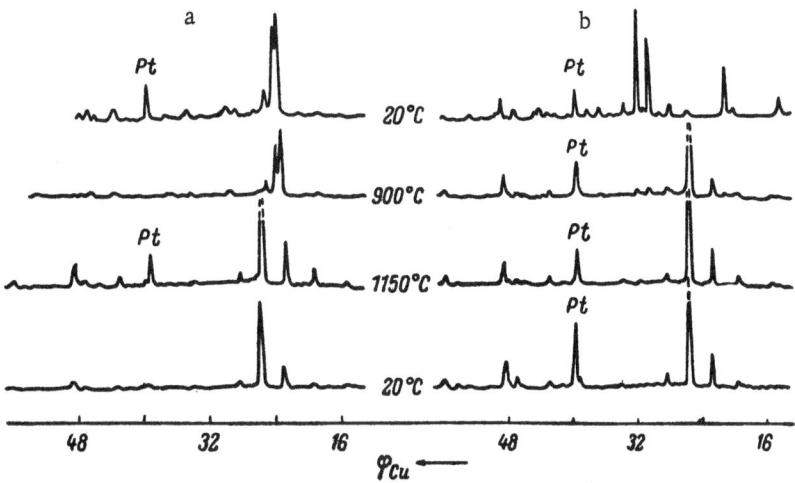

Fig. 17. X-ray diffraction patterns of petalite (a) and spodumene (b) at normal and high temperatures.

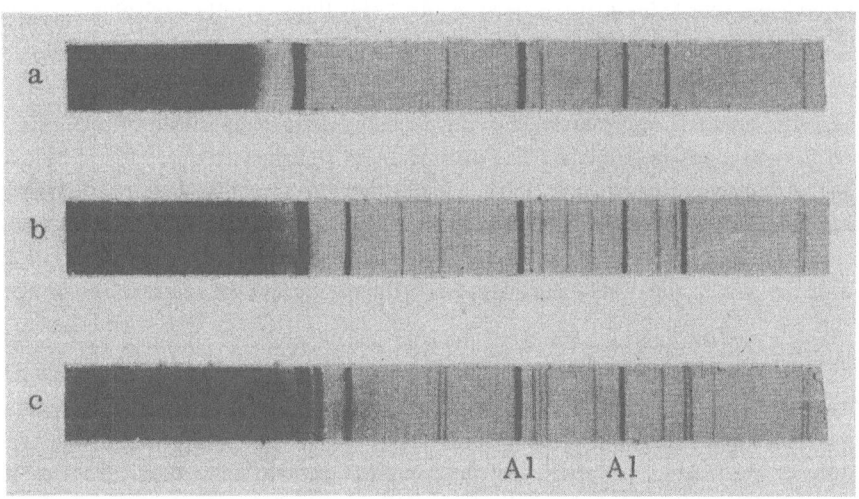

Fig. 18. X-ray diffraction patterns obtained by Guinier's method.
a) β-Eucryptite; b) β-spodumene; c) petalite.

Precise determinations of the parameters of the elementary cell of the tetragonal solid solution were carried out photographically with sodium chloride crystals used as an internal standard.

Natural petalite and spodumene were also converted into β-spodumene by heating for 1.75 h. The dimensions of the elementary cell of β-spodumene obtained in this way are given in Table 8.

The change in the parameters of the elementary cell of the solid solutions examined in relation to the composition is shown in Fig. 16. A detailed x-ray investigation showed that over the whole field of stability of the solid solution of β-spodumene no change is observed in the crystallographic symmetry.

The lattice parameters change symbatically with the composition of the solid solution. Though it was not possible to obtain a solid solution of β-spodumene containing more than 84.3 wt.% SiO_2, the general chemical formula may be written as follows: $Li_xAl_xSi_{1-x}O_2$; when x = 0 we obtain pure SiO_2 in the form of tetragonal keatite, and when x = $\frac{1}{3}$, we obtain β-spodumene. These solid solutions with an SiO_2 content from 84.3 (x = 0.1457) to 100 wt.% are metastable at 1350°C and atmospheric pressure.

Table 9. Lattice Constants of High-Temperature Forms
of Lithium Aluminosilicates, A

Lithium aluminosilicates	a	b	c
$Li_2O : Al_2O_3 : SiO_2$ (1 : 1 : 6).	18.24	10.54	10.5
β-Spodumene	18.38	10.61	10.68
β-Eucryptite	18.15	10.48	11.13

Let us examine the lattice points which may be occupied by lithium and aluminum atoms. As is shown in Fig. 15, the structure of keatite is complex. The possible positions of lithium atoms in the structure of a solid solution of β-spodumene are shown by shaded circles. Samples containing less than 62 wt.% SiO_2 crystallized at 1350°C in the form of a mixture of solid solutions of β-spodumene and solid solutions of β-eucryptite. Thus, a sample containing 60.76 wt.% SiO_2, crystallized for 3 h at 1350°C, consisted of a β-spodumene solid solution containing 62.3 wt.% SiO_2 (a = 7.5576 A, c = 9.1720 A, d_{303} = 1.9438 A) and a β-eucryptite solid solution containing 59.8 wt.% SiO_2 (a = 5.2340 A, d_{212} = 1.6346 A). These and other similar experiments showed that the region of stability of a β-spodumene solid solution lies between compositions corresponding to SiO_2 contents of 62.3 and 82 wt.%.

Saalfeld (1961a) again turned to the problem of the structural interrelations of solid solutions with the general formula $Li_xAl_xSi_{1-x}O_2$. Having confirmed the hexagonal character of the eucryptite solid solutions, Saalfeld made a more detailed x-ray investigation of the β-spodumene solid solutions obtained, in particular, by roasting crystals of natural α-petalite.

X-ray diffraction patterns obtained by Guinier's method showed strong differences between β-eucryptite and β-spodumene. The product from roasting natural petalite had a composition of approximately 1 : 1 : 6 and did not correspond to tetragonal parameters. The displacement from position was found by obtaining crystals (1 : 1 : 6) of a phase with oriented conversion, so that it was possible to obtain texture diagrams. Saalfeld stated that he was able to obtain rhombic crystals. Hexagonal crystals were obtained at 1300°C from the oxides right up to a composition of 1 : 1 : 3.5. Then there are rhombic spodumenes and the orientation of the axes also changes: instead of the a-hexagonal cell there arose a b-rhombic axis (the rearrangement occurred suddenly).

Figure 17 shows x-ray diffraction patterns of petalite and spodumene obtained by Saalfeld at normal and high temperatures. Because of the strong differences in their structures, petalite and spodumene gave very different x-ray diffraction patterns. At 900°C, α-spodumene was rapidly converted into β-spodumene, while no changes were observed in petalite at this temperature. The rate and temperature of the conversion also depended considerably on the fineness of grinding of the samples. In both cases the diffraction patterns obtained at 1150°C showed many lines characteristic of the β-modifications. If we neglect the slight broadening of the interference lines as a result of heating, their position and intensity may be regarded as practically the same. However, it should be noted that reflections obtained with samples of β-petalite which had been quenched were not so clear and sharp as on diffraction patterns obtained directly at high temperature. Stresses arising during the quenching of the samples probably have an effect here.

The shift of the lines due to thermal broadening or contraction is very slight, confirming the small values of the thermal expansion coefficients. High-temperature photographs confirmed the character of the conversion reported previously in the literature, and showed that crystals of the high-temperature β-modifications do not undergo substantial changes during cooling.

Diffraction patterns obtained by Guinier's method are shown in Fig. 18. The similarity of the β-structures of petalite and spodumene is clearly shown. The diffraction patterns differ only in the slightly changed position of the lines as a result of slight differences in the lattice constants of the minerals investigated. The diffraction pattern of β-eucryptite differs sharply from those of the other two.

For a more accurate determination of the symmetry of the elementary cell, Saalfeld (1961b) used the method of oriented conversions, which was used extensively by Taylor (1959) in investigations of the dehydration products of various hydrosilicates. In these cases, the small crystals of the newly formed phase are correctly oriented with respect to the crystallographic axes of the starting (mother) crystal. It was possible to obtain on

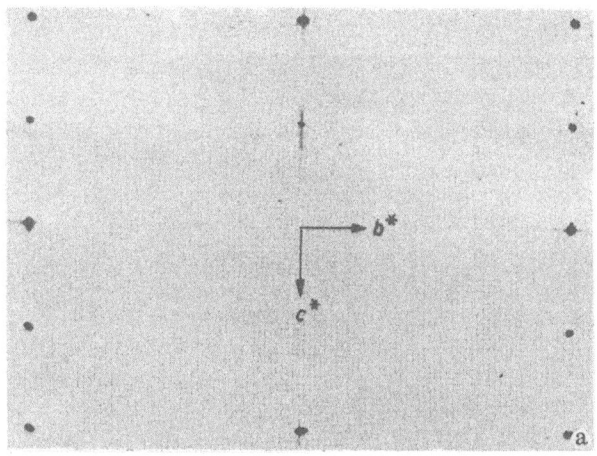

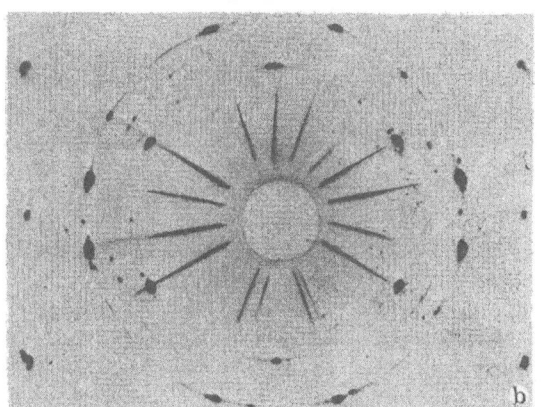

Fig. 19. Precision x-ray diffraction patterns of a petalite crystal. a) Before roasting; b) after roasting, 1200°C (30 min)

x-ray photographs texture diagrams which, though they did not correspond strictly to x-ray diffraction patterns of single-crystal samples, were much more useful for structural analysis than powder patterns. Therefore, a study was made of the products from roasting natural and definitely oriented small crystals of spodumene and petalite. It was found that the right orientation of the small crystals within the primary structure of the starting material was obtained only with petalite, which was investigated in more detail.

The experiments were carried out with plates of clear petalite crystals, cleaved along the (001) plane. These plates were studied by x rays in the original state and after roasting, when they had already been converted into mixed crystals of β-spodumene. Partial vitrification of the starting crystals was observed. On diffraction patterns obtained by Guinier's method, the separation of a certain amount of quartz was also observed (see Fig. 18).

According to Roy, the polymorphic conversion of petalite proceeds congruently, but his work was carried out under hydrothermal conditions. Figure 19 gives precision x-ray diffraction patterns of a petalite crystal. Weissenberg patterns were also obtained for more accurate indexing. The phase obtained after roasting was treated as $Li_2O \cdot Al_2O_3 \cdot 6SiO_2$ (lithium orthoclase). Calculation of the parameters of the elementary cell in the rhombic syngony gave the following results according to Saalfeld (Table 9).

The x-ray diffraction patterns obtained by Guinier's method were indexed on the basis of the values obtained for the constants of the elementary cell and satisfactory agreement was obtained between the observed and calculated values of the parameters of the elementary cell.

The structural ideas obtained from the x-ray data presented confirm the hypothesis put forward by Kolesova (1959), that aluminum is in sixfold coordination in α-spodumene and fourfold coordination in β-spodumene.

BIBLIOGRAPHY

Ballo, R., and E. Dittler. Z. Anorg. Chem. 76: 39 (1912).

Brun, A. Arch. Sci. Phys. 13: 363 (1902).

Endell, K. Z. Anorg. Chem. 74: 33 (1912).

Ginsberg, A. Z. Anorg. Chem. 73: 291 (1912).

Hatch, R. A. Am. Mineralogist 28(9-10): 471 (1943).

Hautefeuille, P. Comptes Rend. 90: 541 (1880).

Hautefeuille, P., and A. Perry. Bull. Soc. Franç. Mineral. 13: 145 (1880).

Heinglein, E. Fortschr. Mineral. 34: 40 (1956).

Hummel, F. A. J. Am. Ceram. Soc. 34: 235 (1951).

Jaeger, F. M., and A. Simek. Verslag. Koninkl. Akad. Wetenschappen 23: 119 (1914).

Keat, P. P. Science 120: 328 (1954).

Kolesova, V. A. Opt. i Spektroskopiya 6(1): 38-44 (1959).

Meissner, F. Z. Anorg. Chem. 110: 187 (1920).

Rindone, G. C., and R. Roy. Z. Krist. 112:409 (1959).

Roy, R., D. H. Roy, and E. F. Osborn. J. Am. Ceram. Soc. 33(5):152 (1950).

Saalfeld, H. Ber. Deutsch. Keram. Gesellschaft 38(7):281 (1961).

Saalfeld, H. Z. Krist. 115(5-6):420 (1961).

Shropshire, I., P. P. Keat, and P. A. Vaughan. Z. Krist. 112:409 (1959).

Skinner, B., and H. I. Ewans. Am. J. Sci., Bradley 258A:312 (1960).

Stein, G. Z. Anorg. Chem. 55:170 (1907).

Tammann, G. Kristallisieren und Schmelzen (1903), p. 114.

Taylor, H. F. W. Mineral Mag. 32:6 (1959); Mag. Concrete Res. 11:151 (1959).

Weyberg, Z. Centralblatt für Mineral. (1905), p. 646.

Winkler, H. G. F. Acta Cryst. 1:27 (1948).

CHAPTER III

STABLE AND METASTABLE PHASE RELATIONS IN THE SYSTEM
MAGNESIUM OXIDE – ALUMINA – SILICA

Investigations of many silicate systems carried out in recent years have again confirmed the great value and validity of the rule of successive conversions or simply the stage rule, first formulated by W. Ostwald. Stable phase states with a minimum energy level may be reached in stages through intermediate metastable states, characterized by higher energy levels.

Of particular interest in this respect are the many investigations of the system $MgO-Al_2O_3-SiO_2$ made by scientists in various countries in connection with the problem of producing ceramic and glass—crystalline articles with exceptionally low or even negative thermal expansion coefficients, the problem of using crystals of cordierite as so-called geological thermometers, order and disorder phenomena in phase transitions in crystalline solids, etc. The formation of metastable phases apparently also occurs frequently in many other silicate, aluminate, and similar systems and, therefore, the results of these experiments deserve a very detailed examination.

Investigations of silicate systems by the classical method of "annealing and quenching" are particularly convenient for such work as here the starting material is for the most part a homogeneous glass, which undergoes crystallization under various temperature conditions. The use of glasses as systems with a high reserve of potential energy which are already metastable creates particularly favorable conditions for investigating metastable structures and the conditions of their interconversions.

The investigation of the system $MgO-Al_2O_3-SiO_2$ in parallel with the system $Li_2O-Al_2O_3-SiO_2$, which has already been examined, is also interesting, in that the ions Mg^{2+} and Li^+ are similar in radius. This makes it possible to seek a more general explanation of the behavior of both cordierite, $2MgO \cdot 2Al_2O_3 \cdot 5SiO_2$, and its derivatives on thermal treatment and catalyzed crystallization. Of particular interest are those compositions which lie along the pseudo-binary system spinel—silica in the same way that the compositions of eucryptite, spodumene, and petalite also lie along the line $Li_2O \cdot Al_2O_3-SiO_2$, which is analogous in the chemical structure of the components. This was examined previously and is characterized by the very complex conditions of the phase conversions, which have not yet been studied fully. One might expect the formation of quartz-like structures in the spinel—silica system, where the spaces of the metastable phases of this system may hold magnesium ions, which are equal in size to lithium ions, in the same arrangement as is found, for example, in β-eucryptite structures.

Synthetic cordierite was first obtained by the Russian mineralogist Morozevich (1897) and was subsequently reproduced by many other investigators. In a physicochemical investigation of the system $MgO-Al_2O_3-SiO_2$, Rankin and Merwin (1918) found, in addition to cordierite, an unstable modification of cordierite (μ-cordierite); however, its composition varied slightly in the range from $MgO:Al_2O_3:2.5SiO_2$ (2:2:5) to $MgO:Al_2O_3:3SiO_2$ (1:1:3). Karkhanavala and Hummel (1953) named this modification μ-cordierite and on the basis of x-ray diffraction data, found that its structure is similar to that of β-spodumene (and subsequently that of keatite). The structure of cordierite, regardless of whether we examine it in an ordered or disordered state, differs strongly from the quartz- and keatite-like structures already described.

In recent years (1957-1961), Schreyer and Schairer (1958) studied anew the crystallization of 46 samples of the cordierite system, points corresponding to which are shown in the diagram in Fig. 20. The phase diagram itself, according to Rankin and Merwin with the additions of Foster (1950) and others, is given in Fig. 21.

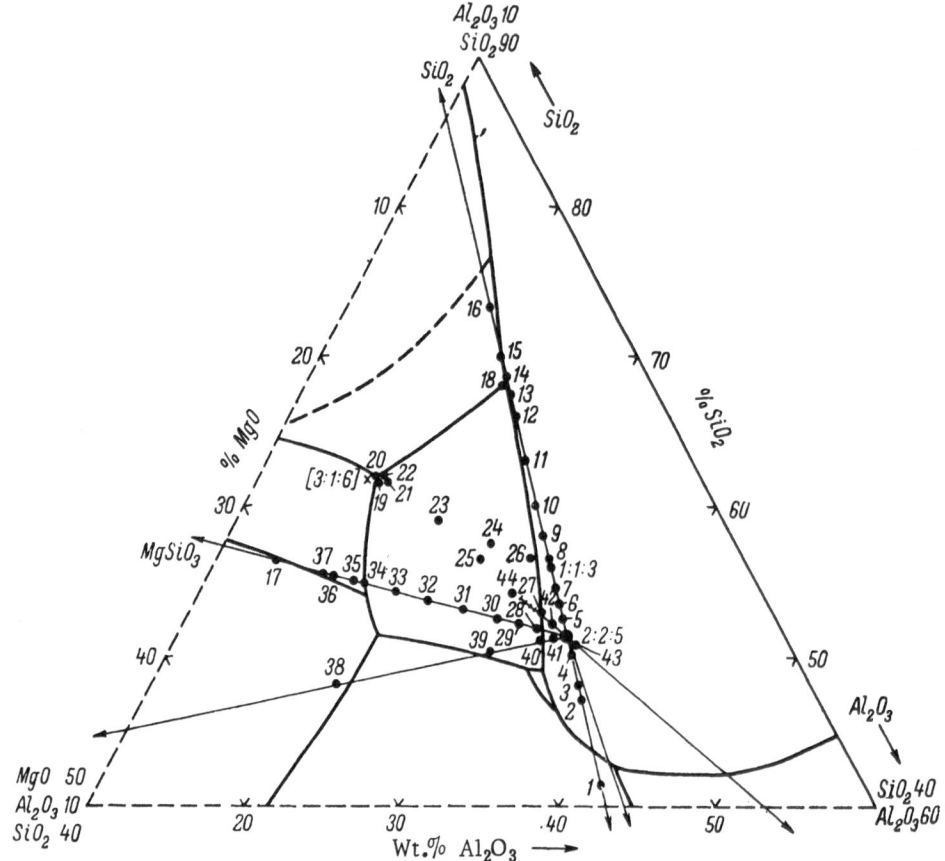

Fig. 20. Enlarged central section of the phase diagram of the system $MgO-Al_2O_3-SiO_2$.
The points indicate compositions studied in the work of Schreyer and Schairer (1958).
The cross marks the composition of Mg beryl $Mg_3Al_2Si_6O_{18}$.

Phases with quartz-like structures were obtained in various amounts as primary and metastable products of low-temperature crystallization of the 46 glasses investigated. In all cases without exception similar phases arose in the first stages of crystallization at the relatively low temperatures of 800-1050°C. For compositions containing approximately 50 wt.% SiO_2, the optimal temperature was 900°C, while no crystallization of these glasses was observed at all below 800°C, even with exposure for two months. Compositions with a high SiO_2 content crystallized more slowly at 900°C, while for those richest in silica the optimal temperature was 1050°C.

At higher annealing temperatures of the order of 1000, 1100, 1200, and 1300°C, there was rapid crystallization of cordierite itself. In this temperature region there arose metastable "high" cordierite, whose chemical composition is constant and crystallographic symmetry hexagonal. With longer heating in this temperature range the metastable "high" cordierite is converted in stages into stable "low" or rhombic cordierite through a series of intermediate states, which were discovered and studied by x-ray diffraction by the Japanese scientist Miyashiro (1956).

Under appropriate thermal treatment, these extreme types of cordierite undergo reversible conversions, but no conversions from cordierite to quartz-like structures could be realized. However, theoretically it may be surmised that some of these phases can still have a region of truly stable existence at high pressures beyond the region of coesite (Fig. 22). Metastable quartz-like phases with inclusions of magnesium ions in the spaces may be obtained not only by crystallization of glasses of appropriate composition, but also by other methods such as rapid heating of the mineral montmorillonite at 1000°C.

On the basis of x-ray diffraction determinations, in 1954 Miyashiro and Iiyama (1954) and, in more detail, A. Miyashiro, Iiyama, Iamasaki, and T. Miyashiro (1955), reported that both synthetic forms of cordierite

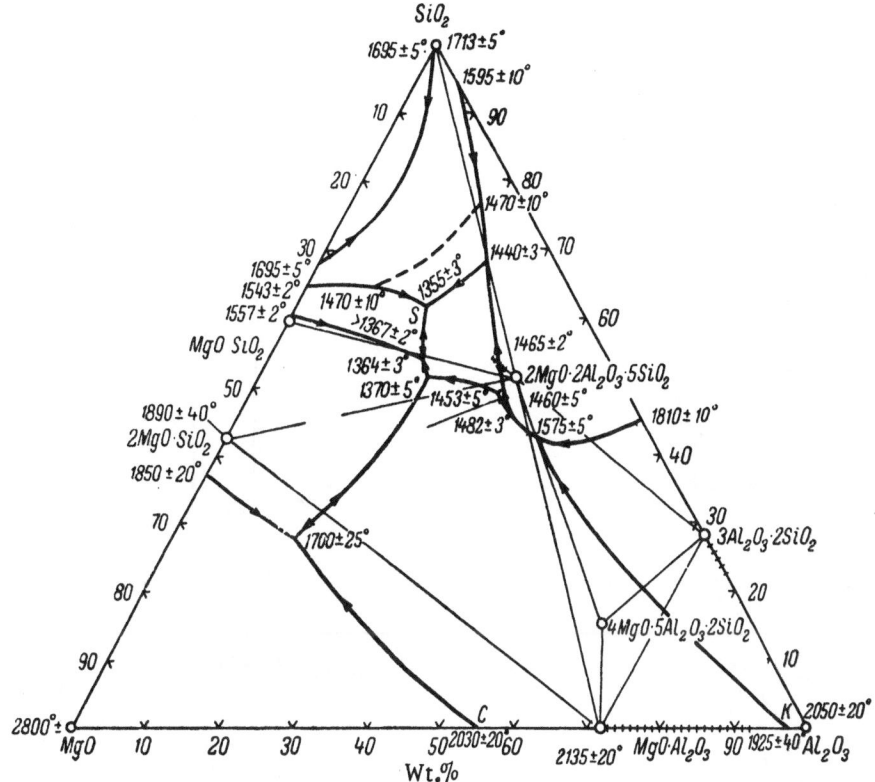

Fig. 21. Phase diagram of the system MgO–Al₂O₃–SiO₂. Without correction for the congruence of the melting of mullite according to the data of Toropov and Galakhov (1958).

show hexagonal symmetry, in contrast to natural forms, which are characterized by rhombic or even, according to Japanese data, monoclinic symmetry.

Only cordierite from fused clay shales of the Vakaro mines in India were found to be structurally similar to α-cordierite and therefore Miyashiro and Komi named all α-cordierites indialites. A. Miyashiro (1957) then developed the cordierite problem in a direction which established the existence of intermediate varieties between indialites of high symmetry and low order and rhombic ordered forms. He confirmed cordierites obtained at low temperatures may be converted into indialites directly around their incongruent melting points, as was established previously by Suguoca and Kuroda (1955).

It should be noted that the terms "high" and "low," introduced by Miyashiro and his co-workers, do not correspond to the true temperature relations of the different forms of cordierite, but in the literature this nomenclature keeps approximately the same meaning as in the description of the different forms of sodium aluminosilicate or albite.

Regions of Separation of Metastable Phases. Quartz-like metastable phases separate from compositions lying along the line cordierite—silica or departing from it by not more than 1 wt.%. In the triangle silica—cordierite—enstatite (see Fig. 21) there separates another metastable phase with the structure of petalite $Li_2O \cdot Al_2O_3 \cdot 8SiO_2$ together with forsterite or enstatite as spontaneously crystallizing phases. Of particular interest is composition 11 (see Fig. 20); though it deviates by 1.5% from the spinel—quartz line, in it there is still the formation of a quartz-like phase in paragenesis with pyroxene.

As was pointed out above, at higher temperatures quartz-like solid solutions disappear and pyroxene is consumed in the formation of a cordierite solid solution (metastable), which then gradually breaks down with the formation of three phases, namely cordierite itself (2:2:5), pyroxene, and cristobalite.

The quartz-like solid solution may contain up to 59 wt.% (MgO + Al₂O₃) until the composition of Mg eucryptite or Mg nepheline is reached.

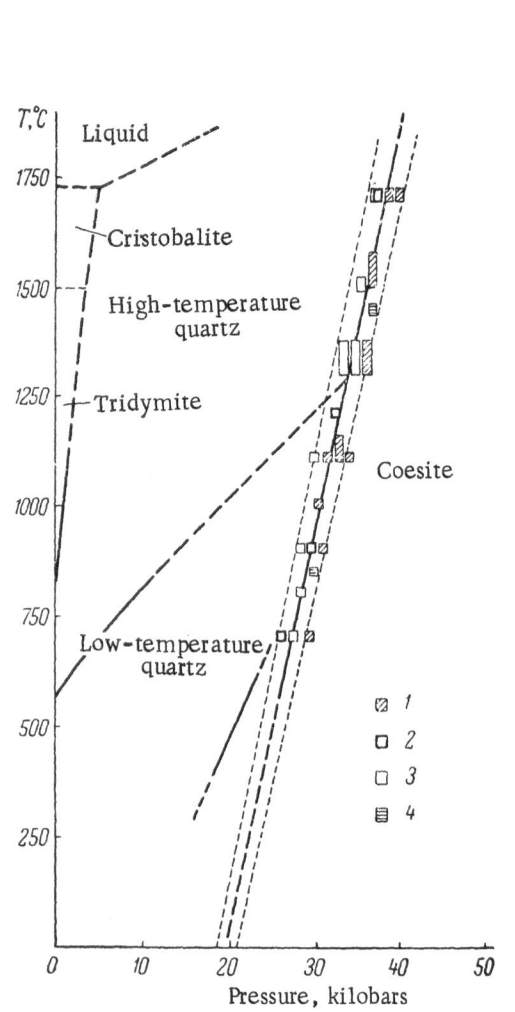

Fig. 22. Region of stability of coesite. 1) Quartz → coesite; 2) coesite → quartz; 3) quartz without change; 4) coesite without change.

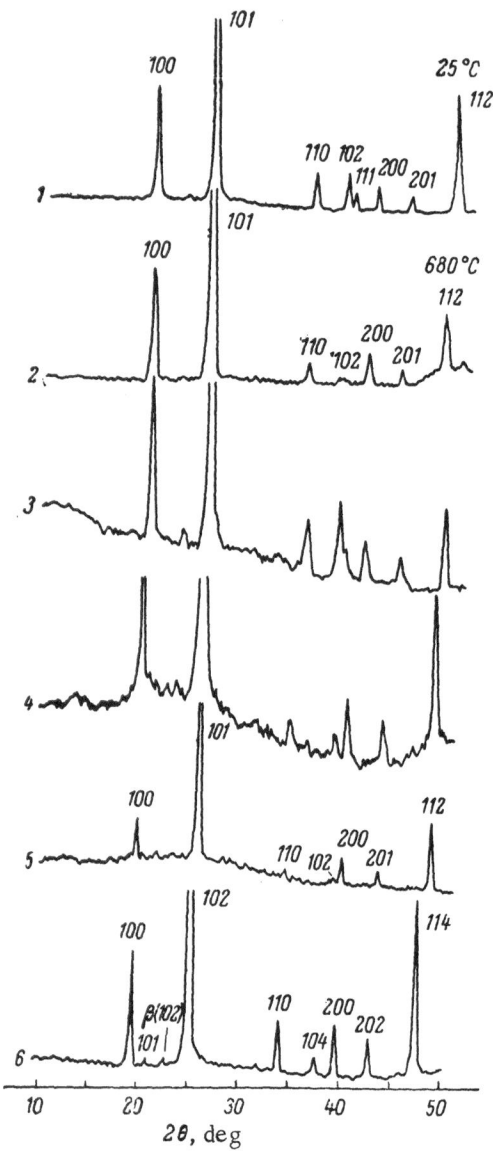

Fig. 23. Ionization curves of quartz-like phases.

X-Ray Diffraction Study of Metastable Quartz-Like Phases. The solid solution obtained at 900°C in 20 h gives on an x-ray diffraction pattern lines which may be indexed well for a hexagonal elementary cell with the parameters a = 5.200 A and c = 5.345. Figure 23 gives ionization diffraction patterns (curves 3-5) of powdered samples lying along the line $SiO_2-MgO \cdot Al_2O_3$ (CuK_α radiation). For comparison we give here diffraction patterns of β-eucryptite (curve 6), high-temperature quartz (curve 2), and low-temperature quartz (curve 1).

Figure 24 shows changes in the lattice parameters of the quartz-like solid solutions in relation to their chemical composition. The anomalous values of the parameters of the elementary cell were obtained with samples in which this phase separated from a glass with 70% SiO_2 at a higher temperature (1250°C).

An examination of Fig. 23, which gives the diffraction patterns for a quartz-like phase from sample 9 (see Fig. 20) at different stages of heating, shows that the behavior of this phase is identical to that of low-temperature quartz. A metastable solution of this type may be fixed even at normal temperature by cooling with retention of the structure of high-temperature and not low-temperature quartz. In these cases, with repeated heatings the transition to the high-temperature form occurs at a lower temperature than for pure quartz (573°C). Thermal analysis gives the position of this transition zone as 400-580°C. For quartz-like solid solutions

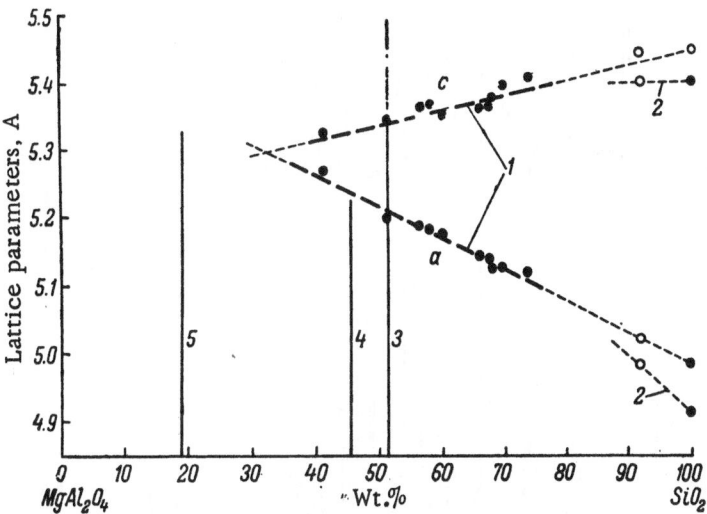

Fig. 24. Change in the lattice parameters of quartz-like solid solutions in relation to their chemical composition. 1) High-temperature quartz; 2) low-temperature quartz; 3) cordierite $Mg_2Al_4Si_5O_{18}$; 4) Mg eucryptite $MgAl_2Si_2O_8$; 5) sapphirine $Mg_2Al_4SiO_{10}$.

the temperature range of the transition zone must be quite clearly marked. An increase in the Mg^{2+} and Al^{3+} content of the high-temperature quartz-like structures increases their refractive indices.

In other words, in this case the relations observed are opposite to those which are found in the β-eucryptite series. This is explained by the different structure of the electron shell of the Mg^{2+} ion as compared with the shell of Li^+. Thus, we arrive at the conclusion that the introduction of the foreign ions Mg^{2+} and Li^+ into the crystal lattice of one of the modifications of silica results in the formation of metastable quartz-like structures at temperatures above the region of stability of quartz, i.e., above 870°C. At the same time, there is a fall in the temperature of the polymorphic conversion α-quartz $\rightleftharpoons$ β-quartz to a much lower temperature region. As the experiments show, at least up to 58.7% SiO$_2$ may be replaced by MgO + Al_2O_3.

With an increase in the content of Mg^{2+} and Al^{3+}, the edge of the elementary cell c decreases, while the edge a and the optical refraction increase. The optical sign of the indicatrices of crystals of the solid solutions change to negative at a composition corresponding to about 73 wt.% SiO$_2$. Samples of the series of solid solutions containing up to 73 wt.% SiO$_2$ may retain the structure of high-temperature quartz on quenching. In high silica compositions, the inversion has a character similar to that of pure quartz. The temperature of this conversion falls as the concentration of the foreign ions introduced increases.

New Metastable Phases of the Osumilite and Petalite Types. In addition to the metastable phases described above with a quartz structure, which are observed in the low-temperature crystallization of a glass of the system MgO · Al_2O_3—SiO$_2$, new metastable phases were found structurally similar to the natural mineral osumilite, which was first described by the Japanese investigator A. Miyashiro (1956) and the lithium aluminosilicate petalite, whose structure was recently investigated satisfactorily by Liebau (1961).

These new phases were observed in the crystallization of the glasses 11 and 19, whose compositions are given in Fig. 20. The formation of an osumilite phase was observed in the crystallization of glass 11 over ten days at 1000°C and over four days at 1250°C. It was also obtained from glass 10 after 6 days at 1050°C and in the form of traces in glass 9 after 20 h at 1000°C.

It is characteristic that glasses whose compositions lie beyond the section 11-9 in Fig. 20 do not form an osumilite phase on crystallization.

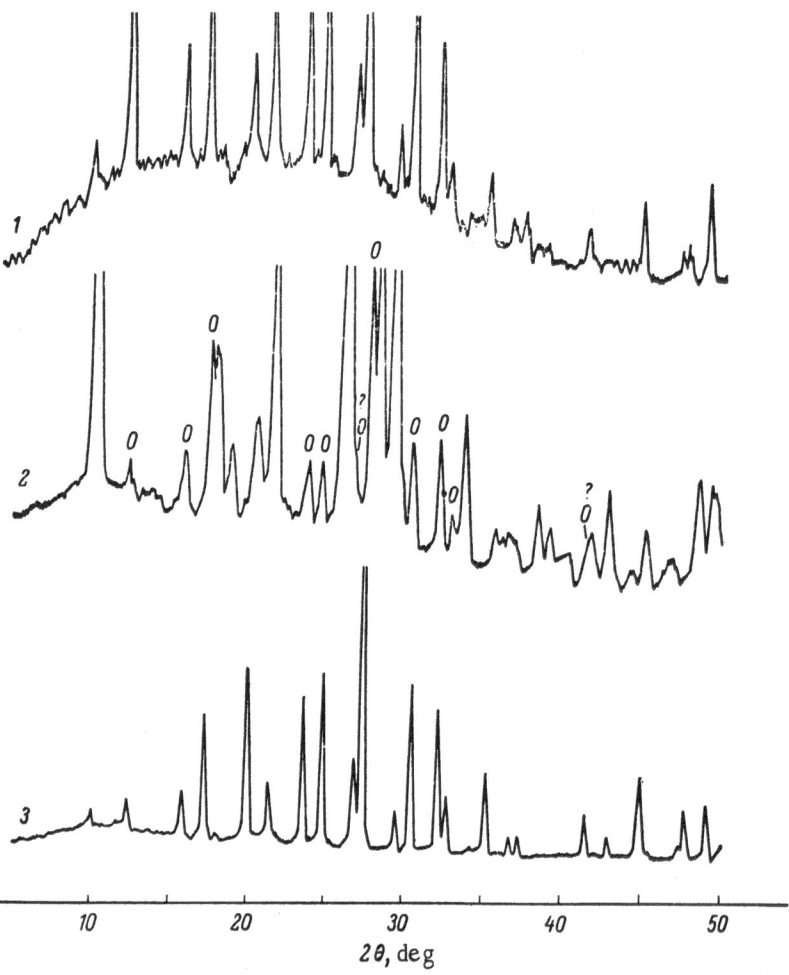

Fig. 25. X-ray diffraction pattern of metastable phase of the osumilite type.
1) Natural osumilite; 2) 9.65% MgO, 24.40% Al_2O_3, 65.95% SiO_2 heated for 10
days at 1000°C and 4 days at 1250°C; 3) synthetic $Na_2O \cdot 5MgO \cdot 12SiO_2$.

Table 10. Interplanar Distances for New Metastable Compounds in the System
$MgO - Al_2O_3 - SiO_2$ (CuK$_\alpha$ radiation)

Osumilite-like phase			Petalite-like phase		
d, Å	2θ, deg	I_{exp}	d, Å	2θ, deg	I_{exp}
7.11	12.45	15	7.20	12.30	8
5.54	16.00	25	4.65	19.10	1
5.01	17.70	55	4.08	21.80	7
3.72	23.91	30	3.79	23.50	20
3.59	24.81	30	3.69	24.10	100
3.20	27.87	100	3.58	24.85	25
2.92	30.62	40	2.99	29.90	8
2.76	32.38	40	2.78	32.20	6
2.71	33.10	15	2.55	35.20	11
			2.47	36.40	8
			2.39	37.70	8
			2.06	44.00	5
			1.95	46.50	6
			1.93	47.00	8
			1.90	47.75	5

Note. The data in Table 10 are incomplete as the samples were heterogeneous.

This phase is always associated with metastable quartz-like solid solutions, usually those varieties of them which are most saturated with silica and also with crystals of cordierite and very small amounts of high-temperature cristobalite. The optimal crystallization range for this phase lies between 1050 and 1250°C. At lower temperatures it is not formed and at higher temperatures it decomposes into cordierite and cristobalite, which are more stable under the given conditions. It is probable that prolonged annealing at the temperatures of its formation also results in the formation of the more stable phase associations. As all three phases which coexist under the conditions described, and also the starting glass belong to the particular system $MgO \cdot Al_2O_3$—SiO_2, the composition of the osumilite-like phase must also lie within this system.

X-ray investigation showed that together with the lines of the other phases accompanying it, this phase gives at least nine sharp characteristic interference maxima (Fig. 25).

It was established that the lattice of osumilite is very similar to that of the recently discovered new synthetic silicates $K_2O \cdot 5MgO \cdot 12SiO_2$ and $Na_2O \cdot 5MgO \cdot 12SiO_2$ (Fig. 25). By using the crystallographic device proposed by Miyashiro for calculating indices, we obtain the dimensions of the elementary cell a = 10.12 A and c = 14.36 A. According to Miyashiro, osumilite is isostructural with milarite $KCa_2(Be, Al, Si)_3(Si, Be)_{12}O_{30}$ $\cdot \frac{1}{2}H_2O$; later, Tennyson (1960) showed that the mineral armenite $BaCa_2Al_3(Al_3Si_9O_{30}) \cdot 2H_2O$ from Norway is also of the same structural type. Thus, at the present time five minerals have already been found which differ strongly among themselves in empirical composition, but are isostructural with osumilite.

A comparison of the atomic structures of such different substances leads to the conclusion that the most probable representative of compounds of the osumilite structure in the system $MgO \cdot Al_2O_3$—SiO_2 must be a silicate with the composition $MgAl_2Si_4O_{12}$ or $MgO \cdot Al_2O_3 \cdot 4SiO_2$. This also corresponds to the above experimental data, which indicate that the composition of the crystals of the phase examined lies between the compositions of cordierite and cristobalite. The composition of osumilite itself, according to Miyashiro, may be represented by the formula

$$(K, Na, Ca)(Mg, Fe^{2+})_2(Al^-, Fe^{3+})(Si, Al)_{12}O_{30}.$$

Osumilite was first observed by the Japanese mineralogist A. Miyashiro in an investigation of volcanic rocks of the deposits at Sakkabira and Kiu Siu (Japan). It is similar in physical properties and conditions of occurrence to cordierite and therefore it had often been described previously as normal cordierite. A detailed structural investigation of natural crystals of osumilite (F. R. Boyd and J. L. England, 1960) showed that they belong to the dihexagonal-bipyramidal class of symmetry: the space group 6m,2m,2m = D_{6h}. The parameters of the elementary cell are a = 10.17 A, c = 14.34 A, a : c = 1 : 1.410. The elementary cell contains two formula units.

Osumilite is isostructural with milarite, whose structure was described by Belov and Tarkhova (1949) and also Ito, Morimoto, and Sadanaga (1952). The structure of osumilite (see above) consists of double hexagonal rings of $(Si, Al)_{12}O_{30}$, bound by the cations Al^{3+}, Fe^{3+}, and Se^{2+} in fourfold and Mg^{2+} and Fe^{2+} in sixfold coordination, and also K^+, Na, and Ca^{2+} in twelvefold coordination. Another metastable phase structurally similar to petalite was obtained from glass 19 (Fig. 20). The optimal temperature of formation of the petalite-like phase is the range from 900 to 1000°C. Like osumilite, this phase is never found as a single devitrification product, but is normally associated with crystals of a different composition. On x-ray diffraction patterns this phase is characterized by a triplet around values of the angle 2θ equal to 24° for CuK_α radiation (Table 10). X-ray diffraction patterns of this type are known for the natural mineral petalite and the synthetic lithium disilicate $Li_2Si_2O_5$.

The very close structural similarity of petalite and lithium disilicate was recently demonstrated by Liebau. In our opinion this is very important as this similarity of structures, on the one hand, makes the formation of solid solutions here and in the lithium aluminosilicate system probable and, on the other hand, indicates the possibility of the existence of metastable magnesium disilicate, which was not found in investigations of equilibria in the system MgO—SiO_2 by Andersen and Bowen (1914) and later by Nikitin (1948).

The discovery of the double silicate $BaMgSi_4O_{10}$ in our work (Toropov and Grebenshchikov, 1962) also makes the existence of the above laminar metasilicate $MgSi_2O_5$ very probable as a component of other series of solid solutions and also an independent, but metastable phase. It seems to us that further investigations in this

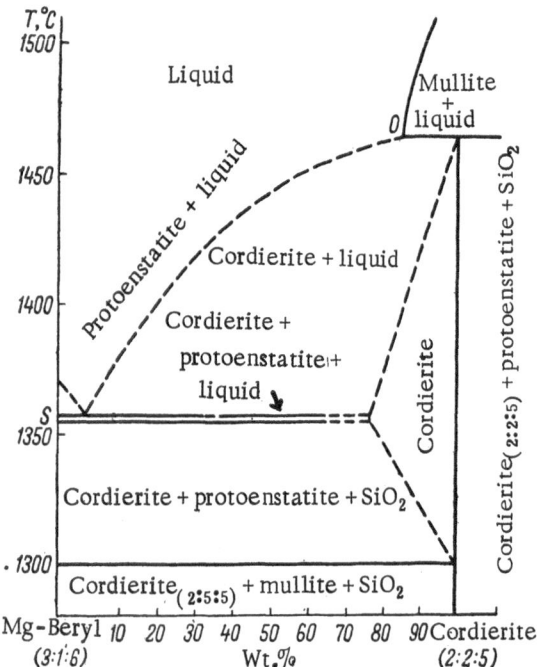

Fig. 26. Enlarged part of the pseudo-binary section cordierite—Mg-beryl. Solid solutions are realized as a result of the heterovalent replacement $Mg^{2+} + Si^{4+} = 2Al^{3+}$.

direction are very timely for the development of the general problem of the structure of sitalls. There is also the possibility of the discovery of the corresponding isomorphous solid solutions, but of a more complex crystallochemical nature, between magnesium and lithium silicates, though the investigations of Hummel indicated the great complexity of the interactions of the phases occurring here.

The magnesium analog of petalite $LiAlSi_4O_{10}$ would be $MgAl_2Si_8O_{20}$ or $MgO \cdot Al_2O_3 \cdot 8SiO_2$. However, no crystallization of a petalite-like phase has been observed in the low-temperature annealing of the corresponding glasses. The composition of the petalite-like phase probably lies in the triangle between the norms of $MgSi_2O_5$ $1:1:8$-$1:1:3$ of this system.

Metastable Solid Solutions of Cordierite. Cordierites separating from glass at low temperatures are hexagonal phases. In actual fact they may be metastable solid solutions based on cordierite. Longer annealing of this system in the low-temperature region results in the breakdown of these solid solutions and the separation of cordierite with the stoichiometric composition $2MgO \cdot 2Al_2O_3 \cdot 5SiO_2$. The facts examined were established by a study of the character of the crystalline phases accompanying cordierite in these samples and also through the appreciable shift in the positions of the diffraction maxima on x-ray patterns of cordierites obtained in the low-temperature region.

The breakdown of the solid solution, which occurs at about 1300°C, is irreversible even on annealing for a year. At lower temperatures there are formed primary solid solutions of cordierite, which may be considered as metastable. In a very small number of cases it was established that at medium temperatures (1250°C) there is the above breakdown of a metastable cordierite solid solution. At very low temperatures (1050°C), these solid solutions remain unchanged for more than a year.

The existence of solid solutions of cordierite at low temperatures was also demonstrated by x-ray diffraction. Thus, on thermal treatment, sample 23 (see Fig. 20) showed a shift of the characteristic peak by 0.2° toward higher angles. At 1050°C, the composition 26 in the same figure crystallized completely to a cordierite solid solution, which showed an appreciable shift of the diffraction maximum. This solution remained stable at 1250°C for three days.

It was pointed out that there is the possibility of the formation of solid solutions under the conditions described along the cordierite—mullite line (though of extremely low concentration) as a result of heterovalent replacements of the type $Mg^{2+} + Si^{4+} \rightarrow 2Al^{3+}$. Moreover, in the particular phase triangle $2MgO \cdot 2Al_2O_3 \cdot 5SiO_2$—$MgO \cdot SiO_2$—$SiO_2$, other solid solutions were detected which were formed by metastable low-temperature cordierite. The formation of solid solutions was also observed at points 1, 2, and 3, where, under the same crystallization conditions there were formed associations of crystals of solid solutions of cristobalite and forsterite; at higher temperatures forsterite disappears and spinel is formed.

As a result of experimental work in the system $MgO-Al_2O_3-SiO_2$ (Fig. 26), Iiyama (1955) discovered solid solutions from cordierite to Mg-beryl (composition $Mg_3Al_2Si_6O_{18}$).

With samples 23, 26, and 32 (see Fig. 20), there was a shift of the interference maxima toward higher values of the angles 2θ, in contrast to the shift toward smaller values of 2θ along the line cordierite—spinel.

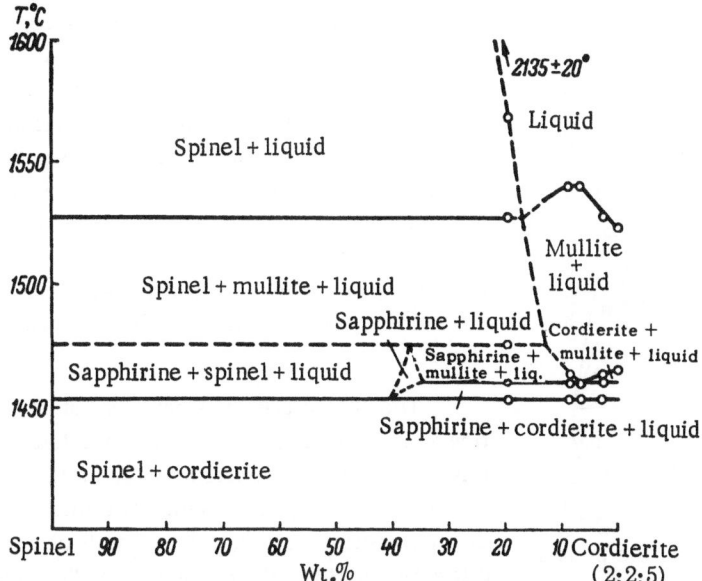

Fig. 27. Pseudo-binary section cordierite−spinel. No formation of solid solutions was observed.

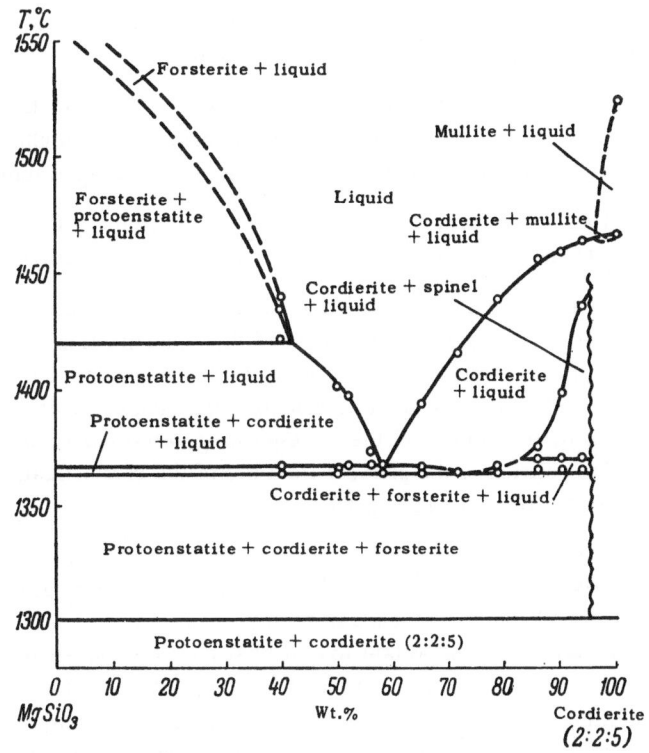

Fig. 28. Pseudo-binary section cordierite−enstatite.

The structural differences between beryl and cordierite mainly reduce to the fact that in beryl the structural element is the group $[Si_6O_{18}]$ while in cordierite the tetrahedral groups $[AlO_4]$ are partly replaced by $[SiO_4]$ in the cyclic radicals and the structural unit is the group $[(Si_3Al)_6O_{18}]$.

It should be emphasized that solid solutions between cordierite and Mg-beryl obtained in the intermediate temperature zone must be regarded as nonequilibrium to a considerable extent. Iiyama worked with very short exposures, namely at 1350°C + 20°, 10 h, and at 1250°C + 20°, 60 h.

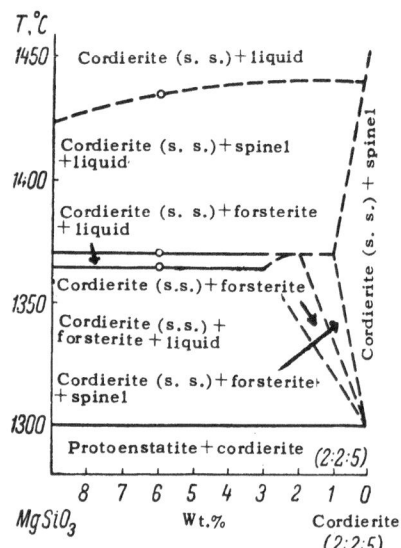

Fig. 29. Enlarged part of the pseudo-
binary section enstatite—cordierite.
(s.s. = solid solution.)

A second very important circumstance is the constancy of the stoichiometric composition $2MgO \cdot 2Al_2O_3 \cdot 5SiO_2$ of the cordierite which is formed by prolonged annealing of all the above series. Prolonged heating of the crystalline material at higher temperatures approximately 50°C below the temperatures of the solidus curves leads to breakdown of the solid solutions described above and a change in the composition of the solid phases. Phase equilibria were attained in experiments lasting for from one week to one month, inclusive. It is also assumed that the breakdown of the solid solutions must occur at lower temperatures, but this proceeds at correspondingly lower rates.

The characteristic associations of phases are: a) the composition 2:2:5 (see Fig. 20) consists completely of homogeneous crystals of cordierite, composition 4 contains a very small amount of spinel crystals; 43, fine needles of mullite; 28, pyroxene; and 7 and the composition 1:1:3, point crystals of cristobalite. On the x-ray diffraction patterns there is a peak at 2θ equal to 28°20', which is characteristic of "high" cordierite. The term "high" is applied to cordierite at the present time in the sense used by Miyashiro, namely, that this is highly symmetrical, randomized hexagonal cordierite. "Low" cordierite is low symmetry, but highly ordered rhombic cordierite. Between the two varieties there exists a whole series of intermediate forms, whose x-ray diffraction characteristics will be examined in more detail below.

Experiments on the annealing of cordierite in the range of 800-1300°C show that here there are no stable solid solutions of cordierite and the formation of these solutions is found only at higher temperatures close to the solidus surface. Only strictly stoichiometric cordierite is stable at low temperatures.

Investigation of Liquidus Curves of System. The study of the crystallization curves shows the presence of definite variations in the composition of cordierite and the formation here of solid solutions of it.

Figures 27-29 give a series of particular pseudo-binary sections of the system spinel—cordierite (Fig. 27), enstatite—cordierite (Fig. 28), and also an enlarged fragment of the concentration region adjacent to the point of cordierite itself (Fig. 29). Figure 26 shows an extremely interesting fragment of the pseudo-binary system Mg-beryl—cordierite. Figures 26-29 make it possible to visualize the conditions of equilibrium crystallization of cordierite close to the liquidus temperatures and the corresponding variations in the composition of the crystalline phases of the system. The position of the primary tie-lines of the crystallization of compositions lying in the field of stability of cordierite are shown in Fig. 30.

Structural Conversions of Cordierite. The interconversions of different forms of cordierites have already been discussed partly above (mainly on the basis of data of Japanese authors) and in this section we make a more detailed examination of the consequence of their great complexity and great technical value in ceramics and the production of sitalls. It is important to note that the interconversions of different forms of cordierite depend not only on temperature, but also on the total composition of the samples studied. The ratio $Al_2O_3 : SiO_2$ in the sample examined is particularly important in this respect. Thus, with a low $Al_2O_3 : SiO_2$ ratio, "high" cordierite is not at all stable in the given system. In compositions with an intermediate $Al_2O_3 : SiO_2$ ratio, "high" cordierite is stable only in the presence of a large amount of liquid phase.

The temperatures of interconversions of individual forms of cordierite are also functions of the total composition of the samples. Crystal optical analysis cannot be used to determine the different polymorphic varieties of cordierite. Only the constant of the angle of the optical axes $2V_\alpha$, which equals 0° for high cordierite and about 90° for low, is sometimes useful here for diagnostic purposes. The best results are obtained from x-ray diffraction data. The variety of cordierite, called "high" cordierite by American authors, was separated out as an independent mineral, namely indialite, which has hexagonal symmetry of the lattice. It was first characterized reliably by Miyashiro et at. by means of the so-called "distortion index," which is denoted by the Greek letter Δ.

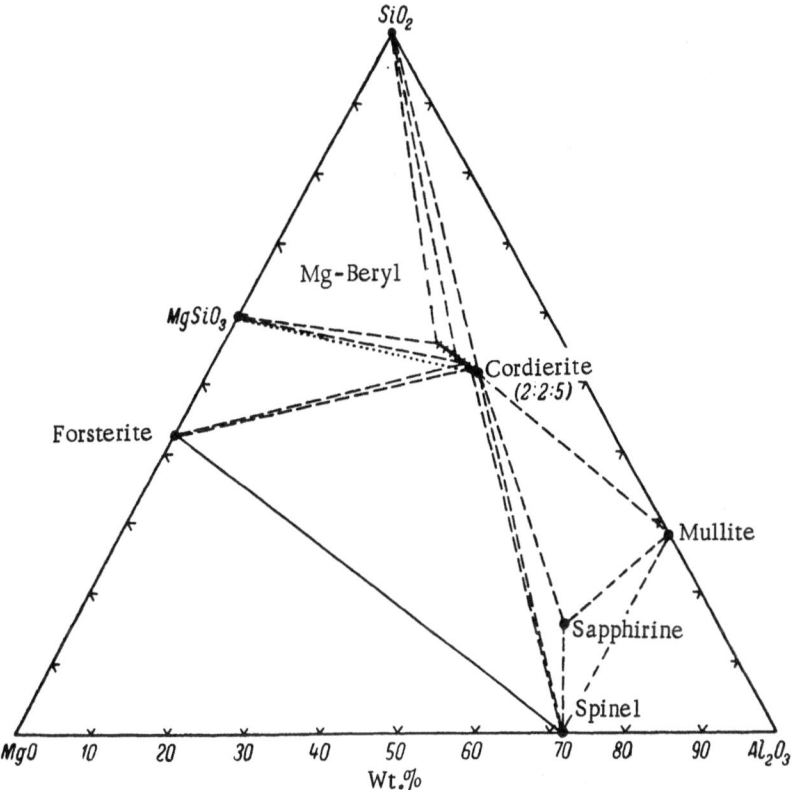

Fig. 30. Tie lines determining the equilibrium crystallization in the
cordierite field.

Cordierite is a widespread mineral in rocks of various origins. Cordierites are found in pegmatites, granulites, and metamorphic schists as well as in contact metamorphic formations, namely paralavas and hornblende slates of various types.

The almost complete optical uniaxial character of some natural cordierites, and also the corresponding synthetic cordierites, which are observed in metallurgical slags or refractories, indicates that the structural relations in cordierite are similar to those for the feldspar group of rock-forming minerals. Only some optically positive natural cordierites are interpreted by the phase diagram of the system MgO$-$Al$_2$O$_3$$-SiO_2$, and then only with great difficulty.

As was pointed out above, Miyashiro and Iiyama made a careful x-ray investigation of various natural cordierites and demonstrated the fundamental differences between the typical hexagonal high-temperature phase, which they named indialite as a new mineral form, and natural true rhombic cordierite of pseudo-hexagonal habit with characteristic complex penetration twinning. A simple x-ray diffraction pattern of indialite (Fig. 31) indicates that it is a randomized phase of cordierite composition, while true (rhombic) cordierite is a more ordered, low-temperature phase.

Indialite phases have structures similar to beryl, but the distribution of the [SiO$_4$] and [AlO$_4$] tetrahedra in the structure is not ordered. The phase relations in the systems examined are complicated by the presence of intermediate order-disorder phases of both indialite and cordierite types.

The disordered synthetic indialite, obtained from melts at the highest temperatures, is also characterized as a component of the solid solutions Mg$_2$Al$_4$Si$_5$O$_{18}$$-Mg_3Al_2Si_6O_{18}$, i.e., Mg-cordierite$-$Mg-beryl with heterovalent isomorphism of the ions Mg^{2+}$+$Si^{4+} $=$ 2Al^{3+} (see p. 32). Many high-temperature cordierites are also observed in effusive rocks of the andesite and basalt type, while low-temperature cordierites are found in deep pegmatites and quartz reefs.

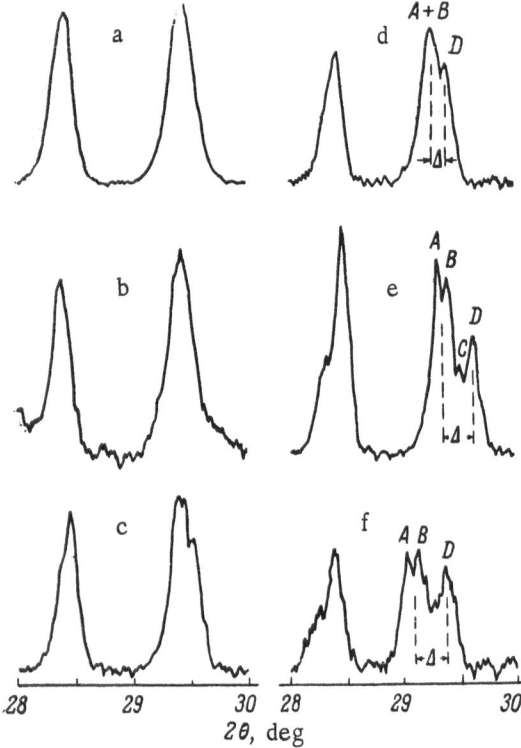

Fig. 31. X-ray diffraction patterns of indialites and cordierites. a) Synthetic Mg-indialite, $\Delta = 0$; b) indialite (Bokaro), $\Delta = 0$; c) cordierite (Khattodama) $\Delta = 0.12$; d) cordierite (Sugama), $\Delta = 0.125$; e) cordierite (Laromi), $\Delta = 0.27$; f) cordierite from blast furnaces, $\Delta = 0.29$. CuK_α radiation.

Some very interesting formations intermediate between indialites and cordierites are also found. These intermediate phases are characterized by a strongly varying angle of the optical axes $2V_\alpha$ from 0 to 80°, and are identified readily by x rays as intermediate between greatest disorder and highest order. As was pointed out above, for determining the degree of order of the structure of these phases, Miyashiro proposed the use of a particular "distortion index."

For recognizing the different varieties and calculating the "distortion index," the Japanese authors proposed the use of the part of the x-ray spectrum lying between $2\theta = 29$ and 30° for copper radiation. In this region (Fig. 32), cordierite shows at least three (and sometimes four or five) well-resolved maxima, while on the diffraction patterns of indialites or "high" cordierites these maxima merge into one because of the disorder of the lattice.

The three most intense peaks of cordierite in the range of 29-30° are denoted by A, B, and D, the letters running from smaller to greater angles. A fourth peak may arise between peaks B and D, and is denoted by C. The peaks A, B, and D probably correspond to reflections from the (511), (421), and (131) planes with CuK_α radiation. Peak C is probably the reflection from (421) for CuK_β radiation. The corresponding single peak on the x-ray diffraction pattern of indialite is the reflection from the (1231) plane of the lattice.

The distances between A, B, and C increase with an increase in the degree of order in the structure of the cordierite. With indialite all these distances equal zero. Therefore, they or their ratios may be used to characterize the degree of distortion of the structure. The most convenient index is Δ.

$$\Delta = 2\theta_D - \frac{2\theta_A + 2\theta_B}{2},$$

where $2\theta_A$, $2\theta_B$, and $2\theta_D$ are the Bragg angles in degrees corresponding to the maxima A, B, and D for CuK_α radiation. For indialite, $\Delta = 0$ and it increases with a fall in the symmetry (increase in the ordering) of the hexagonal lattice. The value of the index Δ in the cases studied of cordierites of various origins varies from 0 to 0.29-0.31°.

The relation of the lattice and the optical indicatrix, expressed in terms of Δ and 2V, is shown in Fig.34. The effect of the conditions of thermal treatment is shown in Fig. 35, where it is clear that there is a tendency for ordering of the lattice with a fall in the temperature of crystallization of cordierites.

Calculation of the parameters of the elementary cell for hexagonal and rhombic cordierites made by Schreyer and Schairer (1962) gave the following results. For hexagonal cordierite, $a = 9.7698$ and $c = 9.3517$ A, which agree well with the values $a = 9.782$ and $c = 9.365$ A obtained by Miyashiro. The space group was determined as D_{6h}^2, the calculated density d = 2.513 g/cm^3, and the refractive index N = 1.524. For rhombic cordierite with a distortion index $\Delta = 0.25°$, the same authors obtained the following parameters: $a = 17.0621$, $b = 9.7208$, and $c = 9.3389$ A; space group D_{2h}^{20}, density (calculated), d = 2.507 g/cm^3, and refractive index N = 1.523.

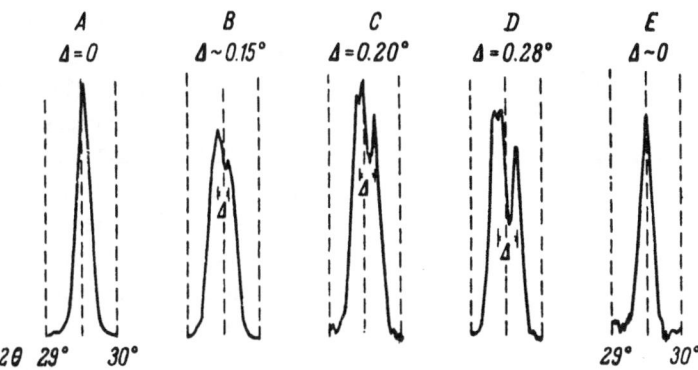

Fig. 32. Form of reflections in the region of 2θ = 29-30° (CuK$_\alpha$ radiation) for cordierites with various values of the index △. All cordierites were obtained from glass with the composition 2:2:5 under the following conditions: A) 1000°C, 301 days; B) 1000°C, 12 days; then 1200°C,11 days;1320°C,17 days;and, finally, 1453°C, 21 h, before the last heating △ = 0.23°; C) 1420°C, 24 h; D) 1300°C, 33 h;1420°C,102 h; and, finally, 1000°C,304 days; E) 1050°C, 4 days; 1200°C, 4 days; 1400°C, 13 days;and finally, 1460°C, 20 h, before the last heating △ = 0.25°.

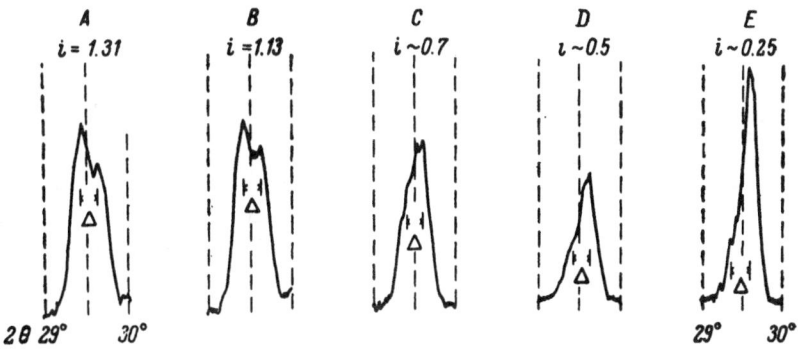

Fig. 33. Form of reflections in the region of 2θ = 29-30° with CuK$_\alpha$ radiation for cordierites with various values of the index i. All cordierites were obtained from glasses with composition 32 (annealing at 1050°C for 8 days and then at 1250° for 1 day). Special (second) thermal treatment: A) 1330°C, 11 days; B) 1330°C, 11 days and 1360°C, 6 days; C) 1370°C, 5 days; D) 1330°C, 11 days and 1385°C, 48 days; E) 1330°C, 11 days and 1400°C, 27 days.

A comparison of the parameters of the cells shows that the rhombic lattice of cordierite is extended in a direction normal to the crystallographic axis c and compressed parallel to the same axis.

The clearest differences between these extreme types of cordierite structure are as follows: with rhombic cordierite close to the angle 2θ = 18° (for CuK$_\alpha$ radiation) there are two or three interference maxima instead of one for the hexagonal, at 28.5-29.5° there are three peaks instead of one (this region is regarded as the most characteristic), at 34, 37, and 48.5° there are also three peaks instead of one. In addition to the appearance of the new peaks, as a result of structural changes there are also some shifts of their positions. The greatest shifts occur at high values of 2θ, such as 69 and 70°.

A more detailed analysis showed that the value of the index △ = 0.20° is the limiting value where there are clear reflections characterizing the presence of the rhombic symmetry of cordierite crystals. For a number of compositions of the system MgO−Al$_2$O$_3$−SiO$_2$ it has been stated that the structure of cordierite is charac-

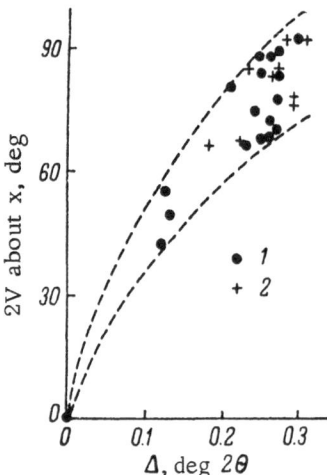

Fig. 34. Relation between the distortion index and the optical angles of cordierite and indialite. 1) Unheated sample; 2) heated sample.

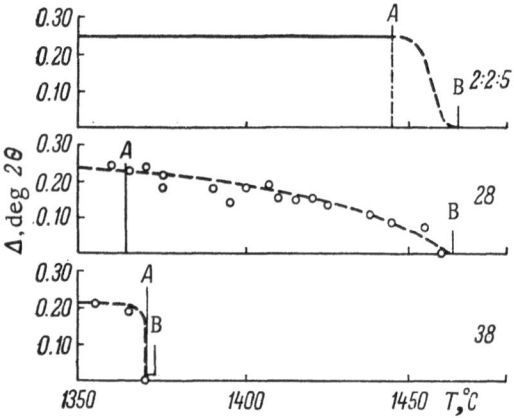

Fig. 35. Relation of the distortion index Δ of cordierites of various compositions to temperature. A) Beginning of fusion; B) disappearance of cordierite. The numbers of the samples correspond to the notation in Fig. 20.

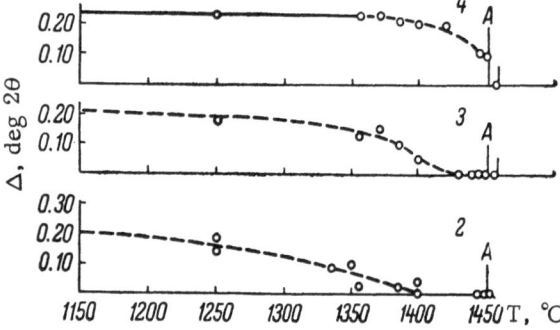

Fig. 36. Relation of distortion index Δ of cordierites to temperature and overall composition of samples 2-4. A) Beginning of fusion.

terized by a fall in the intensity of the peaks of (511) and (421) without an appreciable change in the value of Δ. The typical position of the maxima for these cases is shown in Fig. 32. Thus, these "anomalous" cordierites are up to now rhombic with a constant value of the index Δ, while the intensities of the peaks of (511) and (421) and other corresponding reflections do not equal zero. Such a change to hexagonal cordierite is sudden rather than gradual.

It was also established that the position of the maxima of hexagonal cordierites obtained by such an anomalous conversion is shifted toward higher angles in comparison with normal cordierite (2 : 2 : 5) by approximately 0.10-0.15° in 2θ. This phenomenon is characteristic of a series of solid solutions formed between Mg-cordierite and Mg-beryl. Therefore, for characterizing the polymorphic conversions of rhombic forms of these solid solutions into hexagonal forms, Schreyer and Schairer introduced a new index, which they called the intensity index (Fig. 33):

$$i = \frac{I_{(511)} + I_{(421)}}{2 I_{131}}.$$

As the maxima for (511) and (421) are not always resolved on x-ray diffraction patterns, it is more convenient to use the mean intensity for (511) and (421) and then the intensity index is given in the form

$$i = \frac{I_{(511+421)}}{I_{(131)}}.$$

It will be shown later that the type of conversion of the cordierite tested depends on its chemical composition. Here we indicate only that for hexagonal solid solutions i = 0, while for rhombic solid solutions it varies from 1.15 to 1.35. The change in the position of the interference maxima on x-ray patterns produced by thermal expansion of the crystal lattice of cordierite for maxima lying between values of 2θ from 5 to 50° was found to be insignificant (less than about 0.10° for 2θ) over the whole of the temperature range studied from 25° to 1025°C. This confirms that the thermal expansion of cordierite materials is very very slight.

Hummel and Reid (1951) found a thermal expansion coefficient for cordierite of the order of 20·$\cdot$ 10^{-7} deg^{-1} over the range from 25° to 1000°C. Suguoca and Kuroda also reported that cordierites undergo a slight decrease in volume over the range

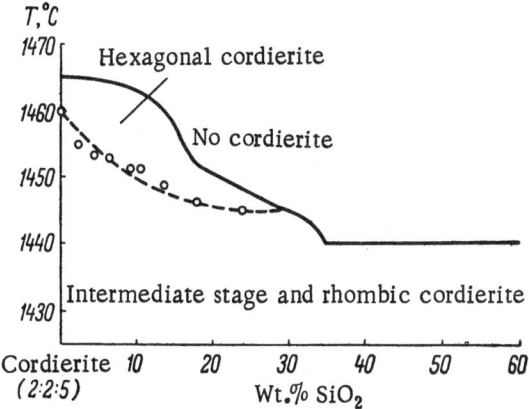

Fig. 37. Relation of the structural state of cordierites with various values of the index Δ to temperature and the total composition of samples along the line cordierite—silica.

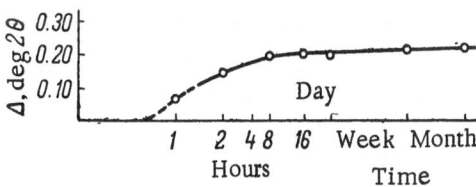

Fig. 38. Increase in index Δ of anhydrous cordierites on annealing at 1400°C. Starting material: glass with the composition $2MgO \cdot 2Al_2O_3 \cdot 5SiO_2$.

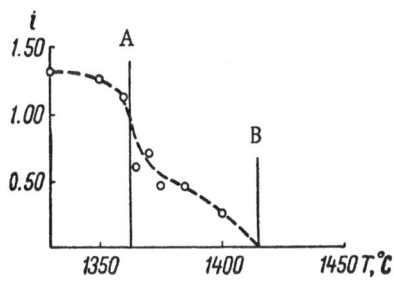

Fig. 39. Temperature dependence of the intensity index i for sample 32. A) Beginning of fusion; B) disappearance of cordierite.

from 20° to 500°C and therefore cordierite is an excellent material for ceramic articles. The density of vitreous cordierite equals 2.62 ± 0.02 and that of purely crystalline material 2.54 g/cm³.

The problem of the optically positive cordierites was solved by the investigations of A. Miyashiro, who described the corresponding crystalline phase as a new mineral called osumilite. As has already been pointed out above, it differs from true cordierite, though it belongs to the same family.

Most natural and synthetic cordierites are intermediate between varieties with Δ = 0 and Δ = 0.29-0.31°, including varieties found in fused argillaceous sedimentary rocks, pegmatites, and quartz reefs.

Low-Temperature Conversions of Metastable Hexagonal Cordierite. Quartz-like, and also osumilite-like and petalite-like unstable crystalline phases, which are formed metastably in the first stages of crystallization, gradually disappear during further thermal treatment and there is then an increase in the content of cordierite and other crystalline phases accompanying it.

In all cases this particular cordierite was determined as hexagonal or "high" cordierite. It is also regarded as metastable since, on further thermal treatment, it is converted into rhombic cordierite (through a series of intermediate states).

In all glasses of the system MgO—Al₂O₃—SiO₂, with the exception of those lying along the line cordierite—spinel, it is precisely the rhombic form which was determined as the stable one at subsolidus temperatures. Metastable hexagonal cordierite forms a series of metastable solid solutions in a series of sections of the system even on prolonged heating. Thereupon there is a change to more stable rhombic cordierite (similarly to the conversion of cristobalite, separated in a metastable form, into more stable tridymite with appropriate thermal treatment).

An example of the use of the index Δ for characterizing phase conversions of this type in the crystallization of a glass with the cordierite composition is presented in Fig. 38. Annealing for an hour yielded a material very similar in structure to true hexagonal cordierite. The maximum on the x-ray pattern at the angle 2θ = 29.5° is somewhat broader than for "high" cordierite with Δ = 0. The index lay between 0.05 < Δ > 0.10. With further heating of the same sample, after less than one day there was obtained cordierite with $\Delta \simeq 0.20°$, which corresponds to the arbitrary boundary between regions of high- and low-symmetry cordierites. At lower temperatures the rate of recrystallization is considerably lower. As will be shown later, the chemical composition of the starting material also affects the rate of the conversion examined; the more it differs from pure $2MgO \cdot 2Al_2O_3 \cdot 5SiO_2$, the slower is the conversion to low-symmetry rhombic cordierite.

Limits of Stability of Hexagonal Cordierites. At temperatures above 1400°C in the presence of the liquid phase there is the reverse conversion of rhombic into hexagonal cordierites. As this conversion is reversible, the hexagonal form may be regarded as the stable modification under these conditions. However, it should be emphasized that these conversions do not proceed at some definite temperature.

Whether or not the conversion will occur at some definite temperature depends on the overall composition of the sample examined. Moreover, the presence of a large amount of liquid phase is a factor stimulating the reversible conversions of cordierite. Thus, for example, the effect of this factor may be seen in Fig. 39, which shows the course of the conversion interesting us in samples with different contents of the liquid phase.

Line A corresponds to the temperature of the beginning of fusion of the test sample, which may be regarded as a heterogeneous system under the given conditions.

Figures 35 and 37 show the conditions of the conversion characterized by changes in the indices Δ or i in samples from different regions of the system. Thus, the relations along the composition line cordierite—silica are shown in a simplified form in Fig. 37. Here the solid line bounds the region where cordierites are generally unstable and the broken line separates the region of stability of hexagonal cordierites from intermediate and low-symmetry, rhombic cordierites.

The relations along the line cordierite—enstatite are somewhat different, as here it is also possible to determine the effect of the index i on the character of the phase conversions examined in connection with the appearance of the heterovalent isomorphous replacement $Mg^{2+} + Si^{4+} \rightarrow 2Al^{3+}$ (see, for example, Fig. 39).

The series cordierite—spinel is extremely interesting. The phase relations here differ sharply from those in all the sections of the system examined above. First of all, the reversible conversion of rhombic into hexagonal cordierite may occur here in the absence of a liquid phase due to the high temperatures of the eutectic points. As Fig. 36 shows, for sample 4 there is a fall in the curve below the melting point by 40°. In the presence of a liquid phase, only hexagonal cordierite is stable above 1453°C. In compositions 3 and 2 the corresponding forms of cordierite were obtained 20° and 50°C below the beginning of melting. This indicates a substantial fall in the temperatures of the conversion in the presence of an increasing amount of spinel crystals. In sample 1 it was impossible to obtain rhombic cordierite even when material that had previously been crystallized at a high temperature was kept at 900°C for 108 days.

Further complications in the structural relations of these compositions rich in aluminum oxide were revealed when sample 2 was heated at two different temperatures. The starting materials were the following products of composition 2:

1) Intermediate cordierite. This was obtained by crystallization of glass at 1150°C for 8 days and then at 1200°C for 7 days. The value $\Delta \simeq -0.15°$.

2) Metastable solid solution of hexagonal cordierite. This was obtained by crystallization of glass at 1000°C for 3 days.

3) Stable hexagonal cordierite obtained from glass at 1350°C for 8 days and at 1380°C for 7 days.

4) Homogeneous glass.

The results of annealing these products at 1250°C show that the stable material at this temperature is a cordierite of intermediate structure with an index Δ between 0.15° and 0.18°. It is obtained in materials of different starting compositions and with different annealing times.

At 1400°C with relative short exposures, cordierites are formed by the route indicated in Fig. 37, i.e., the nature of the cordierite obtained is independent of the character of the starting material. The intermediate cordierite is converted into hexagonal in one hour.

The characteristic changes of the indicator peak of A. Miyashiro for these compositions are shown in Fig. 40. Here, the change in the symmetry of the maximum is particularly noteworthy. There is a gradual broadening of the peaks from B to E and also the "plain" between the x-ray patterns F and G.

In compositions lying on the line cordierite—spinel there was no formation of solid solutions, the amounts of the two phases in the samples were the same, and only the state of the crystal lattice changed. On the basis

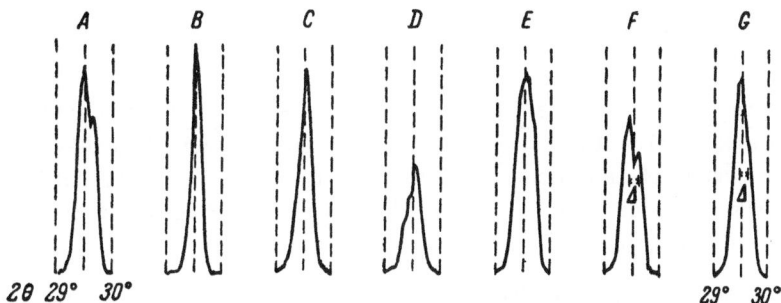

Fig. 40. X-ray patterns of cordierites for 2θ between 20 and 30° (CuK$_\alpha$ radiation) obtained from glass of composition 2 under the following conditions. A) At 1150°C for 8 days, then at 1200°C for 17 days. The structural state of the cordierite was intermediate, $\Delta = 0.16°$; B, C, D, E) material A heated to 1400°C for 1 h, 24 h, 7 days, and 28 days, respectively; F) at 1350°C for 8 days, then at 1380°C for 7 days and, finally, at 1280°C for 28 days, $\Delta = 0.18°$; G) at 1350°C for 8 days, then at 1380°C for 7 days. and, finally, at 1400°C for 28 days.

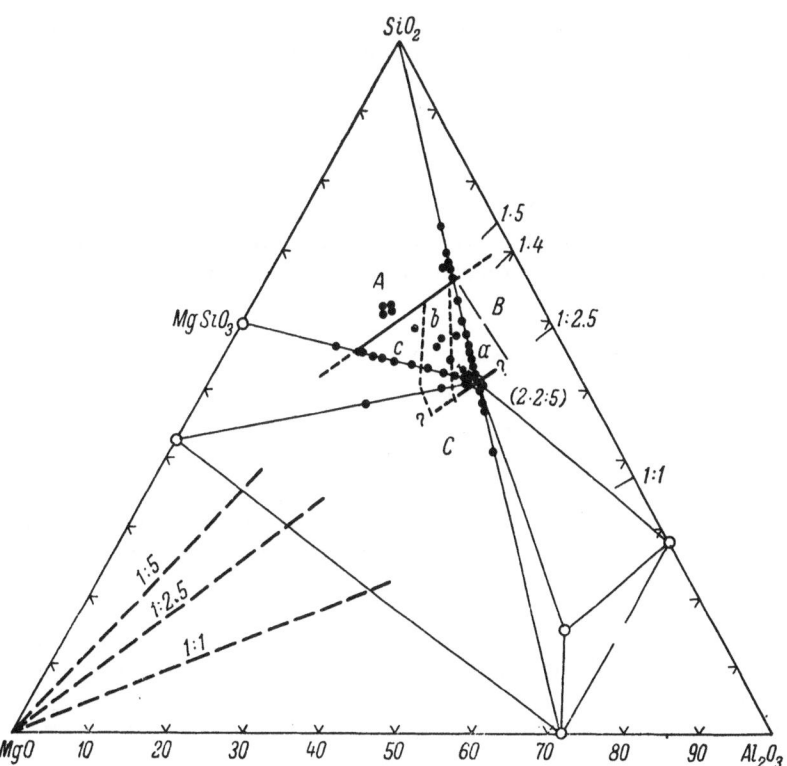

Fig. 41. Position of different structural types of cordierite in the system MgO$-$ $-$Al$_2$O$_3-$SiO$_2$. The heavy points denote compositions studied. The numbers on the sides of the triangle correspond to the molar ratios Al$_2$O$_3$: SiO$_2$.

of the observations described, and taking into account the effect of the character of the initial structures on the route of the conversions of cordierites and, in particular, the tendency of hexagonal cordierites obtained previously from rhombic cordierites for gradual reversible conversions to rhombic cordierites, Schreyer and Schairer (1961) proposed a new characteristic, which they called the "second-order metastability." This concept could be found repeatedly before this in physical metallurgy literature, where it is known under the name of "structural heredity," etc. It is probable that such phenomena may be expressed even more clearly in silicate phases because of the usual complexity of the steric structures of silicates.

No sign of the formation of solid solutions was found in the investigation of cordierite—mullite composi-tions in the high-temperature region.

<u>Structural Changes of Cordierite and Their Relation to the Overall Chemical Composition of the Samples Investigated</u>. X-ray investigations show that the structural conver-sions of cordierite occur without substantial changes in its basic structure. The hexagonal lattice of cordierite is rearranged into the less symmetrical rhombic lattice, which is pseudo-hexagonal, as soon as the conversion process begins. This structural change is not a first-order phase transition.

Miyashiro first proposed that structural changes are produced by the gradual ordering of the distribution of aluminum ions in columns formed by hexagonal or pseudo-hexagonal rings of $[AlSi_5O_{18}]$ probably accompanied by internal redistribution of the ions Al^{3+}, Mg^{2+}, and Fe^{2+} between rings. This hypothesis is confirmed by the fact that in hexagonal cordierites the Al^{3+} ions are arranged randomly, while in rhombic cordierites they have an orderly arrangement. No direct experimental demonstrations of this have yet been obtained. This requires x-ray investigations of single crystals of cordierite.

However, the behavior of cordierites on heating is very reminiscent of phenomena observed in feldspars, where "order—disorder" relations have been studied by Bailey and Taylor (1955) and Ferguson, Traill, and Taylor (1958) by three-dimensional structural analysis, and by Brun, Hartmann, Staub, Hafner, and Laves (1960) by nu-clear magnetic resonance. This confirms the hypothesis of Miyashiro. In particular, anorthite shows behavior very similar to that of cordierite. Its ordered modification is stable up to the melting point, while the meta-stable disordered anorthite may be obtained at subsolidus temperatures only with short exposures. However, up to the present time no data have been obtained on the effect of the accompanying phases on the polymorphic conversions of anorthite in the system $CaO-Al_2O_3-SiO_2$.

In connection with the problem of obtaining glass—crystalline materials from slags based on the system $CaO-Al_2O_3-SiO_2$, it may be surmised that specialists in the field of technical silicate chemistry will pay close attention to phase transitions in anorthite in the very near future. Moreover, corresponding investigations are also necessary in the series of lithium aluminosilicates, which are important components of contemporary pyro-ceram-sitalls.

The effect of the overall chemical composition in the system $MgO-Al_2O_3-SiO_2$ on the character of the phase transitions of cordierite is shown in Fig. 41. Here it is apparent that the composition of the sample has a definite effect on the character of the transitions occurring. Thus, hexagonal cordierites are unstable at any temperature in compositions with a high SiO_2 concentration (region A). Region B, within which hexagonal cor-dierites are stable only in the presence of the liquid phase, corresponds to a medium content of SiO_2. In the third region C, where subsolidus hexagonal cordierites are stable, the SiO_2 content is minimal. It may be as-sumed that the boundaries between the subzones a, b, and c correspond approximately to isotherms lying within the field of primary crystallization of cordierite.

The primary reason for the differences in the phase transitions of cordierite in different zones of the sys-tem is most likely the character of the "chemical environment" regardless of whether the atoms of this environ-ment get into the structure of cordierite or not. Here the most important factor is the ratio $Al_2O_3 : SiO_2$ and, therefore, it may be assumed that the concentration of Mg^{2+} ions does not substantially affect the character of the phase transitions.

The fact that structural changes in cordierite are a function of the $Al_2O_3 : SiO_2$ ratio in the overall com-position of the samples shows that the determining factor in the phase transitions is the "order—disorder" ratio in the arrangement of the Al and Si atoms in the structure. The higher the Al content with respect to Si in the sample, the greater will be the probability of finding Al in the composition of the cyclic radicals of the struc-ture and the formation of disordered cordierites.

<u>Experiments in the System $FeO - Al_2O_3 - SiO_2$</u>. In a separate investigation in the geo-physical laboratory, analogous data were obtained on metastable states in the system $FeO-Al_2O_3-SiO_2$, as the Fe^{2+} ion is similar to Mg^{2+} and Li^+ in geometric size. In particular, it was confirmed that in the system $FeO-$ $-Al_2O_3-SiO_2$ (hercynite—silica), a sample with the composition 23% FeO, 22% Al_2O_3, and 55% SiO_2, heated at

900°C for 24 h, contained hercynite, some fayalite, and the greatest amount of a quartz-like solid solution. Thus, it was established that Fe^{2+} ions are capable of entering the spaces of the quartz structure in considerable amounts. This is not unexpected, as in many natural minerals there is isovalent isomorphism of Fe^{2+} and Mg^{2+}.

The phase relations obtained in the systems examined above raise anew the whole problem of the transitions of silica and, in particular, the quartz varieties. Thus, in the work of Barth and Kvalheim (1944), it was established that tridymite may dissolve in its crystal lattice up to 5.2 wt.% of NaAlSiO$_4$. By thermal analysis, Keith and Tuttle (1952) studied the transitions of 250 quartz samples of different origins and investigated the effect of various additives on the temperature of the $\alpha \rightleftharpoons \beta$-quartz transition. In natural quartz samples the deviations of the transition point from the standard figure of 573°C were no more than 38°C, while with synthetic samples they reached 100°C. Thus, in particular, the addition of Ge^{4+} ions raised the transition point by +40°, while Li$^+$ reduced it by 121° at a very low concentration of additive. Therefore, the study of the corresponding simplest silicate systems at high pressures and temperatures is particularly urgent.

BIBLIOGRAPHY

Andersen, O., and N. L. Bowen. Am. J. Sci. 4:37, 487 (1914).

Bailey, S. W., and W. H. Taylor. Acta Cryst. 8:621 (1955).

Barth, T. E., and A. Kvalheim. Norsk Videns. Akad.,Oslo 22:1 (1944).

Belov, N. V., and T. N. Tarkhova. Dokl. Akad. Nauk SSSR 69:305 (1949).

Boyd, F. R., and J. L. England. J. Geophys. Res. 65(2):749 (1960).

Brun, E. P. Hartmann, H. H. Staub, S. Hafner, and F. Laves. Z. Krist.113:65 (1960).

Ferguson, R. B., B. J. Traill, and W. H. Taylor. Acta Cryst. 11:331 (1958).

Foster, W. R. J. Am. Ceram. Soc. 33(3):73-84 (1950).

Hummel, F.A., and H. W. Reid. J. Am. Ceram. Soc. 34:319 (1951).

Iiyama, F. Proc. Japan Acad. 31:166 (1955).

Ito, T., N. Morimoto, and R. Sadanaga. Acta Cryst. 5:209 (1952).

Karkhanavala, M. D., and F. A. Hummel. J. Am. Ceram. Soc. 36:389 (1953).

Keith, M. L., and O. F. Tuttle. Am. J. Sci., Bowen Vol: 203-280 (1952).

Liebau,F. Acta Cryst. 44:399 (1961).

Miyashiro, A. Am. Mineralogist 41:104 (1956).

Miyashiro, A. Am. J. Sci. 255:43 (1957).

Miyashiro, A., and T. Iiyama. Proc. Japan. Acad. 30:746 (1954).

Miyashiro, A, T. Iiyama, M. Iamasaka, and T. Miyashiro. Am. J. Sci. 253:185 (1955).

Morozevich, I. A. Experiments on the Formation of Minerals in Magma. SPb (1897).

Nikitin, V. D. Izv. Sektora Fiz.-Khim. Analiza, Inst. Obshch. Neorgan. Khim. SSSR 16(3):29-46 (1948).

Rankin, G. A., and H. E. Merwin. Am. J. Sci. 45:301 (1918).

Schreyer, W., and J. F. Schairer. Carnegie Inst. Washington Year Book 57:197 (1958).

Schreyer, W., and J. F. Schairer. Z. Krist. 116(1-2):60 (1961).

Schreyer, W., and J. F. Schairer. Am. Mineralogist 47(1-2):90 (1962).

Schreyer, W., and J. F. Schairer. J. Petrol. (1962).

Suguoca, K., and J. Kuroda. Bull. Inst. Technology, Tokyo 1:36B:1-5 (1955).

Tennyson, C. Neues Jahrb.Mineral., Abhandl. A94:1253 (1960).

Toropov, N. A. and F. Ya. Galakhov. Izv. Akad. Nauk SSSR, Otd. Khim. Nauk,No. 1:8-11 (1958).

Toropov, N. A., and R. G. Grebenshchikov. Zh. Prikl. Khim. 7(2):337-345 (1962).

CHAPTER IV

PHASE DIAGRAMS OF SYSTEMS FORMED BY OXIDES OF ELEMENTS OF VARIABLE VALENCE

At the present time, there is intensive investigation of systems formed by oxides of elements with variable valence. The most important problem in this field is the study of the behavior of iron oxides at high temperature and also the changes underlying the theory of the most important metallurgical processes, which occur under similar conditions in more complex systems. Such systems include oxides of chromium, manganese, and a series of other elements, which are very important for the development of the physicochemical foundations of refractory technology and slag processes.

In the Institute of Silicate Chemistry there have been extensive studies of systems formed by oxides of rare-earth elements, which have variable valence, such as cerium, praseodymium, samarium, neodymium, etc., as part of a general program on the development of a new section of silicate chemistry. All this compels us to pay due attention to the theoretical and experimental material available here.

The investigation of such systems requires careful control of the gaseous medium in which the experimental work is carried out. The required partial pressure of oxygen may be obtained either by vacuum techniques or by mixing oxygen with an inert gas in the range from 1 to 10^{-3} atm. For obtaining more reducing conditions, the required partial pressure of oxygen is obtained by mixing CO_2 and H_2 in various proportions with subsequent heating of this gas mixture at high temperatures. The partial pressure of oxygen realized in this way is calculated from available equilibrium data.

In the description of phase equilibria in such systems the partial pressure of oxygen in the gas phase is an important parameter. Diagrams showing projections of the liquidus surface with liquidus isotherms and fractional separation curves must be supplemented in these cases by curves showing the equilibrium partial pressure of oxygen in the gas phase. New criteria for deviations from routes of equilibrium crystallization must also be established. The compositions of condensed phases change during crystallization in such a way that this change is described by straight lines directed toward the oxygen apex of the nominal model which is usually used for graphical representation of such systems.

Silicates containing iron oxides are the most thoroughly studied of the systems in which there are variations in the degree of oxidation. The systems $Fe-O$, $FeO-Fe_2O_3-SiO_2$, and $MgO-FeO-Fe_2O_3-SiO_2$ may be used to illustrate the main principles for describing the equilibria in such systems (Muan, 1958).

Heterogeneous equilibria are characterized by the transfer of material through phase boundaries. Such equilibria include condensed phases, either solid or liquid or the two simultaneously. The equilibria are described by the Gibbs phase rule. By condensed phases we mean solid and liquid phases.

Until recently, the greatest attention in the study of phase diagrams has been paid to systems of oxides formed by ions with a noble gas structure (SiO_2, Al_2O_3, MgO, CaO). However, there are many other oxides which are no less important, especially in new techniques where materials are produced containing chemical compounds of ions of transition elements, where a change in the degree of oxidation is not only possible, but normal. The study of equilibria in such systems is a more complex problem, as the gas phase plays a substantial part in the equilibria.

1. Methods of Determining Phase Equilibria in Oxide Systems

Normal Systems. In normal silicate systems the gas phase is ignored and the phase rule is applied in the form $P + F = C + 1$, where P is the number of condensed phases present. These systems are investigated in open tempering furnaces under normal atmospheric conditions as, for example, the system $CaO-MgO-SiO_2$, where at 1400°C the partial vapor pressure of CaO is 10^{-10}, MgO, 10^{-9}, and SiO_2, 10^{-9} atm. The conditions for attaining true equilibrium between the gaseous and condensed phases are of no great value. Investigations may be carried out without control of the composition of the gas phase.

Systems in Which There are Changes in the Degree of Oxidation. As an example let us consider an element Me, which forms the two oxides MeO_x and MeO_y. In this case, the element Me is present in two oxidation states Me^{2x} and Me^{2y}. Let us assume that these oxides are dissolved in the condensed phase which is in equilibrium with the gas phase. The following reaction is then of considerable importance:

$$(MeO_x)_{soln} = (MeO_y)_{soln} + \frac{(x-y)}{2} O_{2\,gas}.$$

It is obvious that a change in the oxygen pressure will produce a change in the ratio Me^{2x} : Me^{2y} in the condensed phase, and vice versa. Therefore, in the investigation of systems including such components, the partial pressure of oxygen must also be considered, in addition to the partial pressures of the components of the system itself in the gas phase.

If the initial values of these pressures are very low, then the system may be regarded as practically similar to the system $MgO-CaO-SiO_2$. The partial vapor pressures of oxides as such are a function of temperature and may be calculated beforehand from equations available in the literature. For a large number of the most important oxides such as the iron oxides, these values are very small, right up to the liquidus temperatures.

The equilibrium pressure of oxygen is a function of the ratio MeO_x : MeO_y in the condensed phase and rise rapidly in proportion to an increase in this ratio. If the ratio MeO_x : MeO_y in the condensed phases is such that the equilibrium oxygen pressure in the gas phase is insignificant, the phase equilibria in this system may be studied in vacuum furnaces or in an atmosphere of a purified inert gas. This method is based on the slow exchange of oxygen between the gaseous and condensed phases.

Let us assume, as an example, that the ratio MeO_x : MeO_y in the liquid phase at the experimental temperature corresponds to a partial oxygen pressure of 10^{-9} atm. Let us also assume that an "ideal" inert gas is used, so that there is no oxygen in the gas phase.

The mixture heated in a furnace through which a stream of the inert gas is blown may create an equilibrium oxygen pressure of the order of 10^{-9} atm. The amount of oxygen required to create this partial pressure is very low (if we assume that the inert gas is blown through the system at a moderate rate), namely, the amount of oxygen will be $(10^{-9}/22.4 \cdot 32 = 1.4 \cdot 10^{-9})$ g per liter of gas. Thus, the ratio MeO_x : MeO_y in the condensed phases will change very slightly and for practical purposes the compositions of the condensed phases may be regarded as relatively constant.

If the inert gas is not "ideal," a similar situation is produced. For example, let us assume that the inert gas (or vacuum system) contains small amounts of oxygen corresponding to an O_2 pressure of 10^{-7} atm. For the establishment of equilibrium between the gaseous and condensed phases the latter must absorb oxygen from the gas phase and the difference between O_2^{-9} and O_2^{-7} is very small, while an unusually large volume of gas is required to change appreciably the ratio MeO_x : MeO_y, and, hence, the overall composition of the condensed phases (about $1.4 \cdot 10^{-7}$ g of oxygen per liter of gas). Therefore, for practical purposes, the overall composition of the condensed phases may be regarded as more or less constant.

These initial premises are the foundation of the method used by Schuhmann, Powell, and Michol (1953) for studying part of the system $FeO-Fe_2O_3-SiO_2$. These authors prepared samples of the part of the system poor in Fe_2O_3 by weighing out all three components, SiO_2, Fe_2O_3, and FeO in the required proportions. The mixtures were then calcined in platinum containers in an atmosphere of a neutral gas and the results were used on the assumption that the overall composition of the condensed phases should not change during the time required for equilibrium to be attained between the condensed phases.

Another method of studying such systems under reducing conditions was used by Bowen and Schairer (1932) in the investigation of equilibria in systems with ferrous silicates. In this method the equilibria between the condensed phases are investigated in iron crucibles in an atmosphere of a purified inert gas such as nitrogen. The partial pressure of oxygen in the gas phase under these conditions is determined by the following equilibria:

$$Fe_2O_3 \, _{melt} = 2FeO_{melt} + \tfrac{1}{2}O_2 \, _{gas},$$

$$Fe_2O_3 \, _{melt} + Fe_{solid} = 3FeO_{melt}.$$

Under these conditions, at any given temperature in the binary system Fe—O there is only one degree of freedom and in the presence of the phases listed (liquid, metallic iron, and gas) the oxygen pressure is a function of temperature alone.

In a ternary system such as $FeO-Fe_2O_3-SiO_2$, with the given association of phases there is one additional degree of freedom and the equilibrium pressure of oxygen is already a function of the temperature and the concentration of one of the components of the liquid phase. If wustite FeO is one of the phases present, the partial pressure of O_2 will be equivalent to the equilibrium pressure at which FeO and Fe coexist at the given temperature in equilibrium in the binary system Fe—O.

If wustite is not present as an individual phase at temperatures below its melting point (1371°C), the partial pressure of O_2 will be below the pressure corresponding to the equilibrium of FeO : Fe.

At relatively high O_2 pressures, the equilibrium between the oxygen of the gas phase and the condensed phases may be reached within the time for an experiment. Thus, for example, 1 liter of gaseous oxygen at a pressure of 1 atm contains 1.43 g of oxygen. Therefore, in the given case with a moderate gas flow rate through the furnace a sufficient amount of it will be transferred to change the ratio $MeO_x : MeO_y$ during the time used in practice. Therefore, it is possible to investigate equilibria in systems containing oxides in the presence of a gas phase with an O_2 pressure of 1 atm or in air ($P_{O_2} = 0.21$ atm).

If the oxygen pressure in the gas phase is reduced, for example by the use of vacuum or by mixing with an inert gas, the volume of gas required to reach the equilibrium composition by the condensed phases is increased sharply (in proportion to $1/P_{O_2}$). A partial oxygen pressure of approximately 10^{-3} atm is the lower limit in practice for studying equilibria in which oxygen participates solely in the form of O_2 molecules in the gas phase.

In addition, it is possible to reach equilibrium between the gas and condensed phases also at lower partial pressures of O_2 by using indirect methods to reach the required values of the partial pressure of oxygen. The most convenient method is the use of an oxygen-containing gas which decomposes to another gas with the liberation of free oxygen on heating in the furnace. Such gases as H_2O and CO_2 are suitable for these purposes and on heating they dissociate in accordance with the equations

$$2H_2O = 2H_2 + O_2,$$

$$2CO_2 = 2CO + O_2.$$

The equilibrium constants of the reactions as functions of temperature are well known. If CO_2 or H_2O is used in a pure form and the total pressure is kept constant, the partial pressure of oxygen depends solely on the temperature. The latter means that the oxygen pressure and temperature may not be changed independently, and this is a great experimental hindrance.

If we mix a certain amount of H_2 with H_2O or CO with CO_2, the degree of dissociation of the H_2O or CO_2 is partly suppressed to an extent which is controlled by the proportions in the gas mixtures used. The partial pressure of oxygen is then a function of temperature and the ratio $H_2O : H_2$ or $CO_2 : CO$ in the gas mixtures.

The oxygen pressure and temperature may be changed independently. The advantage of using a mixed gas atmosphere over vacuum and an inert gas lies in the fact that oxygen is present here not only in the form of O_2, but also in other forms (H_2O or CO_2). At sufficiently high temperatures for achieving high rates of gas reactions, it is possible to attain a reaction mechanism in which the exchange of O_2 between the gas and condensed phases gives practically applicable rates of reaction at low partial pressures of O_2.

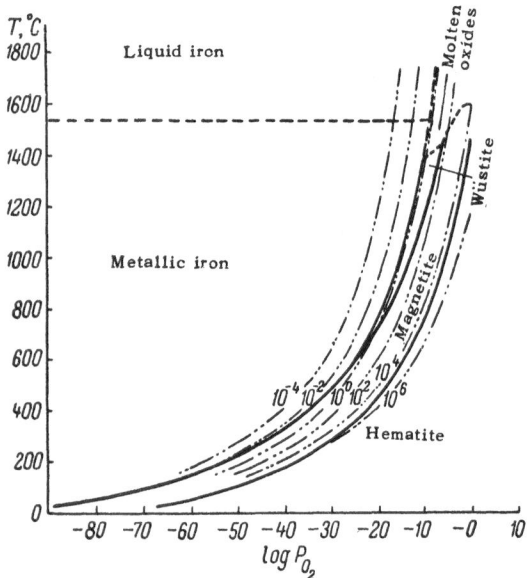

Fig. 42. Diagram of equilibrium phase relations between hematite Fe_2O_3, magnetite $FeO \cdot Fe_2O_3$, wüstite FeO, a melt of iron oxides, and liquid iron in relation to the temperature and partial pressure of O_2. The thick solid lines are boundary curves separating the regions of different phases. The thin broken lines are lines of equal $P_{HO} : P_{H_2}$ ratios in the gas phase with a total pressure of 1 atm.

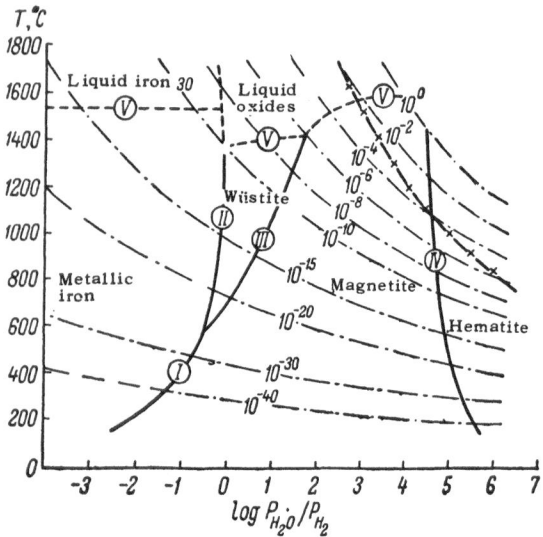

Fig. 43. Diagram of equilibrium phase relations of iron oxides and metallic iron in relation to temperature and the ratio $H_2O : H_2$ in the gaseous medium in equilibrium with the condensed phases. The thick solid lines represent the boundary curves separating regions of different phases. The fine broken lines are oxygen isobars. The thick cross-dash line shows the equilibrium ratio $P_{H_2O} : P_{H_2}$ of pure water vapor in relation to temperature at a total pressure of 1 atm.

As the O_2 is removed from the gas phase and transferred to the condensed phases, the equilibrium is rapidly restored as a result of the following decomposition reactions:

$$2H_2O = 2H_2 + O_2,$$

$$2CO_2 = 2CO + O_2,$$

which occur at high temperatures. Such methods of studying equilibria may be used only at temperatures at which gas reactions proceed at sufficient rates.

2. Methods of Constructing Phase Diagrams

A detailed idea of the phase equilibria in the system Fe−O is very important for further discussion of more complex multicomponent systems containing iron silicates. Figure 42 gives the phase equilibrium relations in the system Fe−O in relation to the temperature and partial pressure of oxygen. The following rules should be applied in using this diagram.

In regions where the condensed phase exists in equilibrium with gas, the system has two degrees of freedom. Therefore, within the limits of the field of stability of one phase such as wüstite, both temperature and pressure may be changed without a change in the number of phases present. Along the boundary curves where two condensed phases coexist in equilibrium with a gas phase, the system has only one degree of freedom. Thus, for example, if the system is examined at a definite temperature, then it must be characterized by a quite definite oxygen pressure, and vice versa. The partial pressure of oxygen at low temperatures is so low that it has very little physical meaning. Of greater physical importance is the ratio $P_{H_2O} : P_{H_2}$ or $P_{CO_2} : P_{CO}$ in the gas phase in equilibrium with the different oxides.

The phase relations of the system Fe−O relative to the given relations are shown in Figs. 43 and 44.

The diagram in Fig. 43 was constructed from the following data: the boundary curve along which metallic iron and wüstite or liquid oxides exist in equilibrium with the gas phase (curve II), exactly like the part of the curve where magnetite, wüstite, and gas exist together in equilibrium (curve III), were constructed from data obtained by direct measurements of the ratio $H_2O : H_2$ in the gas in equilibrium with the given phases according to data available in the literature.

The part of curve II (Fig. 43) referring to low temperatures (600-1000°C) was constructed from the data of Emmett and Schultz(1930), while the high-

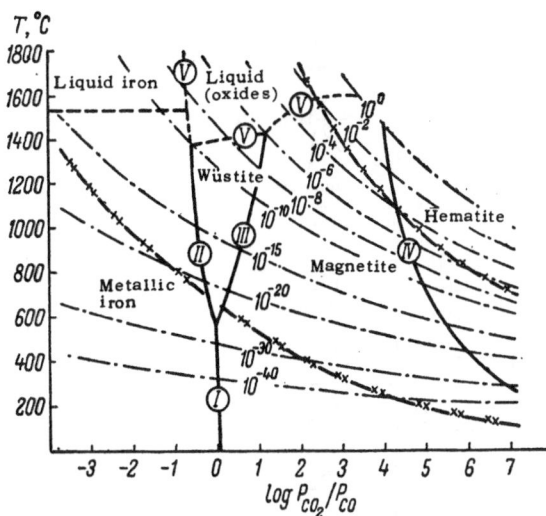

Fig. 44. Diagram of equilibrium phase relations of iron oxides and metallic iron in relation to temperature and the ratio $CO_2:CO$ in the gas phase in equilibrium with the condensed phases. The thick solid lines and the thin broken lines mean the same as in Fig. 43. The thick double cross-dash line shows the equilibrium ratio $P_{CO_2}:P_{CO}$ in equilibrium with carbon (graphite) in relation to temperature at a total pressure of 1 atm.

temperature part (1200-1515°C) was from the data of Chipman and Marshall (1940). The low-temperature part of curve III (600-800°C) was constructed from the data of Emmett and Schultz, and the upper part of this curve (1096-1388°C) from the data of Darken and Gurry (1945, 1946). These authors used $CO_2:CO$ and $CO_2:H_2O$ mixtures to obtain the data used for localization of the above lines in Fig. 44.

The calculations were based on the experimental data of Coughlin (1954) obtained for the reaction

$$2CO_2 = 2CO + O_2,$$

$$2H_2O = 2H_2 + O_2.$$

Curve I and part of curve III (below 1000°C) were obtained by calculation from the free energy values (Fig. 43).

The position of the thick cross-dash line (Fig. 43) relative to the boundary lines determines the region of stability of the various iron oxides in pure water vapor. It should be noted that at temperatures below approximately 1070°C, hematite is the phase stable in an atmosphere of water vapor, while above 1595°C under the same conditions the stable phase is a melt of iron oxides. It should also be pointed out that neither wustite nor metallic iron is in equilibrium with water vapor at any temperature and a pressure of 1 atm.

For investigating equilibria in regions of the diagram below the dissociation curve of pure water, it is necessary to use a mixture of H_2 (or another reducing agent) with H_2O in proportions which may be calculated from Fig. 43. One of the practically important consequences is the fact that the values of the ratio $H_2O:H_2$ given in Fig. 43 are equilibrium values at any given temperature.

For an experimental study of the phase equilibria of iron oxides, it is necessary to use a definite mixture of the two gases prepared at room temperature and then to heat this mixture to the given temperature at which the equilibrium between the condensed phases is being studied.

These two ratios (the initial ratio in the mixture at room temperature and the equilibrium ratio under the experimental conditions) are not equal. If $P_{H_2O}(i)$ and $P_{H_2}(i)$ are the partial pressures of H_2O and H_2 in the initial gas mixture at room temperature, on heating the H_2O will decompose according to the equation

$$2H_2O = 2H_2 + O_2,$$

and under equilibrium conditions we will have

$$K = \frac{P_{O_2}(e) P_{H_2}(e)}{P_{H_2O}(e)^2},$$

where $P_{O_2}(e)$ is the partial pressure of the gases at high temperature, etc., and K is the equilibrium constant. By means of this equation it is possible to derive a simplified formula for determining the conditions of mixing the gases.

Figure 43 shows that some of the boundary curves have positive and others negative slopes, while one (the boundary of Fe-wüstite and Fe-liquid oxides) is practically vertical. The latter case illustrates the interesting situation when the ratio $P_{H_2O}:P_{H_2}$ in the gas in equilibrium with two condensed phases remains constant over a

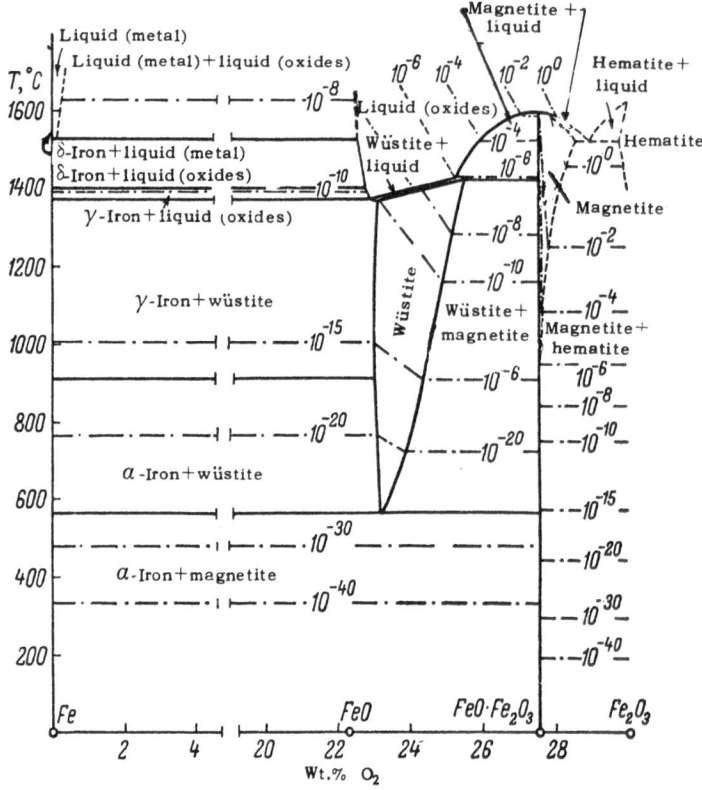

Fig. 45. Diagram of equilibrium phase relations of the system Fe—O based mainly on the data of Darken and Gurry (1945, 1946). Thick solid lines are boundary curves separating regions of different phases. The thin dot-dash lines are oxygen isobars (in atm). The broken curves in the upper right-hand part of the system are provisional boundary curves in a region where there are no experimental data ($P_{O_2} > 1$ atm).

wide range of temperatures (1000-1600°C). The reason for this is the similarity in the heats of reduction of ferrous oxide to Fe and oxidation of hydrogen to H_2O

$$FeO + H_2 = Fe + H_2O.$$

Curves with a positive slope (for example the boundary of $Fe-Fe_3O_4$) arise in cases where the heat of reduction of the oxide from one oxidation state to a lower one, and to metallic iron, is greater than the heat of oxidation of H_2 which participates in the reaction.

The diagram illustrated in Fig. 44 is constructed from the same data as Fig. 43. It is readily seen that Figs. 43 and 44 are similar to each other, and this is caused by the similarity in the dissociation constants of H_2O and CO_2.

The additional line (double cross-dash) showing the ratio $P_{CO_2} : P_{CO}$ in equilibrium with graphite in accordance with the following equation is of great importance:

$$2CO_{gas} = C_{solid} + CO_{2\,gas}.$$

It should be noted that this line intersects the Fe_3O_4-FeO and $FeO-Fe$ boundaries at 660 and 720°C, respectively. These temperatures are the minimal values at which wüstite and metallic iron (respectively) may be in equilibrium with gases of the C—O system. These ratios are of fundamental importance for the theory of the blast-furnace process and have a series of important applications in geochemistry, such as the stability of siderite under oxidation at high pressures.

The diagrams presented in Figs. 43 and 44 have one substantial drawback: they do not show the compositions of the condensed phases. However, the phase diagram of the Fe−O system may be represented in such a way that the compositions of the condensed phases will be shown on it.

The most complete diagram of this system in this form was published by Darken and Gurry. It was in the form of a normal diagram of a binary system in the coordinates temperature−concentration. However, for the type of system examined, such diagrams do not, unfortunately, show the composition of the gas phase in equilibrium with the condensed phases. The diagram shown in Fig. 45 was constructed in a way which makes it possible to eliminate the drawbacks of the methods just given for representing the phase states of binary systems. This diagram largely reproduces the diagram of Darken and Gurry, constructed in the coordinates temperature−concentration. However, in addition to the boundary curves, this diagram also shows a series of curves representing identical values of the partial pressure of oxygen in the gas phase in equilibrium with the condensed phases. It is also possible to construct an analogous type of diagram for a binary system showing curves of equal ratios $P_{H_2O} : P_{H_2}$ or $P_{CO_2} : P_{CO}$ in the equilibrium gas phase.

The relation between the different methods of representing phase equilibria may be seen by comparing Figs. 43-45. In Figs. 43 and 44, each region corresponds to conditions under which one condensed phase exists in equilibrium with the gas phase. The boundary curves on these diagrams indicate the conditions for coexistence of a gas and two condensed phases. In Fig. 45 the separate regions of the diagram may correspond to

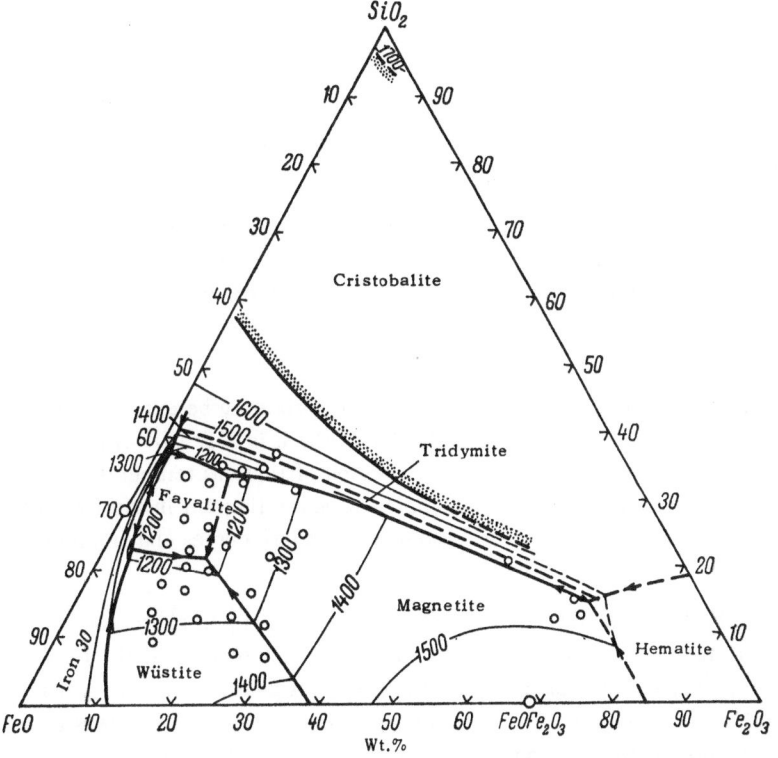

Fig. 46. Diagram of equilibrium phase relations of the system FeO−
−Fe₂O₃−SiO₂ according to Muan (1955). Fine lines are liquidus isotherms
at intervals of 100°C and the thick lines are boundary curves, the arrows
on which indicate the direction of falling temperature. The tridymite−
−cristobalite boundary curve is shown by a thick broken line. The lines
shaded on one side are the boundaries of regions of two immiscible liquids.

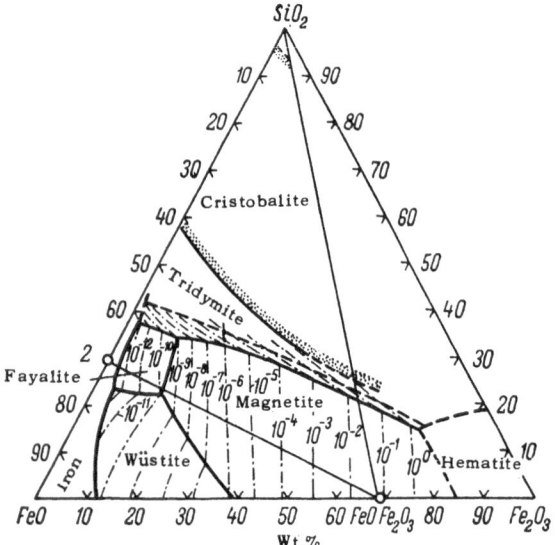

Fig. 47. Diagram of equilibrium phase relations of the system $FeO-Fe_2O_3-SiO_2$. Thick lines are bound-ary curves, fine dot-dash lines are O_2 pressure iso-bars for points on the liquidus surface. The lines shaded on one side bound regions of two immiscible liquids (diagram according to Muan, 1955).

conditions under which one or two condensed phases exist simultaneously in equilibrium with the gas phase. These conclusions are not contradictory to each other.

Figures 43 and 44 have no coordinate axes on which the compositions of the condensed phases could be shown. Two condensed phases of different composi-tions in equilibrium along the boundary curve are rep-resented on these diagrams by one point, namely that which corresponds to the combination of temperature and the ratio $H_2O : H_2$ ($CO_2 : CO$), where the two phases are in equilibrium. In Fig. 45 the axes are selected in such a way that the compositions of the condensed phases may be determined directly from the graph. As the compositions of the phases in equilibrium differ, the two phases correspond to two different points on the diagram. The construction is such that the bound-ary curves of Figs. 43 and 44 are used for the forma-tion of regions in Fig. 45.

In Figs. 43 and 44, regions of adjacent phases are separated by boundary curves, while in Fig. 45 each single-phase region (one condensed phase) is bounded by regions in which two condensed phases coexist in equilibrium with a gas phase. As the system has two degrees of freedom in those cases where one condensed phase exists in equilibrium with the gas phase, both the temperature and the oxygen pressure or the ratio $H_2O : H_2$ ($CO_2 : CO$) may be changed within the limits of one-phase regions (Figs. 43-45). Therefore, the lines corresponding to the O_2 isobars may pass through these regions at a slope in Fig. 45.

In the two-phase regions (two condensed phases) there is only one degree of freedom and P_{O_2}, like the ratio $H_2O : H_2$ ($CO_2 : CO$) in the gas phase, is fixed at a given temperature. Therefore, the O_2 isobars must be horizontal lines parallel to the concentration axis in the two-phase region.

3. Ternary Systems Containing Iron Oxides

The method examined above for describing phase equilibria may be extended to more complex systems. Let us examine, for example, the system $FeO-Fe_2O_3-SiO_2$, which is only part of the more general system $Fe-$ $-Si-O$. Below we shall mainly be concerned with the region of liquidus temperatures of the system examined. Thus, in Fig. 46 we give the usual type of three-component phase diagram in the form of a projection onto the concentration base of the liquidus surface of the system of boundary curves and liquidus isotherms according to Muan.

From the phase rule it follows that if one crystalline and one liquid phase coexist in a three-component system with a gas phase, then the system will have two degrees of freedom. If the temperature and composition of the liquid phase are given, then the oxygen pressure in the gas phase must have a strictly defined value. Therefore, it is possible to draw a series of curves corresponding to equal partial pressures of oxygen (O_2 isobars) in the gas phase in equilibrium with the condensed phases at the liquidus temperature. In the $FeO-Fe_2O_3-SiO_2$ triangle these lines are an extension of the system of lines starting from the points obtained by the intersection of the O_2 isobars with the liquidus curve in the system $Fe-O$ (Fig. 45). An example of the construction of iso-baric lines is given in Fig. 47.

Analogously, it is possible to draw lines corresponding to constant ratios $P_{H_2O} : P_{H_2}$ or $P_{CO_2} : P_{CO}$ shown in Figs. 43 and 44. Therefore, on the phase diagram it is possible to draw a series of curves of equal partial pressures of oxygen (O_2 isobars) in the gas phase in equilibrium with the condensed phases on the liquidus surface. These lines begin at the points of intersection of the oxygen isobars and the liquidus curve in the system $Fe-O$

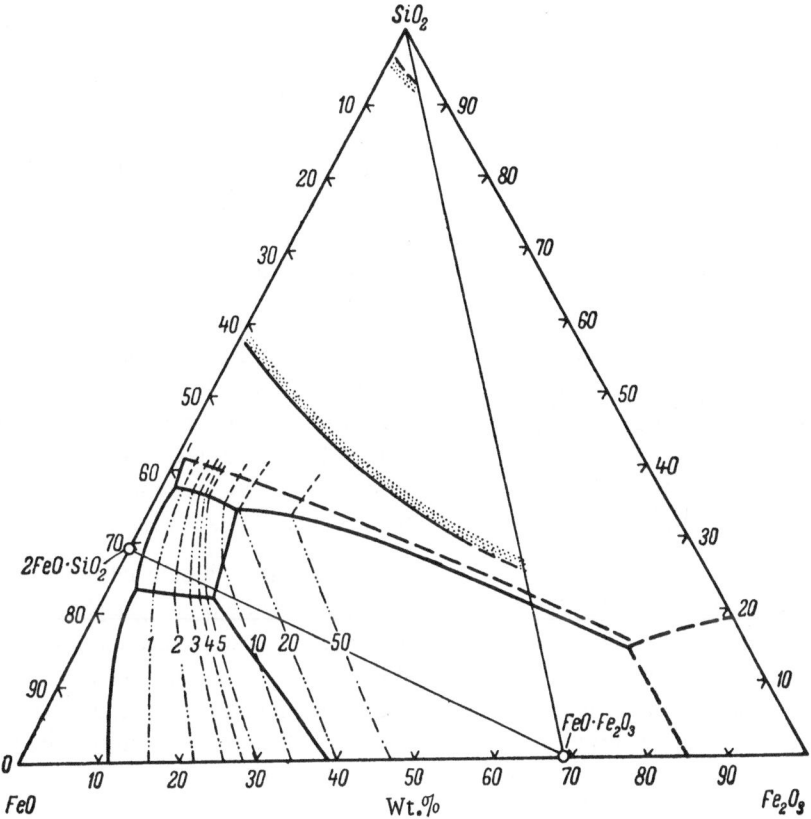

Fig. 48. Diagram of equilibrium phase relations of the system $FeO-Fe_2O_3-$
$-SiO_2$. Thick lines are boundary curves and fine dot-dash lines are lines of
equal ratios $P_{CO_2} : P_{H_2O}$ in the gas phase in equilibrium with the melt at the
liquidus temperature. The lines shaded on one side bound regions of two immiscible liquid phases (diagram according to Muan, 1955).

and continue into the field of the system $FeO-Fe_2O_3-SiO_2$. It should be noted that the lines of equal ratios
$P_{CO_2} : P_{H_2}$ (Fig. 48) are approximately parallel to the boundary curve between the fields of wüstite and metallic iron, i.e., almost as in Fig. 47.

In exactly the same way it is possible to draw the corresponding isobars for other temperatures lying above
or below the liquidus surface. To demonstrate the relations of the compositions of the crystalline phases in
equilibrium with the melt and the gas phase, Fig. 49 shows the fractional crystallization curves. The fractional
crystallization curves may be constructed from the tie lines (conodes), as these are tangential to the fractional
crystallization curves. The compositions of the crystalline phase in the melt with the liquid may be obtained
on the diagram in Fig. 49 by drawing the tangent to the fractional crystallization curve at the point representing
the composition of the liquid phase on the line $FeO-Fe_2O_3$.

4. Equilibrium Crystallization Routes

The set of lines showing changes in the composition of the liquid phase during crystallization on the diagram of phase equilibria are usually called crystallization routes. If one solid phase of constant composition
separates, the crystallization route lies along a straight line, while if one solid phase of varying composition
(solid solution) separates, the change in composition of the residual liquid phase will pass along some curve.

In systems where there is a change in the degree of oxidation of the elements whose oxides form the solid
phases, the composition of the latter will naturally change with a change in temperature as a result of the introduction or removal of oxygen from these phases, and this will lead to a change in the ratio $MeO_x : MeO_y$. Therefore, the overall composition of the condensed phases will change continuously during crystallization from the

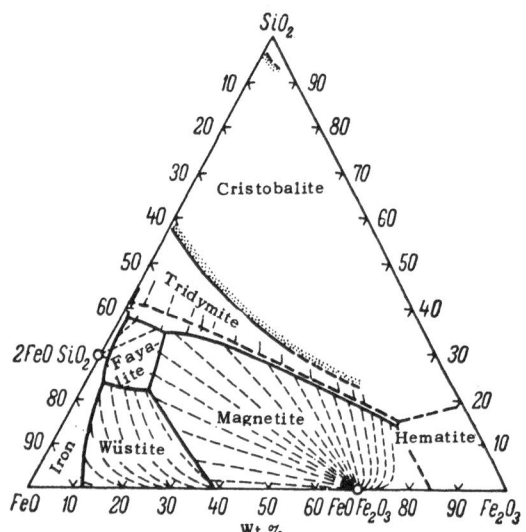

Fig. 49. Diagram of the system $FeO-Fe_2O_3-SiO_2$ illustrating the compositions of the liquid and solid phases in equilibrium with each other (according to Muan, 1955). The thick lines show boundary curves while the broken lines are fractional crystallization curves. The lines shaded on one side are the boundaries of regions of two immiscible liquids.

liquid. This change may be described by a straight line passing through the oxygen corner of the diagram of the system examined.

By combining these straight lines with the liquidus isotherms, fractional crystallization curves, O_2 isobars, or lines of equal $P_{CO_2}:P_{H_2}$ ratios, it is possible to determine the crystallization routes for various idealized conditions which approach the practical conditions to various extents.

Let us briefly examine this on the example of the systems $Fe-O$ and $FeO-Fe_2O_3-SiO_2$ for the following conditions: 1) the condensed phases have constant compositions; 2) the O_2 pressure is constant; 3) there is a constant $P_{CO_2}:P_{H_2}$ ratio in the gas phase; 4) crystallization occurs in contact with metallic iron.

The first of these conditions is realized in cases where the volume of the gas phase in a closed system is insignificant. The second and third conditions are fulfilled in many experimental studies of recent years. Finally, the fourth condition is fulfilled in the work of Bowen and Schairer, who investigated various systems containing iron silicates.

The System Fe − O. In binary systems the relations of the phases are not normally too complex and crystallization routes are rarely used here. Phase changes with a constant composition of the condensed phases (first condition) may be established directly on the diagram in Fig. 45. With a constant O_2 pressure (second condition), the equilibrium crystallization route is connected with the isobar corresponding to the O_2 pressure adopted.

Let us examine crystallization at a constant O_2 pressure of 10^{-8} atm. Above 1626°C the condensed phase will be only metallic iron. At 1626°C, metallic liquid and molten oxides will coexist in equilibrium with the gas phase. Between 1626 and 1397°C, only liquid oxides are present as the condensed phase under equilibrium conditions. At 1397°C, liquid, wüstite crystals, and the gas phase participate in the equilibrium. In the temperature range of 1397-1280°C, wüstite is in equilibrium with the gas, and at 1280°C, wüstite and magnetite coexist with the gas phase. From 1280 to 840°C, magnetite is stable in an atmosphere with the partial pressure of oxygen given, at 840°C, magnetite and hematite coexist, and below 840°C, hematite exists.

We should note one substantial difference from the previous case. Here two condensed phases may coexist in equilibrium only at one constant temperature: one degree of freedom is lost at a constant partial pressure of oxygen. In the first case there is no such limitation and two condensed phases may be in equilibrium over a certain range of temperatures.

In the third case (crystallization with a constant ratio $P_{CO_2}:P_{H_2}$ in the gas phase), the picture obtained is very similar to that just examined, and under these conditions the lines of equal ratios $P_{CO_2}:P_{H_2}$ are the crystallization routes. If crystallization occurs in contact with metallic iron, the changes in phase associations may be followed from Fig. 45 along the boundary curve between the regions of metallic iron and the lowest stable oxide from high to low temperatures.

Between 1535 and 1524°C, solid iron is in equilibrium with liquid metal and the gas phase. At 1534°C, solid iron, metallic liquid, and liquid oxides coexist with the gas phase. Between 1534 and 1371°C, liquid oxides are in equilibrium with iron and the gas phase. Between 1371 and 560°C, wüstite is in equilibrium with iron and below 560°C, magnetite coexists with iron. On the other hand, changes in the partial pressure of

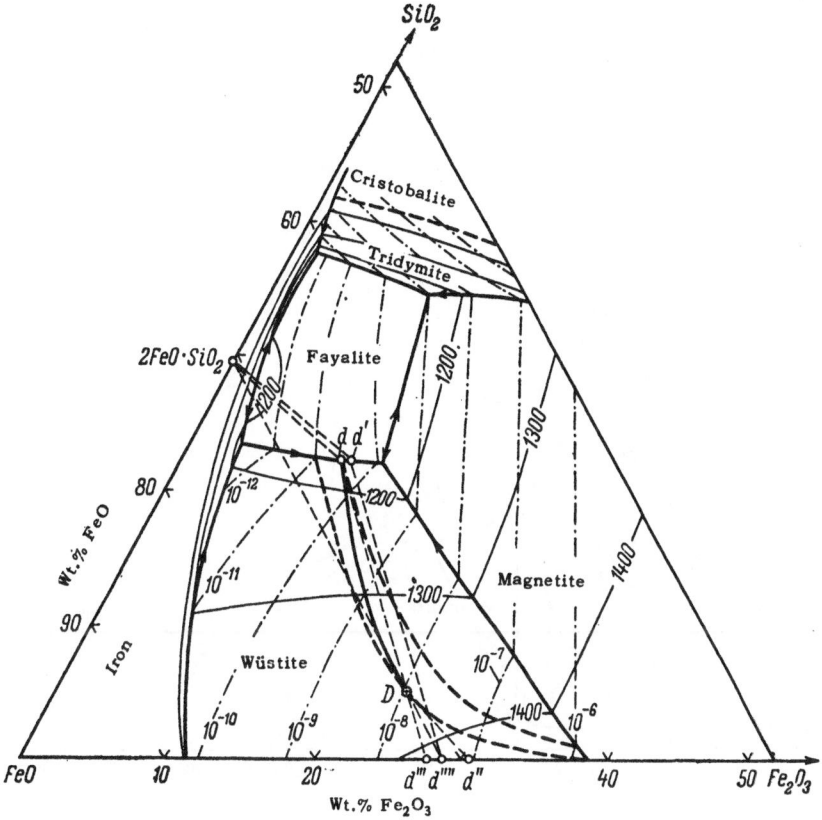

Fig. 50. Diagram of part of the system $FeO-Fe_2O_3-SiO_2$, illustrating the equilibrium crystallization routes under conditions of constant overall composition of the condensed phases. The thick lines show boundary curves, lines of medium thickness are equilibrium crystallization curves, and broken lines of medium thickness are fractional crystallization curves. The fine solid lines are liquidus surface isotherms, the fine dot-dash lines are O_2 isobars for points lying on the liquidus surface, and the fine broken lines are tie lines (diagram according to Muan, 1955).

oxygen during crystallization are determined by intersections of the O_2 isobars and the boundary curve along which iron is one of the equilibrium phases.

The System $FeO-Fe_2O_3-SiO_2$. The crystallization routes under the first of the above conditions may be demonstrated by means of liquidus surface isotherms (Fig. 46), oxygen isobars at the liquidus surface (Fig. 47), and fractional crystallization curves (Fig. 49) with the assumption that the overall composition of the condensed phases remains unchanged during the crystallization process. On the basis of these graphs, the crystallization routes in part of the system are drawn in Fig. 50.

Let us examine mixture D to illustrate the crystallization routes of melts from the wüstite field. At approximately 1370°C, wüstite begins to crystallize and the composition of the liquid phase therefore changes along the equilibrium crystallization curve D—d (solid line of medium thickness). The composition of wüstite in equilibrium with the liquid is determined by tangents to the fractional crystallization curve drawn through points representing the composition of the liquid phases and it changes gradually from d″ (31 wt.% Fe_2O_3) to d‴ (28 wt.% Fe_2O_3). Fayalite begins to separate together with wüstite from the liquid at point d and the composition of the liquid phase changes along the wüstite—fayalite boundary toward point d' along the next part of the crystallization curve. Beginning with point d, where the boundary curve is reached, reenrichment of the wüstite in iron oxide is observed.

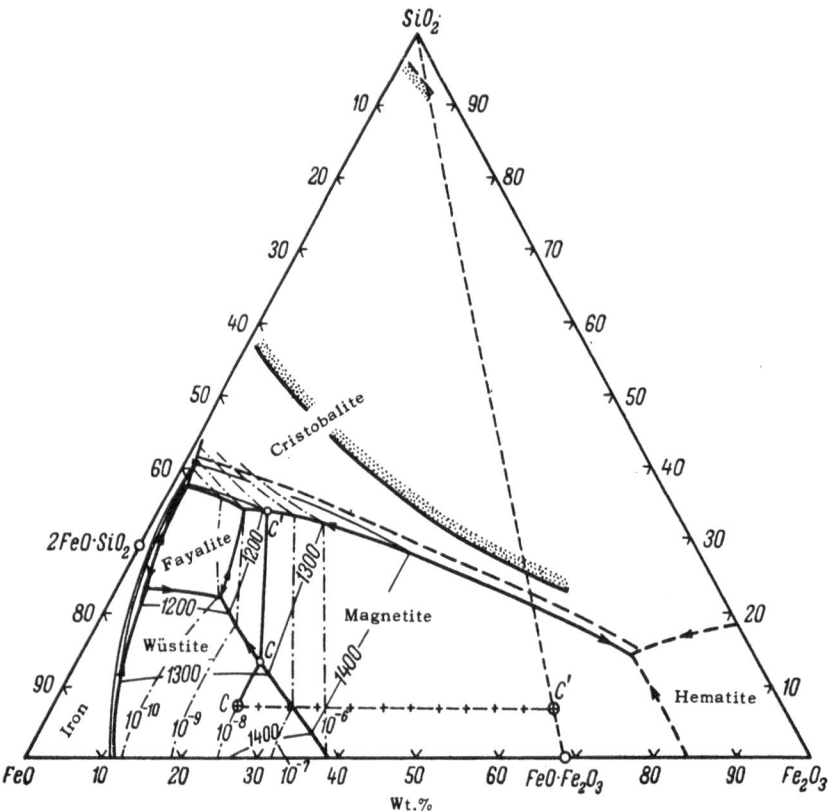

Fig. 51. Diagram of the system $FeO-Fe_2O_3-SiO_2$ showing the equilibrium crystallization routes at constant O_2 pressure. In addition to the system of lines given in Fig. 50, fine cross-dash lines are used to show the change in overall composition of the condensed phases during the crystallization process.

The liquid curve disappears at point d and the final product is a mixture of fayalite and wüstite. The composition d'''' of wüstite at this point is determined by the conode obtained by continuing the line from fayalite $2FeO \cdot SiO_2$ through point D. During crystallization the O_2 pressure changes continuously. Firstly it falls along the section $D-d$ from approximately 10^{-8} atm at point D to $10^{-10.5}$ atm at point d.

When the partial oxygen pressure is kept constant, the composition of the condensed phases in general must change during crystallization. As a result of the reaction with the gas phase there is a change in the ratio $Fe_2O_3 : FeO$ in the solid phases, while the amount of the third component SiO_2 remains constant. These changes in composition are described by a straight line passing through the O-corner of the triangle of the system $Fe-Si-O$ of which the system $FeO-Fe_2O_3-SiO_2$ is only part. According to Muan, these lines should be called o x y g e n   r e a c t i o n   l i n e s. An example of a crystallization route at constant O_2 pressure is shown in Fig. 51.

In the mixture C, wüstite begins to separate at about 1350°C and the composition of the liquid phase changes along the liquidus surface of wüstite along the isobar for 10^{-8} atm. The composition of wüstite changes continuously along the route shown by the fractionation curve. This curve is not shown in Fig. 51. When the composition of the liquid phase reaches point C, the wüstite disappears and magnetite begins to separate at constant temperature. When the last traces of wüstite have disappeared, the composition of the liquid phase begins to change along the isobar for 10^{-8} atm with further continuous separation of magnetite. When the magnetite—tridymite boundary is reached at point C', the liquid phase disappears and the final product will consist of magnetite and tridymite. During the process the overall composition changes along the straight line from C to C".

The sharp sudden change in the overall composition of the condensed phases begins when the liquid phase reaches the compositions c and c' in Fig. 51. This change is analogous to that which occurs in the binary system

Fe—O along the horizontal line which cuts through the region of two condensed phases. A similar situation arises as a result of the need for the disappearance of one of the phases before the subsequent change of the process from an isothermal to a polythermal one.

The isobaric sections for O_2 through the ternary system $FeO—Fe_2O_3—SiO_2$ are not true binary systems as the compositions of all the coexisting phases cannot be represented by means of the components selected. With the exception of this condition, the system may be regarded as binary. The compositions of the liquid phases may be represented by points along the oxygen isobar, regardless of the nature of the crystalline phase separating, which may be a chemical compound or a solid solution.

The oxygen of the atmosphere acts as a regulator for the composition of the liquid phase as regards its equilibrium composition in cases when the oxygen concentration in the atmosphere falls as a result of the crystallization process. Therefore, the lowest temperature at which the liquid may exist under such conditions is the lowest temperature of the liquidus along the O_2 isobar corresponding to the composition of the atmosphere selected. In the case of crystallization at a constant ratio $P_{CO_2}:P_{H_2}$ (third condition), the relations are identical to those obtained on replacement of O_2 isobars by lines of equal ratios $P_{CO_2}:P_{H_2}$ (Fig. 48).

Crystallization in contact with metallic iron corresponds to the experimental conditions in the work of Bowen and Schairer. Under these conditions the crystallization routes in the system $FeO—Fe_2O_3—SiO_2$ are the boundary curves between the fields of metallic iron and in parallel with an increasing SiO_2 content, wüstite, fayalite, tridymite, and cristobalite. Mixtures whose compositions lie along the boundary curves between the fields of metallic iron and wüstite crystallize in the following way: above the liquidus temperature (of the oxide phase) the liquid phase is in equilibrium with iron; with a fall in temperature wüstite will separate and the composition of the liquid phase will change along the boundary curve; at 1117°C the liquid will be already saturated with fayalite and therefore this phase begins to crystallize; with the disappearance of the last traces of the liquid phase, the sample will consist of wüstite, fayalite, and metallic iron (from 1177 to 560°C). The crystallization curves of other compositions may be analyzed similarly.

The system iron oxides—oxygen in contact with metallic iron is not a true binary system, but the crystallization routes may be treated as above.

5. Latest Research on the Manganese — Oxygen System

As manganese forms a series of oxides with various oxygen contents, many problems encountered in the system Fe—O may be studied under other thermodynamic conditions in the system Mn—O. Despite the fact that it would be of undoubted interest to study this system over quite a wide range of partial oxygen pressures (approximately from 10^{-20} to 1 atm), it is easier experimentally to work at the upper end of this range of oxygen pressures.

In the most complete investigation of recent years (Hahn and Muan, 1960), the range chosen was from 1 to $10^{-3.66}$ atm. The temperatures were limited to 1550°C by the stability of the crucibles (platinum + platinum — 20% rhodium) and at the lower limit, by the rate at which equilibrium was attained. The authors concentrated on the investigation of the equilibria

$$6Mn_2O_3 = 4Mn_3O_4 + O_2,$$

$$2Mn_3O_4 = 6MnO + O_2.$$

The solidus regions were studied for compositions lying between MnO and Mn_3O_4.

For the investigation of subsolidus equilibrium relations, the samples were heated at definite temperatures and partial oxygen pressures. After being heated under definite conditions, the mixtures of Mn_2O_3 and Mn_3O_4 or Mn_3O_4 and MnO were quenched and the phases formed were investigated by x rays. Additional experiments were also carried out by thermogravimetric analysis. In addition, samples were examined as polished sections.

The composition of the gaseous medium was controlled in various ways. Samples in which the equilibria were studied at a partial pressure of O_2 of 1 atm were heated in furnaces through which oxygen was blown from tanks. Samples studied at a pressure of 0.21 atm of O_2 were obtained in furnaces with free circulation of air.

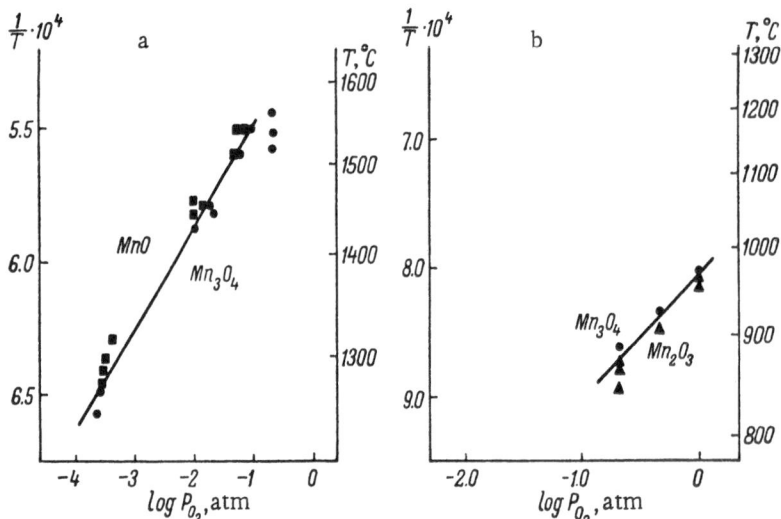

Fig. 52. Diagram of equilibrium phase relations of MnO and Mn_3O_4 (a) and Mn_2O_3 and Mn_3O_4 (b) in relation to O_2 pressure and temperature. The black squares and circles (a) and circles and triangles (b) show the stable coordinates of the existence of MnO and Mn_3O_4 (a) and Mn_2O_3 and Mn_3O_4 (b), respectively. The solid line is given for demarcation of the data.

Samples at other partial pressures of oxygen (from 1 to $10^{-2.00}$ atm) were heated in mixtures of O_2 and CO_2 of the required composition. In these cases the partial pressure of oxygen under the experimental conditions corresponded to that in the mixtures at normal temperature as the dissociation of CO_2 could be neglected at the given O_2 concentrations. Preparations for experiments with an even lower partial pressure of oxygen were obtained in an atmosphere of CO_2 or a mixture of CO_2 and H_2O. The partial pressures of oxygen in these cases were calculated from Coughlin's bulletin.

The gases used had the following degrees of purity: O_2, 99.5%; CO_2, 99.956%; and H_2, 99.5%. For the removal of traces of oxygen, the carbon dioxide was passed over copper turnings heated to 500°C. The gas flow rate was fixed with a special apparatus. The starting oxides Mn_2O_3 and Mn_3O_4 were obtained by calcination of manganese dioxide of high purity at 800°C and 1100°C, respectively. The MnO was obtained by heating MnO_2 at approximately 1100°C in an atmosphere of CO_2 and H_2 in a ratio of 10:1.

Special attention was paid to the determination of "excess oxygen," i.e., the amount of oxygen excess relative to the stoichiometric ratio Mn:O = 1:1. Approximately 0.1 g of the oxide analyzed and 0.75 g of $Fe(NH_4)_2(SO_4)_26H_2O$ were added to a solution consisting of 25 ml of water, 10 ml of sulfuric acid (1:4), and 3 ml of 49% hydrofluoric acid; the mixture was then boiled for 10 min for complete solution of the sample. After the solution had cooled to room temperature, to it were added 25 ml of distilled water and 15 ml of 2% boric acid solution. The solution was then titrated with 0.5 N $KMnO_4$ solution. The amount of excess oxygen was calculated as the difference between the total amount of ferrous iron added and the amount of it consumed in titration with permanganate.

According to the data of Hahn and Muan, equilibrium was reached rapidly when Mn_3O_4 and MnO were used as the starting material for experiments in the range 1100°C to 1800°C. Therefore, only Mn_3O_4 was used subsequently as the starting material.

It was established that air quenching does not give satisfactory results because of the interaction between the samples and atmospheric oxygen and, therefore, the samples were subsequently quenched with water. The phase conversions in Mn_3O_4 proceed quite rapidly and no high-temperature modification could be kept under normal conditions. It was established that although the structure of Mn_3O_4 changes during quenching, its composition does not undergo any appreciable changes. The compositions of Mn_3O_4 obtained as a result of the experiments lay within the range from $MnO_{1.326}$ to $MnO_{1.334}$, and no changes were observed in relation to temperature and pressure.

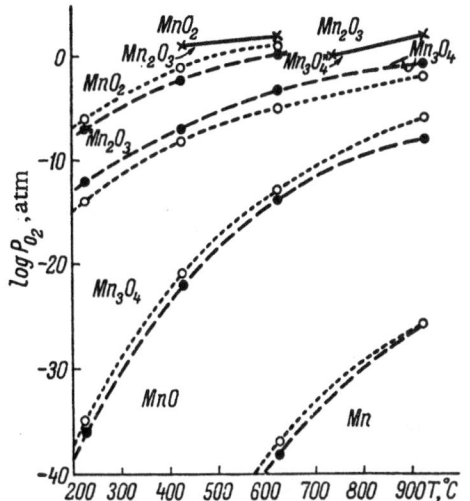

Fig. 53. Equilibrium oxygen dissociation pressures of oxides of the system Mn—O. White circles, calculations based on data from Coughlin's tables; black circles, according to the data of Tatevskaya, Chufarov, and Antonov (1948); crosses, according to the data of Klingsberg and Roy. Mn_3O_4 refers to phases formed in the presence of water. The solid line (indicated by an arrow) shows the equilibrium $Mn_2O_3 \rightarrow Mn_3O_4$ in dry air.

The equilibrium between Mn_3O_4 and Mn_2O_3 (Fig. 52b) was investigated in air at 1 atm and at other pressures. The heat of the reaction

$$6Mn_2O_3 = 4Mn_3O_4 + O_2$$

was calculated as equal to 46,200 ± 3300 kcal. The equilibrium temperatures of the dissociation of Mn_2O_3 into Mn_3O_4 were determined as 968 ± 5°C in a pure oxygen atmosphere and 877 ± 8°C in air. Figure 52a shows the conditions of equilibrium between Mn_3O_4 and MnO.

The lower temperature limit for the investigation examined was the temperature of the equilibrium Mn_3O_4- −MnO in an atmosphere of pure carbon dioxide. The ratios $CO_2 : H_2$, required to obtain the necessary partial pressures of oxygen at lower temperatures, reached values greater than 500 : 1. Under these circumstances serious difficulties arose with the gas measuring apparatus. The heat of reaction for the process

$$2Mn_3O_4 = 6MnO + O_2$$

was calculated as 118,500 ± 2500 cal.

Experiments carried out in evacuated (80% platinum and 20% rhodium) crucibles made it possible to determine the minimal liquidus temperature in the compositions studied. The eutectic between Mn_3O_4 and MnO lies at 1540°C with the curve of the MnO−Mn_3O_4 equilibrium (Fig. 52a); this value equaled approximately 0.1 atm. Moreover, the authors observed that when MnO was used as the starting material in the region of compositions and pressures realized in the field of stability of the phase MnO, in particular close to the boundary curve Mn_3O_4−MnO, quenching to normal temperatures gave mixtures of Mn_3O_4 and MnO as a result of the complex conversion of MnO into other phases with higher oxygen contents. The authors believe that this process cannot be reduced to elementary separation of Mn_3O_4 from the MnO phase during quenching.

Monovariant equilibria at various conditions of temperature and pressure for the same system, but at higher partial pressures of oxygen, were studied in 1960 by Klingsberg and Roy. They determined the monovariant curves describing the equilibria between β -MnO_2, Mn_2O_3, and O_2 at P_{O_2} = 200 atm. The data obtained differed substantially from the results of thermodynamic calculations. The equilibrium curves are given in Fig. 53. Separate experiments were carried out to study the catalytic effect of small amounts of water on the phase conversions examined.

Investigations of phase equilibria at high pressures under conditions where oxidation−reduction processes occur are of particularly great importance for the further development of the physical chemistry of silicates, as here it is possible to accelerate considerably slow low-temperature reactions as a result of the catalytic effect of oxygen which is present at high pressure. This new and important section of heterogeneous equilibria in silicate systems forms part of the general program of work of the Institute of Silicate Chemistry.

The investigations of silicate systems containing manganese oxides are less complete than those of systems containing iron despite their substantial value in metallurgy, refractory technology, and theoretical petrography. The great discrepancies in the phase diagrams published earlier for the system MnO−SiO_2 by Herty (1930), White, Howat, and Hay (1934), Glasser (1958), and Doerinkel (1911) are explained to a considerable extent by the lack of control over the partial pressure of oxygen in the gas phase at the experimental temperatures.

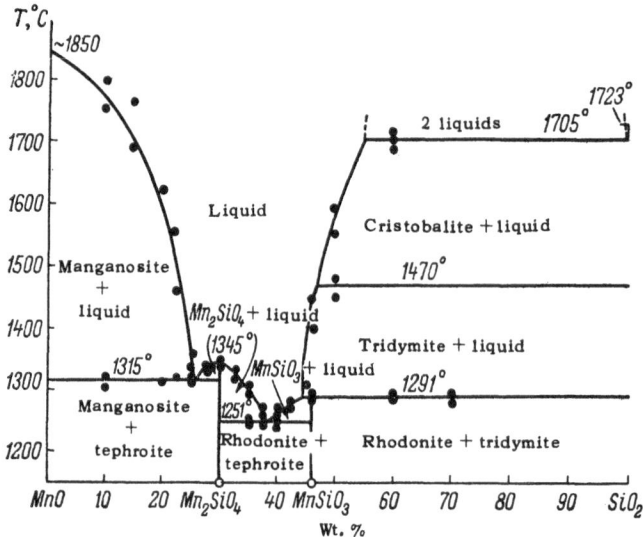

Fig. 54. Phase diagram of the system $MnO-SiO_2$ (according to Glasser's data).

In later work, Glasser took precautions to allow for this important factor. Fusions were carried out in platinum crucibles under strongly reducing conditions in an atmosphere of a gas mixture $CO_2 : H_2 = 5.1$. The gas contained 83.3 mol.% CO_2 and 16.7 mol.% H_2. The partial pressure of oxygen equaled 10^{-10} atm at 1130°C, 10^{-8} atm at 1370°C, and 10^{-6} atm at 1550°C. Moreover, a series of precautions were taken during the preparation of the samples for the experiment and a careful chemical analysis was carried out to determine the presence of high degrees of oxidation of the manganese after quenching experiments. The phase diagram of the system $MnO-SiO_2$, constructed in this way, is given in Fig. 54. In cases where the heating was carried out in air, there was strong enrichment of the samples in oxygen, which increased systematically with an increase in the basicity ($MnO : SiO_2$) of the crystalline products.

To elucidate the behavior of the systems manganese oxides—silica at high temperatures under normal conditions (heating in air or at $P_{O_2} = 0.21$ atm), Muan (1959) carried out appropriate experiments and constructed the diagram of $Mn_3O_4-SiO_2$ analogous to the system $Fe_3O_4-SiO_2$.

The examples of the systems Fe—Si—O and Mn—Si—O examined above characterize one of the most important trends in modern silicate chemistry, namely the change to the study of more general systems in which the components are not separate oxides and silica, but simpler, namely the chemical elements. However, work in this direction is as yet only in the initial stage.

BIBLIOGRAPHY

Bowen, N. L., and J. F. Schairer. Am. J. Sci. 24:177 (1932).
Chipman, J., and S. Marshall. J. Am. Chem. Soc. 62:299 (1940).
Coughlin, J. P. U. S. Bureau of Mines, Bulletin 542 (1954).
Darken, L. S., and R. W. Gurry. J. Am. Chem. Soc. 67:1398 (1945).
Darken, L. S., and R. W. Gurry. J. Am. Chem. Soc. 68:798 (1946).
Doerinkel, F. Metall und Erz 8:201 (1911).
Emmett, P. H., and J. F. Schultz. J. Am. Chem. Soc. 52:4268 (1930).
Glasser, F. P. Am. J. Sci. 256:398 (1958).
Hahn, W. C., and A. Muan. Am. J. Sci. 258:66 (1960).
Herty, C. Metals and Alloys 1:883 (1930).
Klingsberg, C., and R. Roy. J. Am. Ceram. Soc. 43(12):620 (1960).
Muan, A. Trans. AIME 203:965-976 (1955).
Muan, A. Am. J. Sci. 256:171 (1958).

Muan, A. Am. J. Sci. 257 : 297 (1959).

Schuhmann, R., K. G. Powell, and E. J. Michol. J. Metals (Sept. 1953), p. 197; Trans. AIME (1953), p. 1097.

Tatevskaya, E. P., G. I. Chufarov, and V. K. Antonov. Izv. Akad. Nauk SSSR, Otd. Tekhn. Nauk (1948), pp. 371-383.

White, J., D. D. Howat, and R. Hay. Roy. Tech. Coll. J., Glasgow 3: 231 (1934).

CHAPTER V

THERMODYNAMIC INVESTIGATIONS OF BINARY OXIDE SYSTEMS

The free energy of formation of silicates from oxides, i.e., of the process oxide + silica = silicate, is known for a limited number of these reactions. Equally few data are available for the free energies of formation of ferrites, aluminates, chromites, etc.

The heat of formation is often determined experimentally (by the solution method), while the entropy is usually obtained on the basis of various assumptions or calculated from structural data or the heat capacity. By precisely this method, Mchedlov-Petrosyan (1950, 1952, 1956), Glushkova (1957), and Glushkova and Keler (1956) calculated the free energy of formation of silicates and spinels for processes occurring in the solid state.

A more direct determination of the free energy is possible in those cases where the formation of the compound involves gaseous substances or where it is possible to investigate several oxidation—reduction processes (essentially involving gaseous materials) as a result of which we find the required free energy of formation of a compound between two oxides. It should be pointed out that the accuracy of data obtained by studying oxidation—reduction reactions is not always satisfactory.

Methods of obtaining thermodynamic characteristics of reactions between oxides by determining the electromotive forces (emf) of galvanic cells constructed with solid electrolytes are beginning to assume great importance. Work in this direction has developed only very recently and there are ever-increasing data demonstrating the promise of electrochemical investigations of solid electrolytes. By this method it has been possible to study systems of oxides of one element in different valence states and also to construct cells which make it possible to find the free energies of formation of complex compounds such as silicates, aluminates, etc.

Methods of studying oxidation—reduction equilibria involving gaseous compounds and electrochemical methods have been used in recent years for thermodynamic investigations of solid solutions of oxide systems. Some new work in this field will be reported in this chapter.

Thermodynamic methods of investigating processes have not yet been applied adequately to the chemistry of solids and the data presented below show in actual fact only searches for accurate methods of obtaining data required for describing the energy characteristics of reactions.

1. Determination of the Free Energy of Reactions in Oxide Systems by Measuring the Electromotive Forces of Galvanic Cells with Solid Electrolytes

The possibility of obtaining thermodynamic data by measuring the emf of galvanic cells containing solid electrolytes was demonstrated by Reinhold (1934), Treadwell, Amman, and Zurrer (1936), Groatto and Bruno (1947), Rose, Davis, and Ellingham (1948), and Sator (1952).

In recent years, measurements with galvanic cells containing solid electrolytes have been made by Kiukkola and Wagner (1957), other co-workers of Wagner, and also Corresponding Member, Academy of Sciences of the USSR Gerasimov and Rezukhina with a group of co-workers.

By studying galvanic cells with solid electrolytes it is possible to determine the thermodynamic characteristics of reactions of the type

$$Me' + Me''O \rightleftarrows Me'' + Me'O,$$
$$Me'O + Me''O \rightleftarrows Me' \cdot Me''O_2.$$

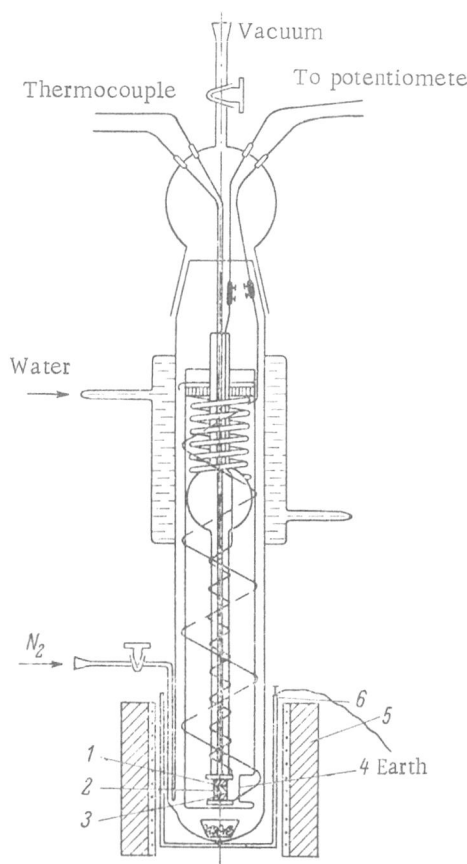

Of greatest interest from the point of view of silicate chemistry are reactions involving the combination of two oxides and also the combination of oxides with silicates,etc. Only in very recent years have there been electrochemical investigations of such reactions.

The most refined galvanic cell for the investigation of emf with solid electrolytes was developed in 1961 by Rezukhina, Lavrent'ev, Levitskii, and Kuznetsov (1961). The apparatus proposed by Kiukkola and Wagner is quite bulky, inconvenient in assembling the cell, and intended only for work in a stream of an inert gas. The first variant of cell design used in the department of Ya. I. Gerasimov, Corresponding Member of the Academy of Sciences of the USSR (Gerasimov and Vasil'eva, 1960), with a metal clamp and quartz washers as insulators did not eliminate the possibility of spontaneous short-circuiting of the cell during an experiment.

Figure 55 shows the design of the cell of Rezukhina et al. This apparatus is convenient in operation, gives a reliable contact between the parts of the cell, and can operate both in vacuum and in an atmosphere of an inert gas. The construction of the cell eliminates spontaneous short-circuiting of the cell on heating as the insulators are in contact only in the cold part of the apparatus.

Tablets of compressed powders are arranged in the form of a column one on top of another with them clamped at the top by a quartz glass tube. A small crucible of titanium powder was placed at the bottom of the outer quartz tube to avoid oxidation of the electrodes on heating of the cell by traces of oxygen remaining in the apparatus in an adsorbed state. A block of heat-resistant steel was used to obtain a more uniform temperature in the cell, and this also protected the cell from induced emf's.

Fig. 55. Cell for measuring the emf of solid electrolytes. 1-3) Tablets forming the cell; 4) side window; 5) resistance furnace; 6) heat-resistant steel block.

To check the cell, Rezukhina, Lavrent'ev, and their co-workers determined the thermodynamic properties of calcium tungstate $CaWO_4$, which are known from the study of the equilibrium with hydrogen (Rezukhina, Gerasimov, and Simanov, 1949; Yakovleva and Rezukhina, 1961). The equilibrium of the reduction of $CaWO_4$ by hydrogen involves one stage

$$CaWO_4 + 3H_2 \rightarrow CaO + W + 3H_2O,$$

and, therefore, for determining the thermodynamic properties of $CaWO_4$, it is necessary to have the cell

$$Pt\,|\,Fe_{0.95}O,\ Fe\,|\ \text{Solid electrolyte}\,|\,CaWO_4,\ CaO,\ W\,|\,Pt.$$

The reference electrode ($Fe_{0.95}O + Fe$) and the mixture of $CaWO_4 + CaO + W$ were prepared by reduction of Fe_2O_3 and $CaWO_4$, respectively, with H_2 or CO. From the reduction products were pressed tablets 5 mm in diameter and 1-3 mm thick, which were heated in vacuum at 900-1000°C for 100-120 h.

The solid electrolyte (mixed crystals in the system $CaO-ZrO_2$) was prepared by the method described below, which was proposed by Kiukkola and Wagner, but annealed finally at 1900-2000°C and not at 1450°C. This made the electrolyte more compact and hence more indifferent to the tablets in contact. Heating above 1900°C prevented decomposition of the solid solution to $CaZrO_3$ and monoclinic ZrO_2.

The equilibrium emf's were established quite rapidly and after from 3 to 8 h the variations in the emf's did not exceed 0.001-0.002 V.

The equilibrium emf's of the cell constructed correspond to the change in free energy* of the reaction ($\Delta F = -6$FarE).

$$3Fe_{0.95}O + CaO + W = 3 \cdot 0.95Fe + CaWO_4.$$

From the measured emf of this reaction it was found that $AH^0_{298} = -242.9$ kcal and $\Delta F_{298} = -221.7$ kcal. Calculation of these values from the equilibrium constants of the reaction of $CaWO_4$ with hydrogen measured at 1220-1373°K gives (with an accuracy of $\pm 2\%$), $\Delta H_{298} = -245.9$ kcal and $\Delta F_{298} = -225.3$ kcal, i.e., the discrepancies between the values does not exceed 2%. Hence, a galvanic cell in the form proposed by Rezukhina and her co-workers is quite suitable for obtaining thermodynamic data.

For valid measurements it is necessary that the "solid electrolyte" (see the galvanic cell given above) should have purely ionic conductivity. At the present time oxide systems are known which are largely solid solutions and which have purely ionic, oxygen conductivity. Such substances are solid solutions of zirconium dioxide with calcium oxide and solid solutions of thorium dioxide with oxides of rare-earth elements. These solid solutions form a fluorite structure with complete cationic and defect anionic lattices. In the work of the school of Professor Ya. I. Gerasimov, a solid solution with the composition $0.85ThO_2 + 0.15LaO_{1.5}$ is used preferentially, while Kiukkola and Wagner used a solid solution containing $0.85ZrO_2$ and 0.15 CaO. It is recommended that these solid solutions are obtained from nitrate solutions by precipitation of the mixed oxides with ammonia with subsequent calcination. A method of preparing solid solutions has also been described by Peters and Mobius (1958).

Kiukkola and Wagner made special investigations of the electrical conductivity of zirconium and thorium solid solutions, and also measured the emf's of galvanic cells with iron oxides

$$Fe, \ Fe_xO \mid (0.85ZrO_2 + 0.15CaO) \mid Fe_yO, \ Fe_3O_4.$$

Kuznetsov (1961) and Lavrent'ev (1961) studied a similar cell with a solid solution of 85 mol.% ThO_2 and 15 mol.% $LaO_{1.5}$ as the solid electrolyte. The emf of this cell obtained was compared with the emf calculated from data on the reduction of iron oxides by carbon monoxide

$$E = \frac{RT}{2\text{Far}} \ln \frac{(P_{CO_2}/P_{CO})_{II}}{(P_{CO_2}/P_{CO})_I}.$$

The good agreement of the two values of the emf showed that the solid electrolytes proposed were suitable and that their conductivity was purely ionic.

Measurements of the emf's of galvanic cells with solid electrolytes for some systems characterize the electrochemical process so accurately that it is possible to determine, for example, the nature of the electrical conductivity in the oxide systems.

By making measurements on cells in which one of the half cells contained tablets consisting of three oxides (CeO_2, ZrO_2, and CaO), Pal'guev, Karpachev, Neuimin, and Volchenkova (1960) were able to determine the fractions of ionic and electronic conductivity in the system CeO_2-ZrO_2-CaO. For the separate binary systems it is known that solid solutions of ZrO_2-CeO_2 are practically electronic conductors, while solid solutions of the oxides ZrO_2-CaO have purely ionic conductivity. For estimation of the fraction of electronic (and ionic) conductivity, the emf was measured of a cell which included the solid electrolyte investigated (melts of the system $CaO-CeO_2-ZrO_2$).

---

*By the free energy here and subsequently we mean the function $\Delta F = \Delta H - T\Delta S$. In Soviet scientific literature this function is often called the isobaric thermodynamic potential.

The authors studied cells of two types:

$$\text{Pt, } O_2(P_1) \,|\, \text{Solid electrolyte} \,|\, O_2(P_2)\text{Pt,} \qquad\qquad\qquad \text{I}$$

$$\text{Me}', \text{Me}_2'O \,|\, \text{Solid electrolyte} \,|\, \text{Me}''O, \text{Me}''. \qquad\qquad \text{II}$$

In the case of a cell of type II, the left-hand half-cell was the system Cu, Cu_2O, and the right-hand, Fe, FeO. The "solid electrolyte" in the cells studied had a variable composition. The ratio of CeO_2 to ZrO_2 was always constant, corresponding to the composition $0.75CeO_2 + 0.25ZrO_2$. To this solid solution was added from 5 to 80 moles of CaO.

In these cells the emf is determined by the oxygen pressure at the electrodes; the thermodynamic value of the emf of these cells may be calculated from the relation

$$E_0 = \frac{RT}{4\,\text{Far}} \ln \frac{P_2}{P_1},$$

where P_1 and P_2 are the oxygen pressures in the left- and right-hand sides of the cell, respectively. In cell I, the oxygen pressure is set directly, while in cell II the partial pressures of oxygen over the electrodes are determined by the dissociation pressures of the corresponding oxides, i.e., FeO and Cu_2O. The construction of cell II makes no provision for a supply of gas (oxygen) to the electrodes and its functions in vacuum.

The following considerations were taken into account in the determination of the fraction of electronic conductivity. It is known (Wagner, 1933; Karpachev and Pal'guev, 1960) that the emf of a cell whose electrolyte has ionic and also electronic conductivity (regardless of whether n- or p-type) is determined by the equation

$$E = [1 - (\bar{t}_e - \bar{t}_0)]E_0,$$

where E is the measured emf of the cell, $\bar{t}_e$ and $\bar{t}_0$ are the mean transport numbers of electrons and holes through the electrolyte, and E_0 is the thermodynamic emf of the cell (i.e., in the absence of electronic and hole conductivity in the solid electrolyte). In order to estimate the value of $(\bar{t}_e - \bar{t}_0)$, i.e., the fractions of electronic conductivity, it is sufficient to measure the emf of a cell which includes the electrolyte investigated, and to compare it with the emf calculated thermodynamically. The overall conductivity was determined by special measurements in the temperature range of 500-1000°C. The emf of the cell was determined in the same temperature range.

This investigation showed that for the system studied we are dealing with an interesting case of a gradual change of electronic into ionic conductivity and not with a relative increase in the ionic conductivity with a constant electronic component. With a content of 40 mol.% CaO at 750-800°C there is 100% ionic conductivity. Karpachev and Pal'guev drew interesting conclusions on the structure of the ternary solid solutions in the system $CaO-CeO_2-ZrO_2$. As this example shows, the emf method with solid electrolytes makes it possible to solve important problems concerning the nature of oxide systems.

The method of determining the fractions of ionic and electronic conductivity proposed by Pal'guev, Karpachev, and their co-workers supplements well and, in some cases corrects, the electrolysis which is normally used for this purpose. As was pointed out by Ol'shanskii (1959), the electrolysis method may lead to incorrect conclusions, as was the case with Tubandt, Eggert, and Schibbe (1921), who studied the character of the conductivity of solid argentine Ag_2S and chalcosine Cu_2S. As Ol'shanskii showed, in a mixed electron−ion conductor there is so-called internal electrolysis as a result of the formation of a galvanic cell which is short circuited by the electronic conductor. Tubandt came to the incorrect conclusion that these substances have purely ionic conductivity at high temperature, as he did not take into account the fact that transfer of material occurred through internal electrolysis as well as external electrolysis. The emf method is naturally free from this drawback.

The accuracy of the electrochemical method was confirmed for a series of cells which were studied by comparison of the results obtained by studying the equilibrium of the corresponding oxides with hydrogen or carbon monoxide.

Table 11. Thermodynamic Characteristics of the Reaction Nb + ½O₂ = NbO

ΔH^0_{298}, kcal	ΔS^0_{298}, e.u.	Author
−108.8	−21.57	Morozova and Getskina (1959)
−99.9	—	Morozova and Stolyarova (1960)
−97.7	−20.2	Kusenko and Gel'd (1960)
−98.4	−20.19	Lavrent'ev (1961), from emf data

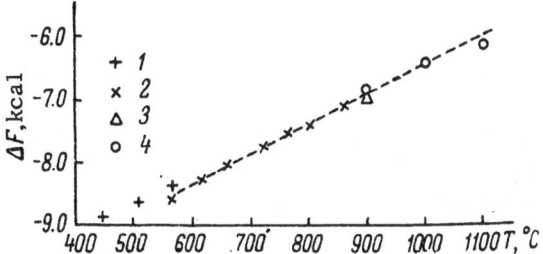

Fig. 56. Change in free energy in the reaction $CoO_{solid} + CO_{gas} = Co_{solid} + CO_{2\,gas}$. From data on the investigation of equilibria: 1) Emmett and Schultz; 2) Watanabe; 3) Schenck and Wesselkock; from measurements on emf's: 4) Kiukkola and Wagner.

Thus, Kiukkola and Wagner studied the cells

$$Fe,\ Fe_xO\,|\,(0.85ZrO_2 + 0.15CaO)\,|\,Co, CoO,$$

$$Fe,\ Fe_xO\,|\,(0.85ZrO_2 + 0.15CaO)\,|\,Ni, NiO,$$

$$Fe,\ Fe_xO\,|\,(0.85ZrO_2 + 0.15CaO)\,|\,Cu, Cu_2O,$$

and found the free energies of the reactions

$$Co\ (solid) + \tfrac{1}{2}O_2\ (gas) = CoO\ (solid),$$

$$Ni\ (solid) + \tfrac{1}{2}O_2\ (gas) = NiO\ (solid),$$

$$2Cu\ (solid) + \tfrac{1}{2}O_2\ (gas) = Cu_2O\ (solid),$$

which were compared with the free energies of these reactions found by studying gas equilibria. The good agreement of the free energy values obtained by completely independent methods is shown by Fig. 56. The data on ΔF are taken from the work of Emmett and Schultz (1930), Watanabe (1933), and Schenck and Wesselkock (1929).

In the work of Lavrent'ev (1961), emf measurements were made on the cells

$$Pt\,|\,Fe,\ Fe_{0.950}O\,|\,Solid\ electrolyte\,|\,Nb, NbO\,|\,Pt,$$
$$Pt\,|\,NbO,\ NbO_2\,|\,Solid\ electrolyte\,|\,Nb, NbO\,|\,Pt.$$

The free energies of the processes occurring in these cells,

$$Nb + \tfrac{1}{2}O_2 = NbO,$$
$$Nb + NbO_2 = 2NbO$$

could also be calculated from data on gas equilibria (equilibria with hydrogen), using the values of the heat capacity, etc.

Table 11 gives the values of ΔH^0_{298} and ΔS^0_{298} for the reaction

$$Nb + \tfrac{1}{2}O_2 = NbO,$$

obtained by studying gas equilibria and emf.

As Table 11 shows, the data of thermal investigations and emf measurements with solid electrolytes agree well.

The most detailed studies have been made of galvanic cells in which the reactions may be reduced to the transfer of oxygen from one element to another. As a result of these investigations, it was possible to determine the thermodynamic properties of various oxides. In this way, Ya. I. Gerasimov, Corresponding Member of the Academy of Sciences of the USSR, and his co-workers carried out thermodynamic investigations of oxides which form homogeneous phases of variable composition.

In the work of Kuznetsov (1961), the emf method with a solid electrolyte was used to study the thermodynamic properties of cerium oxides over the range of compositions of $Ce_2O_3 - CeO_2$. The investigation was carried out with cells of the following type:

| Pt | CeO_x x varies from 2 to 1.5 | Solid electrolyte | Fe + Wüstite | Pt. |

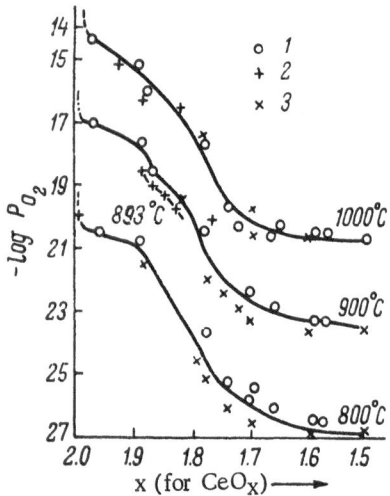

Fig. 57. Curves of the relation of the dissociation pressure of cerium oxides to the composition. 1) From emf measurements; 2) from data from the study of equilibria; 3) dynamic method (Brauer's data).

Table 12. Comparison of Values for the Free Energy ΔF of the Reaction $\frac{1}{2}W + \frac{1}{2}O_2 = \frac{1}{2}WO_2$ Obtained by Thermal and Electrochemical Methods, kcal

Temp., °K	Emf method	Study of equilibrium reduction of tungsten oxides
973	48.8 ± 1.5	49.5
1073	47.0 ± 1.5	47.5
1173	45.2 ± 1.5	45.4
1273	43.3 ± 1.5	43.3

The current-forming reaction here will be

$$\frac{1}{\delta}CeO_x + Fe_{0.947}O \rightarrow \frac{1}{\delta}CeO_{x+\delta} + 0.947Fe.$$

The free energy of this reaction, $\Delta F = -2Far \cdot E$, where Far is the Faraday number and E, the measured electromotive force.

For determining the thermodynamic parameters of the reaction

$$\frac{1}{\delta}CeO_{x+\delta} \rightarrow \frac{1}{\delta}CeO_x + \frac{1}{2}O_2,$$

which is interesting from the point of view of studying the properties of cerium oxides, it is necessary to know the free energy of formation of wüstite, i.e., $0.947Fe + \frac{1}{2}O_2 \rightarrow Fe_{0.947}O$. This value may be taken from the work of Darken and Gurry (1945) and Peters and Mobius.

By measuring the emf at various temperatures, it is also possible to determine the enthalpy and entropy of the dissociation of cerium oxide from the equations

$$\Delta H = 2Far\left[E - T\frac{dE}{dT}\right],$$

$$\Delta S = 2Far \cdot \frac{dE}{dT}.$$

The investigations of Kuznetsov described with galvanic cells containing cerium oxides gave results which could be compared with data from investigations of the reduction of cerium oxides with hydrogen and also data on the dissociation pressure of cerium oxides obtained by the dynamic method.

Figure 57 gives curves of the dissociation pressures of cerium oxides in relation to the oxygen content of the cerium oxide. The data obtained from the emf and from measurement of the reduction constants of cerium oxides by hydrogen are those of F. A. Kuznetsov. The data obtained by the dynamic method were obtained by Brauer, Gingirich, and Holtschmidt (1960). The discrepancies between $\log P_{O_2}$ and the data of Kuznetsov and Brauer do not exceed 1.5-2%. Taking into account the difference in the methods, this agreement is good.

In thermodynamic investigations of the lower oxides of tungsten, Gerasimov, Vasil'eva, Chusova, Geiderikh, and Timofeeva (1960) used the cell:

$$WO_x \,|\, 0.85ZrO_2 + 0.15CaO \,|\, Fe_{0.95}O, \; Fe,$$

where x = 2.719, 2.66, 2.39, 1.90, 1.69, and 1.45. The values for the free energy of the reaction

$$^1/_2W + {}^1/_2O_2 = {}^1/_2WO_2,$$

obtained from electrochemical measurements and from equilibrium data on the reduction of tungsten oxides by hydrogen according to Gerasimov and Vasil'eva (1959) are given in Table 12.

Table 12 shows that the agreement of the data obtained by the different methods is good for the system of tungsten oxides examined.

Galvanic cells with solid electrolytes were used by Aronson and Belle (1958) for thermodynamic investigation of the region of homogeneity of oxygen compounds of uranium $UO_2-UO_{2.20}$.

The authors constructed the galvanic cell:

$$\text{Fe, Wüstite} \mid (0.85ZrO_2 + 0.15CaO) \mid UO_{2+x} \mid Pt,$$

where x varied from 0.01 to 0.20. Uranium oxides of variable composition were obtained by oxidation or reduction (with oxygen or hydrogen) of a commercial preparation with the formula $UO_{2.03}$ and also by mixing uranium dioxide with U_3O_8 and calcining the mixture.

For the cell with uranium oxide corresponding to the formula $UO_{2.20}$, the temperature dependence of the emf showed an inflection at 940°C, indicating a phase conversion of this oxide at the given temperature.

By studying the temperature dependence of the emf of cells containing various uranium oxides, the authors were able to calculate the partial molar energies, entropies, and enthalpies of solution of oxygen in UO_{2+x}. The calculation was carried out by the formulas

$$\bar{F}_{O_2} = RT \ln P_{O_2},$$
$$\bar{S}_{O_2} = -\frac{\partial \bar{F}_{O_2}}{\partial T},$$
$$\bar{H}_{O_2} = \bar{F}_{O_2} + T \bar{S}_{O_2}.$$

The relation of the oxygen pressure to temperature and the composition of the oxide is represented by the formula

$$P_{O_2} \text{ (atm)} = 76 \exp(-33\,000/T°K) \exp(31x/1-x).$$

The authors derived theoretical formulas for calculating the free energy, entropy, and enthalpy of solution of oxygen in UO_{2+x}. In these formulas these thermodynamic values were related to the degree of nonstoichiometry and it was assumed that uranium oxides have a fluorite lattice and that the dissolved oxygen ions in it lie at interstitial points (interstitial solid solution). The fact that only "semiquantitative" agreement with the experimental data was obtained demonstrates the arbitrary nature of the initial assumptions, especially the assumption that a solution of oxygen in the oxide is ideal.

The emf method with solid electrolytes was used by Kiukkola (1962) to determine the thermodynamic properties of the system $UO_2-U_3O_8$. He used the normal galvanic cell

$$\text{Fe, Fe}_x\text{O} \mid (0.85ZrO_2 + 0.15CaO) \mid UO_{2+x},$$

and measurements with this were made in the temperature range 800-1200°C; x, which characterizes the ratio O/U, was varied from 0.0 to 0.66.

For compounds corresponding to the formulas $UO_{2.189}$, $UO_{2.219}$, and $UO_{2.239}$, the curves of the temperature dependence of the emf of the cell studied showed inflections corresponding to polymorphic conversions of these three oxides.

As the electrolyte used had purely anionic (oxygen) conductivity, it seemed possible to carry out a direct determination of the relative partial molar free energy of oxygen in uranium oxide as a function of the atomic ratio O/U and temperature. The emf of the galvanic cell studied may be represented in the following way:

$$E = \frac{(\bar{F}''_{O_2} - \bar{F}^\circ_{O_2}) - (\bar{F}'_{O_2} - \bar{F}^\circ_{O_2})}{4\,\text{Far}},$$

where $\bar{F}'_{O_2}$ and $\bar{F}''_{O_2}$ are the partial molar free energies of oxygen in the left- and right-hand sides of the galvanic cell studied, $\bar{F}^\circ_{O_2}$ is the standard molar free energy of oxygen, and Far is the Faraday number. The difference

$(\bar{F}'_{O_2} - \bar{F}^{\circ}_{O_2})$ is the relative partial free energy of oxygen over the system iron—wüstite for the temperature range 800-1200°C, which is known from the work of Kiukkola and Wagner. Thus, it is possible to determine $(\bar{F}''_{O_2} - \bar{F}^{\circ}_{O_2})$, i.e., the relative partial molar free energy of oxygen over the system $UO_2-U_3O_8$. Knowing this difference it is possible to calculate the partial pressure of oxygen over uranium oxides from the formula

$$F''_{O_2} - \bar{F}^{\circ}_{O_2} = RT \ln P_{O_2}.$$

The values of P_{O_2} obtained were found to be in good agreement with the tensimetric measurements of Biltz and Müller (1927).

As a result of his electrochemical measurements, Kiukkola was able to determine the enthalpy, free energy, and entropy of the uranium oxides studied and also the same thermodynamic functions for the formation of higher uranium oxides from UO_2 and O_2. A comparison of the high temperature enthalpy $H_{1000°K} - H_{298°K}$ obtained by Kiukkola for U_3O_8 with the corresponding data given by Popov, Gal'chenko, and Senin (1958) and Westrum and Gronvold (1959) shows that there is still a considerable discrepancy between the electrochemical (13.8 kcal/mole) and thermochemical (18.0 kcal/mole) results.

The work of Kiukkola examined demonstrates the promise of galvanic cells with solid electrolytes for studying oxides with variable oxygen contents.

The possibility of using the electrochemical method for determining the thermodynamic properties of systems with iron and manganese oxides of various valences was demonstrated by the work of Blumenthal and Whitmore (1961). These authors studied the two following galvanic cells with solid electrolytes:

$$\text{Pt} \mid \text{Fe, wüstite} \mid 0.85ZrO_2 + 0.15CaO \mid \text{Magnetite} (Fe_3O_4), \text{ hematite} (Fe_2O_3) \mid \text{Pt,} \qquad \text{I}$$

$$\text{Pt} \mid \text{Fe, wüstite} \mid 0.85ZrO_2 + 0.15CaO \mid \text{Manganosite} (MnO), \text{ hausmannite} (Mn_3O_4) \mid \text{Pt.} \qquad \text{II}$$

The magnetite—hematite and manganosite—hausmannite electrodes were prepared by prolonged heating of the mixtures $Fe + Fe_2O_3$ and $Mn + MnO_2$, respectively, in a sealed evacuated quartz tube.

The purely anionic mechanism of the electrical conductivity of the electrolyte with 85 mol.% ZrO_2 + + 15 mol.% CaO with a transport number of the oxygen ion equal to unity makes it possible to regard the current-forming process in the given cells as the transfer of oxygen from the right to the left of the cell. Then,

$$4 \, \text{Far} \, E = \Delta \bar{F}''_{O_2} - \Delta \bar{F}'_{O_2},$$

where $\Delta \bar{F}''_{O_2} = \bar{F}''_{O_2} - \bar{F}^{\circ}_{O_2}$ and $\Delta \bar{F}'_{O_2} = \bar{F}'_{O_2} - \bar{F}^{\circ}_{O_2}$. Here, $\bar{F}'_{O_2}$ and $\bar{F}''_{O_2}$ are the partial molar free energies of oxygen on the left and right sides of the cell, and $\bar{F}^{\circ}_{O_2}$ is the standard molar free energy of oxygen.

$\Delta \bar{F}'_{O_2}$, i.e., the relative partial molar free energy of oxygen in the system Fe—wüstite, is known for a wide temperature range from the numerous investigations of gas equilibria (Darken and Gurry, 1945).

As the mean of the data of different authors, Blumenthal and Whitmore give $\Delta \bar{F}'_{O_2} = -117,740 + 31.01T$ (°C) (600-1360°). Then $\Delta \bar{F}''_{O_2}$ (the relative partial molar free energy of oxygen for the mixtures magnetite—hematite or manganosite—hausmannite) will be given by

$$\Delta \bar{F}''_{O_2} = 4 \, \text{Far} \, E - 117 \, 740 + 31.01T \; (°C) \; (600—1360° \, C).$$

Having carried out several series of measurements and obtained very close results, the authors concluded that the data they obtained was of high accuracy. Thus, $\Delta \bar{F}''_{O_2}$ for the reaction

$$2Fe_3O_4 + \frac{1}{2} O_2 \rightleftarrows 3Fe_2O_3,$$

according to Blumenthal and Whitmore is given by the expression $\Delta \bar{F}''_{O_2} = -103,800 + 70.78T$ (°C), and the accuracy was estimated as ± 850 cal. The accuracy of the data obtained for the given reaction by studying gas equilibria and by thermochemical measurements is considerably lower. Thus, Richardson and Jeffes (1948), on the basis of equilibrium measurements and the heat of formation of hematite of Roth and Wienert (1934) estimated the accuracy of the determination of $\Delta\bar{F}''_{O_2}$ as 10 kcal. Coughlin (1954) believes that the data given in his summary for the standard free energy of the oxidation of magnetite have the still lower accuracy of $\pm 15,000$ cal.

According to the statement of Blumenthal and Whitmore, the relative partial molar free energy of oxygen for the mixture manganosite—hausmannite was also obtained with a high accuracy. They give for cell II, with which three series of measurements were made, $\Delta\bar{F}''_{O_2} = -92,790 + 53.62T$ (°C) and estimate the accuracy as ± 600 cal.

Thermodynamic data for the reaction

$$3MnO + \frac{1}{2}O_2 = Mn_3O_4$$

are given in Coughlin's summary (1954). The values $\Delta\bar{F}''_{O_2}$ [$-92,830 + 62.56T$ (°C)] obtained from these data differ from those obtained by the electrochemical method and Blumenthal and Whitmore explained this by the nonstoichiometry of the manganese oxides they used.

The work given above with galvanic cells with solid electrolytes give important information on the properties of individual oxides. In a study of reactions between oxides, it is naturally necessary to make a thorough study of the oxides themselves. We considered it necessary to present in detail the results of these investigations, especially as this made it possible to assess the emf method by comparison with thermochemical data. In general, it may be concluded that this method is quite reliable. We will next report the few studies in which the thermodynamic characteristics of reactions between oxides were determined by the emf method.

Several spinel-forming reactions were studied by Schmalzried (1961). He investigated reactions of the general type

$$AO + B_2O_3 = AB_2O_4,$$

for which the following cells were constructed:

$$Pt \mid A, B_2O_3, AB_2O_4 \mid (0.85ZrO_2 + 0.15CaO) \mid A, AO \mid Pt.$$

As a result of work with galvanic cells with solid electrolytes, the free energies of the following reactions were obtained:

$$NiO + Al_2O_3 = NiAl_2O_4 \quad \Delta F^0_{1000°C} = -5.2 \pm 0.2 \text{ kcal/mole},$$

$$CoO + Al_2O_3 = CoAl_2O_4 \quad \Delta F^0_{1000°C} = -4.9 \pm 0.2 \text{ kcal/mole},$$

$$NiO + Cr_2O_3 = NiCr_2O_4 \quad \Delta F^0_{1000°C} = -6.0 \pm 0.5 \text{ kcal/mole}.$$

In addition, studies were made of the following reactions, which lead to the formation of substances with a spinel structure:

$$Cu_2O + Al_2O_3 = Cu_2Al_2O_4 \quad \Delta F^0_{1000°C} = -4.0 \pm 0.4 \text{ kcal/mole},$$

$$CoO + CoTiO_3 = Co_2TiO_4 \quad \Delta F^0_{1200°C} = -2.0 \pm 0.4 \text{ kcal/mole}.$$

Some galvanic cells were found whose emf values changed with time. Thus, varying emf values were found for the cells

$$Pt \mid Co, Cr_2O_3, CoCr_2O_4 \mid (0.85ZrO_2 + 0.15CaO) \mid Ni, NiO \mid Pt,$$

$$Pt \mid Co, TiO_2, CoTiO_3 \mid (0.85ZrO_2 + 0.15CaO) \mid Ni, NiO \mid Pt.$$

Traces of oxygen were found to have a harmful effect in cells with iron oxides, despite the purification of the gas entering the cell. This oxygen oxidized the iron to FeO as a result of which measurement of the emf of the cell

$$Pt\,|\,Fe,\ Al_2O_3,\ FeAl_2O_4\,|\,(0.85ZrO_2+0.15CaO)\,|\,Ni,\ NiO\,|\,Pt$$

actually gave the emf of the cell

$$Pt\,|\,Fe,\ FeO\,|\,(0.85ZrO_2+0.15CaO)\,|\,Ni,\ NiO\,|\,Pt,$$

as the reaction which could have eliminated the ferrous oxide ($FeO + Al_2O_3 = FeAl_2O_4$) proceeds relatively slowly. Schmalzried emphasized that the main difficulty in carrying out experiments with galvanic cells in which spinel-forming reactions occur lies in the provision of a sufficiently inert atmosphere as only then is it possible to avoid a mixing potential.

Unsatisfactory results were also obtained by Fischer (1958) with a galvanic cell of solids

$$Pt\,|\,FeO,\ Al_2O_3\,|\,Al_2O_3\,|\,Pt.$$

For the free energy of the reaction occurring in this cell

$$FeO + Al_2O_3 = FeAl_2O_4$$

the values obtained lay over a very wide range (from -7.5 to -14 kcal). To obtain satisfactory results with Fischer's cell, it is necessary to know the degree of participation in current transfer of the cations and anions, and also it is necessary to establish a strictly defined partial pressure of oxygen in the surrounding atmosphere. The experimental design of the galvanic cell of Fischer evidently did not satisfy these conditions.

In experiments with cells, in one of whose half-cells equilibrium was established in the system $Ni-Cr-O$ (i.e., there was the equilibrium of the three phases Ni, Cr_2O_3, and $NiCr_2O_4$) there was a strong dependence of the emf on the flow rate of the inert gas. Schmalzried (1961) suggested that the reason for this was volatilization of the components, especially Cr_2O_3. As a result of this, experiments were carried out in an inert gas atmosphere under stationary conditions. The corresponding cell for the system $Co-Cr-O$

$$Pt\,|\,Co,\ Cr_2O_3,\ CoCr_2O_4\,|\,(0.85ZrO_2+0.15CaO)\,|\,Ni,\ NiO\,|\,Pt,$$

did not give a constant emf despite very careful experiments. The values actually measured here corresponded to mixing potentials of the cell

$$Pt\,|\,Co,\ CoO\,|\,(0.85ZrO_2+0.15CaO)\,|\,Ni,\ NiO\,|\,Pt.$$

As a reference electrode Schmalzried used the half-cell Ni, NiO/Pt, which was studied by Kiukkola and Wagner. In a study of copper−aluminum spinel the following cell was used:

$$Pt\,|\,Cu,\ Al_2O_3,\ Cu_2Al_2O_4\,|\,(0.85ZrO_2+0.15CaO)\,|\,Cu,\ Cu_2O\,|\,Pt.$$

A strong variation (10% of the mean value) was observed in the determination of the emf of this cell.

From the free energy values given above it is obvious that for spinel formation ΔF is in the range of 5-6 kcal. For the formation of $Cu_2Al_2O_4$ and Co_2TiO_4, the free energy values are lower (p. 71).

In general, it should be recognized that obtaining the free energy of spinel formation by the emf method with solid electrolytes still presents considerable difficulties. Further development of the method is required to eliminate the existing drawbacks.

Up to the present time there have been thermodynamic investigations of silicate formation in only two systems, namely $PbO-SiO_2$ and $CaO-SiO_2$.

In the work of Benz and Schmalzried (1961), which was devoted to the determination of the free energy of formation of lead silicates, the authors used two types of galvanic cell. The first type was similar to that used by Schmalzried in the study of spinels:

$$Ir\,|\,Pb,\ PbO\,|\,(ZrO_2,\ CaO)\,|\,Pb,\ PbO-SiO_2\,|\,Ir.$$

The second type coincides with the cells used by Benz and Wagner (1961), who studied calcium silicates:

$$\text{Pt, } O_2 | PbO | PbF_2 | PbO-SiO_2 | \text{Pt, } O_2. \qquad\qquad \text{II}$$

With oxygen at atmospheric pressure and at a temperature somewhat lower than 500°C, PbO changes into Pb_3O_4. It is not possible to carry out the experiments at a temperature above 500°C as the eutectic point in the system $PbO-PbF_2$ is only 494°C. For these reasons it was necessary to give up the use of the above cell, II.

It might have been possible to replace PbO by Pb_3O_4 in the left-hand half-cell and to use for the calculations thermodynamic data for the reaction

$$3PbO + \frac{1}{2}O_2 \rightleftarrows Pb_3O_4.$$

However, at the required low temperatures, the conversion of these oxides proceeds so slowly and the final product is so heterogeneous (Eberius, 1931) that it is worthwhile to turn to somewhat different cells with lead silicates:

$$\text{Pt, } O_2 | Pb_2SiO_4 + Pb_4SiO_6 | PbF_2 | Pb_2SiO_4 + PbSiO_3 | \text{Pt, } O_2, \qquad\qquad \text{III}$$

$$\text{Pt, } O_2 | PbSiO_3 + Pb_2SiO_4 | PbF_2 | PbSiO_3 + SiO_2 | \text{Pt, } O_2. \qquad\qquad \text{IV}$$

Measurements with these cells were made at 325°C, and those with cell I at 640°C.

The following cell was also studied for test experiments:

$$\text{Pt, } O_2 | Pb_4SiO_6 + Pb_2SiO_4 | PbF_2 | PbSiO_3 + SiO_2 | \text{Pt, } O_2.$$

The emf of cell I was readily reproducible and remained constant for a whole day.

The emf of the galvanic cells studied may be regarded as the difference in the chemical potentials of lead oxide. The corresponding difference in chemical potentials for silica was calculated by the Gibbs–Duhem method. For cell I we may write

$$\mu_{PbO} - \mu^0_{PbO} = -2E\,\text{Far} = F^\mu_1$$

where E is the electromotive force of the cell, Far is the Faraday constant, μ^0_{PbO} is the chemical potential of pure PbO, and μ_{PbO} is the chemical potential of lead oxide in the system $PbO-SiO_2$.

The equation given holds in the absence of electronic conductivity and, according to the data of Tubandt (1932), it is actually valid for the oxides used. Moreover, mixing potentials must be absent.

The emf of cell I was determined with the following compositions in the right-hand half-cell: $0.80PbO \cdot 0.20SiO_2$ (Pb_4SiO_6), $0.67PbO \cdot 0.33SiO_2$ (Pb_2SiO_4), and $0.50PbO \cdot 0.50SiO_2$ ($PbSiO_3$).

The values obtained for the free energy $F^\mu_1 = \mu_{PbO} - \mu^0_{PbO} = -2E\,\text{Far}$, are given in Table 13. Here we also give the values for silica $F^\mu_2 = \mu_{SiO_2} - \mu^0_{SiO_2}$ and the integral molar free energy of mixing F^μ, corresponding to the formation of the three silicates examined.

Table 13 gives the data of Richardson and Webb (1955), who calculated the free energy of mixing in the system $PbO-SiO_2$ in the liquid state using the concentration of oxygen in liquid lead in equilibrium with the system $PbO-SiO_2$. Benz and Schmalzried (1961) considered that their data were in good agreement with the values of Richardson and Webb, which were obtained at a higher temperature. The temperature dependence of cell I was not studied because of fusion of the mixture of lead silicates and an increase in the resistance of the solid electrolyte ZrO_2(+CaO). A microscopic study of the electrolyte ZrO_2(+CaO) after an experiment showed that neither PbO nor the mixture of lead silicates reacted with ZrO_2.

The results of determining the emf of cell III made it possible to find the difference in the chemical potentials of lead oxide in the mixed phases $Pb_2SiO_4 + PbSiO_3$ (μ'_{PbO}) and $Pb_2SiO_4 + Pb_4SiO_6$ (μ''_{PbO}). The difference $\mu'_{PbO} - \mu''_{PbO} = -2.3$ kcal found is close to the value -2.0 kcal obtained in the investigation of the cell I at 640°C and the data of Richardson and Webb for 718°C (-2.2 kcal).

Table 13. Free Energy of Mixing for the System $PbO-SiO_2$

x_{PbO}	Cell I (640°C)			According to the data of Richardson and Webb (718°C)		
	$-F_1^\mu$	$-F_2^\mu$	$-F^\mu$	$-F_1^\mu$	$-F_2^\mu$	$-F^\mu$
0.80	0.4	8.2	1.7	0.80	10.3	2.1
0.67	2.4	6.4	2.5	3.0	7.3	3.0
0.50	5.2	2.9	2.6	5.9	2.9	3.0

The investigation of galvanic cells by Benz and Schmalzried made it possible to find the change in free energy for the formation of lead silicates from the oxides in the solid state:

$$PbO + SiO_2 = PbSiO_3 \quad \Delta F^0 = -5,2 \text{ kcal/mole},$$

$$2PbO + SiO_2 = Pb_2SiO_4 \quad \Delta F^0 = -7,5 \text{ kcal/mole},$$

$$4PbO + SiO_2 = Pb_4SiO_6 \quad \Delta F^0 = -8,5 \text{ kcal/mole}.$$

Of great interest is the thermodynamic investigation by Benz and Wagner (1961) of the reaction

$$CaO + SiO_2 = CaSiO_3,$$

which was carried out by means of galvanic cells with solid electrolytes. The authors studied a cell of the following form:

$$Pt, O_2 (gas) | CaO (solid) | CaF_2 (solid) | CaSiO_3 (solid), SiO_2 (solid) | Pt, O_2 (gas) .$$

In the production of 2Far of electricity the reaction on the left-hand side will be*

$$CaO (solid) + 2Far^- \left(\overset{2F^-}{\longleftarrow}\right) = CaF_2 (solid) + {}^1/_2 O_2 (gas) + 2e^-$$

and, on the right-hand side

$$SiO (solid) + {}^1/_2 O_2 (gas) + 2e^- \left(\overset{Ca^{2+}}{\longleftarrow}\right) = CaSiO_3 (solid) + 2Far^-.$$

The sum of these reactions leads to the overall process

$$CaO + SiO_2 = CaSiO_3.$$

Direct measurements in this form are difficult because of the very low electrical conductivity of calcium oxide, silica, and calcium silicates, and also the slowness with which the equilibrium state is established.

To eliminate these drawbacks, the authors added to the right and left half-cells a mixture of the fluorides CaF_2, KF, and NaF, which formed a eutectic at 710°C. The amount of fluorides added was small and could not have a substantial effect on the solid single crystal of CaF_2, which remained as the diaphragm separating the two half-cells. Moreover, it was observed that at the experimental temperatures the electrochemical equilibrium of the oxygen electrodes was reached very slowly and then only with the use of "catalysts," which consisted of oxygen compounds of elements of variable valence. Cr_2O_3, $K_2Cr_2O_7$, and PbO were tested. These "catalysts" (0.03-0.05 mole per mole of CaO) were added to each of the half-cells adjacent to the platinum electrode.

As the solid electrolyte separating the two half-cells, the authors used a single crystal of CaF_2, 5 × 5 × 15 mm. CaF_2 is an ionic conductor in which fluorine ions migrate. Ure (1957) considers that the fluorine ion moves through interstitial points.

* The arrows show the direction of movement of the ions F^- and Ca^{2+}.

Table 14. Emf of the Cell Pt, O_2| CaO| CaF_2| $CaSiO_3 \cdot SiO_2$| Pt, O_2
with Various "Catalysts"

T, °C	"Catalyst" content in moles per mole of CaO	Time, h	Emf, mV
785	0.03 Cr_2O_3	7	462±2
625	0.03 $K_2Cr_2O_7$	24	463±2
665	0.03 $K_2Cr_2O_7$	5	465±2
670	0.03 $K_2Cr_2O_7$	3	463±2
750	0.03 $K_2Cr_2O_7$	5	463±1
875	0.03 $K_2Cr_2O_7$	1	463±2
662	0.05 PbO	24	460±1
700	0.05 PbO	6	460±1
730	0.05 PbO	2	460±1
763	0.05 PbO	2	461±1
810	0.05 PbO	5	460±2
830	0.05 PbO	3	458±1

Mean value 461±4

Table 15. Partial Molar Free Energies of Mixing of CaO and SiO_2
in the System $CaO-SiO_2$ at 700°C

Phases	Mole fraction of SiO_2	F_1, kcal	F_2, kcal
CaO (I) + Ca_2SiO_4	0 to 0.333	0.0	—32.3±0.5
Ca_2SiO_4 + $Ca_3Si_2O_7$	0.333 to 0.400	— 2.8±0.2	—26.7±0.8
$Ca_3Si_2O_7$ + $CaSiO_3$	0.400 to 0.500	—19.3±0.3	— 2.0±0.5
$CaSiO_3$ + SiO_2(V)	0.500 to 1.000	—21.3±0.20	0.0

In addition to cells of the above form, the following cells were studied:

Pt, O_2 (gas)| CaO (solid)| CaF_2 (solid)| $Ca_3Si_2O_7$ (solid), $CaSiO_3$ (solid)| Pt, O_2 (gas),

Pt, O_2 (gas)| CaO (solid)| CaF_2 (solid)| Ca_2SiO_4 (solid), $Ca_3Si_2O_7$ (solid)| Pt, O_2 (gas).

The authors studied the temperature dependence of the emf of the galvanic cells they constructed. It was found that this dependence was so slight that it could be neglected. This insignificant temperature dependence is understandable if we take into account the fact that the entropy changes are small in reactions between solids.

Symmetrical cells in which the only differences lay in the "catalysts," namely, chromium compounds of different valences, were used for determining the reproducibility. The emf of these cells was very small (1-2 mV), while the lowest emf of the cells studied in which silicate formation occurred was 60 mV.

To give an idea of the reproducibility of the results of measuring the emf with solid silicates, Table 14 gives data for the cell

Pt, O_2| CaO| CaF_2| $CaSiO_3$, SiO_2| Pt, O_2.

Regarding the free energy obtained by means of the emf as the difference in chemical potentials of CaO in the left and right parts of the galvanic cell, we have

$$F_1 = \mu_{CaO} - \mu^0_{CaO},$$

where μ^0_{CaO} is the chemical potential of pure calcium oxide and μ_{CaO} is the chemical potential of CaO in the corresponding phase.

Table 16. Values of the Integral Free Energy of Mixing
for the System $CaO-SiO_2$ at 700°C

Phases	X_1	F, kcal	
		Emf method	Calorimetric method
II (Ca_2SiO_4)	0.333	-10.8 ± 0.1	-10.5 ± 0.1
III ($Ca_3Si_2O_7$)	0.40	-12.4 ± 0.15	—
IV ($CaSiO_3$)	0.50	-10.6 ± 0.15	-10.6 ± 0.1

The three galvanic cells studied made it possible to determine the partial molar free energies of mixing of CaO and SiO_2 in the system $CaO-SiO_2$. Table 15 gives the corresponding data for 700°C.

To determine the partial molar free energy of mixing of SiO_2 as the second component it is possible to use the Gibbs−Duhem equation. As the values of F_1 and F_2 are constant in each two-phase region, integration of the Gibbs−Duhem equation leads to the following algebraic formulas:

$$F_2(III+IV) = -[F_1(III+IV) - F_1(IV+V)],$$

$$F_2(II+III) = -^3/_2 F_1(II+III) + ^1/_2 F_1(III+IV) + F_1(IV+V),$$

$$F_2(I+II) = +^1/_2 F_1(II+III) + ^1/_2 F_1(III+IV) + F_1(IV+V).$$

The integral molar free energy of mixing, i.e., the free energy of formation of silicates, may be calculated from the formula

$$F = x_2 \int_{x_2}^{1} \left(\frac{F_1}{x_2^2}\right) dx_2,$$

where x_2 is the mole fraction of SiO_2. As F_1 is constant within the limits of each two-phase field, we obtain the following equations:

$$F(x_2=0.5) = 0.5F_1(IV+V),$$

$$F(x_2=0.4) = 0.4F_1(IV+V) + 0.2F_1(III+IV),$$

$$F(x_2=0.333) = 0.333F_1(IV+V) + 0.167F_1(III+IV) + 0.167F_1(II+III).$$

The values of F from calorimetric data given in Table 16 were obtained by calculations from the values of ΔH^0 for the reactions

$$2CaO \text{ (solid)} + SiO_2 \text{ (quartz)} = Ca_2SiO_4 \text{ (solid)},$$

$$CaO \text{ (solid)} + SiO_2 \text{ (quartz)} = CaSiO_3 \text{ (wallastomite)}.$$

For the first reaction, King (1957) gives the value $\Delta H_{298}^0 = -30.19 \pm 0.23$ kcal and for the second reaction, according to Torgeson and Sahama (1948), $\Delta H_{298} = -21.25 \pm 0.13$ kcal. The entropies of these reactions may be taken from Circular No. 500 of the National Bureau of Standards (1952). For the first reaction, $\Delta S_{298} = +1.5$ en. units, and for the second, $\Delta S_{298} = +0.1$ en. units.

To reach 700°C (973°K) it is possible to use Kelley's (1959) tables. As the table shows, the values of the free energy of these two reactions obtained from electrochemical and calorimetric measurements agree well.

The emf method with solid electrolytes was used by Rezukhina, Levitskii, and Kazimirova (1961) for studying the thermodynamics of the formation of magnesium molybdate $MgMoO_4$.

The authors used three galvanic cells:

$$\text{Pt} \mid \text{MgMoO}_4, \text{MgMoO}_3 \mid \begin{array}{c}\text{Solid}\\\text{electrolyte}\end{array} \mid \text{Wüstite, Fe} \mid \text{Pt,} \qquad \text{(I)}$$

$$\text{Pt} \mid \text{MgMoO}_3, \text{MgO, Mg} \mid \begin{array}{c}\text{Solid}\\\text{electrolyte}\end{array} \mid \text{Wüstite, Fe} \mid \text{Pt,} \qquad \text{(II)}$$

$$\text{Pt} \mid \text{MgMoO}_3, \text{MgMoO}_4 \mid \begin{array}{c}\text{Solid}\\\text{electrolyte}\end{array} \mid \text{MgO, Mo, MgMoO}_3 \mid \text{Pt.} \qquad \text{(III)}$$

All the electrodes of these cells were made from phases in equilibrium with each other in the form of tablets. The tablets were heated in evacuated quartz ampoules on an indifferent backing at 900-1000°C. The phase composition of the tablets after heating and after the emf measurements was checked by x-ray diffraction.

Solid solutions in the system $\text{ThO}_2-\text{La}_2\text{O}_3$ or ZrO_2-CaO were used as the "solid electrolyte." The absence of electronic conductivity from these systems was checked by measurements of the emf of the cell

$$\text{Pt} \mid \text{Wüstite, Fe} \mid \text{Solid electrolyte} \mid \text{Fe}_3\text{O}_4, \text{wüstite} \mid \text{Pt.}$$

The data thus obtained were in good agreement with the data presented by Kiukkola and Wagner (1957). Measurements with the galvanic cells were made in the range of 1100-1300°K. The current-forming reaction in cell I will be

$$0.947\text{Fe} + \text{MgMoO}_4 = \text{Fe}_{0.947}\text{O} + \text{MgMoO}_3$$

and, in cell II,

$$\text{Fe}_{0.947}\text{O} + {}^1\!/_2\text{MgO} + {}^1\!/_2\text{Mo} = {}^1\!/_2\text{MgMoO}_3 + 0.947\text{Fe}$$

The current-forming reaction in cell III,

$$\text{MgMoO}_4 + {}^1\!/_2\text{MgO} + {}^1\!/_2\text{Mo} = {}^3\!/_2\text{MgMoO}_3$$

is the overall reaction of cells I and II. Thus, cell III served as a check on the results obtained with cells I and II. Below we give the emf values $E_I + E_{II}$ and E_{III} for two temperatures.

$$1239°\text{K} \quad E_{III} = 0.245 \text{ V}, \quad E_I + E_{II} = 0.241 \text{ V},$$
$$1283°\text{K} \quad E_{III} = 0.277 \text{ V}, \quad E_I + E_{II} = 0.270 \text{ V}.$$

The agreement between these values is quite satisfactory. It is important to note that the study of gaseous reducing reactions for magnesium−molybdenum systems could not give reliable thermodynamic data because of the high values of the equilibrium constants. In this case, the electrochemical method has the advantage over thermal methods.

From the measurements of the emf of cells I and II we obtain the change in free energy ΔF of the reaction

$$3\text{Fe}_{0.947}\text{O} + \text{MgO} + \text{Mo} = \text{MgMoO}_4 + 3.0.947\text{Fe}.$$

By using thermodynamic data from the literature for the system

$$0.947\text{Fe} + {}^1\!/_2\text{O}_2 = \text{Fe}_{0.947}\text{O},$$

it is possible to find the value of ΔF of the reaction

$$\text{MgO} + \text{MoO}_3 = \text{MgMoO}_4.$$

To obtain the standard values of the thermodynamic functions of the reaction given, Rezukhina and her co-workers used data from handbooks for the heat capacities, entropies, and enthalpies of MgO, Mo, and O_2 and their own values for the heat capacity of MgMoO_4. These data made it possible to determine the enthalpy of the reaction

$$MgO + MoO_3 = MgMoO_4,$$

which was found to equal $\Delta H_{298}^0 = -19$ cal. Obtaining thermodynamic values for reactions between oxides is known to present great difficulties because of the slowness of the reactions and the smallness of the heat effects. Determination of the heat effect of the reaction between MgO and MoO_3, which was carried out by Tamman and Westerhold by means of the heating curve of a mixture of MgO and MoO_3, gave a value which differed strongly from the data of Rezukhina, Levitskii, and Kazimirova. Tamman gave $\Delta H_{298} = 7$ kcal, which cannot be regarded as correct at all. No other determinations of the thermal effect of the reaction between magnesium oxide and molybdenum trioxide have been reported in the literature.

Though in this work galvanic cells did not give direct data for the thermodynamic characterization of the reaction

$$MgO + MoO_3 = MgMoO_4,$$

with the aid of supplementary calculations, using the required constants, it was possible to obtain the thermodynamic characteristics of a reaction between oxides.

2. Some Oxidation — Reduction Equilibria Used for Obtaining Thermodynamic Characteristics of Reactions Between Oxides

A series of reactions have been studied and the results have made it possible to obtain values for the free energy of spinel formation

$$R'O + R_2''O_3 = R'R_2''O_4.$$

Fricke and Weitbrecht (1942) determined the equilibrium $CO-CO_2$ over a three-phase mixture of Ni, Al_2O_3, and $NiAl_2O_4$, i.e., they studied the equilibrium of the reaction

$$NiAl_2O_4 + CO \rightleftarrows Ni + Al_2O_3 + CO_2.$$

The thermodynamic data obtained for this reaction made it possible to find the free energy of the reaction

$$NiO + Al_2O_3 = NiAl_2O_4,$$

which was found to equal -5.0 kcal (for 1000°C).

Boricke and Bangert (1945) investigated the equilibrium of the reduction reaction

$$FeCr_2O_4 + H_2 \rightleftarrows Fe + Cr_2O_3 + H_2O.$$

The data obtained by Boricke and Bangert made it possible to determine the free energy of the reaction

$$FeO + Cr_2O_3 = FeCrO_4,$$

which was found to equal (for 1000°C), -8.2 kcal.

Chen and Chipman (1947) studied the following reduction reactions of iron and chromium:

$$FeCr_2O_4 + 4H_2 \rightleftarrows Fe \ (liq) + 2Cr + 4H_2O,$$
$$Cr_2O_3 + 3H_2 \rightleftarrows 2Cr + 3H_2O.$$

The free energy $\Delta F \ (1870°) = -9.5$ kcal was obtained for the spinel-forming reaction

$$FeO \ (liq) + Cr_2O_3 = FeCr_2O_4.$$

This value is more negative than the value obtained for this reaction by Boricke and Bangert as, according to the latter authors, ΔF increases with a rise in temperature.

From data for the reactions

$$3Fe_2O_3 = 2Fe_3O_4 + \frac{1}{2}O_2,$$

$$Fe_3O_4 + CO = 3FeO + CO_2 \quad \text{and} \quad CO + \frac{1}{2}O_2 = CO_2,$$

which were obtained by Schmahl (1941) and Darken and Gurry (1945), Wagner calculated the free energy of the reaction

$$FeO + Fe_2O_3 = Fe_3O_4$$

and obtained for 1000°C, $\Delta F° = -7.5$ kcal.

The thermal characterization of mullite and a detailed thermodynamic treatment of the formation of $3Al_2O_3 \cdot 2SiO_2$ from the oxides are of great interest for elucidating the stability of mullite under various conditions and the interaction of mullite with other compounds of alumina and silica, namely, sillimanite, andalusite, and kyanite.

Attention is attracted by the high value of the free energies of formation of sillimanite, andalusite, and kyanite (from −50 to −25 kcal per mole of the corresponding oxide). For other silicate compounds the free energy of formation is much lower. Thus, for calcium and magnesium silicates, the free energies of formation ΔF are −10 and −4 kcal, respectively.

Flood and Knapp (1957) give all the known free energy values for the systems $CaO-Al_2O_3-SiO_2$, $MgO-Al_2O_3-SiO_2$, and $FeO-Al_2O_3-SiO_2$. In general, the free energy of formation of compounds in these systems is between 0 and −10 kcal.

Some conclusions on the free energy of mullite have been drawn on the basis of approximate calculations and indirect considerations. Clark, Robertson, and Birch (1957) studied the equilibrium between minerals of the sillimanite group at high pressures. The heat of conversion of one mineral of the sillimanite group into another was estimated as −(500-1000) cal. ΔF^0 will be of the same order since ΔS^0 is very small.

The first experimental investigation of oxidation−reduction equilibria making it possible to find the thermodynamic properties of mullite, which was carried out by British scientists, is of great interest. For determining the free energy of mullite, Cooper, Kay, and Taylor (1961) used a reaction which had already been used to determine the activity of silica in slag solutions. In the latter case, a study was made of the equilibrium pressures for the reaction

$$SiO_2 + 3C = SiC + 2CO . \tag{1}$$

If P° is the pressure of carbon monoxide formed by this reaction (the starting substance is pure SiO_2) and P is the pressure over the slag, the activity of silica in the slag will be given by the expression $a_{SiO_2} = (P/P^0)^2$.

To find the free energy for formation of mullite, in addition to the reaction given above, Cooper and his co-workers studied the equilibrium

$$3Al_2O_3 \cdot 2SiO_2 + 6C = 3Al_2O_3 + 2SiC + 4CO.$$

In the two reactions examined, all the substances apart from CO are in the solid state and for determining the free energy of formation of mullite by the reaction

$$3Al_2O_3 + 2SiO_2 = 3Al_2O_3 \cdot 2SiO_2$$

it is only necessary to determine the pressure of carbon dioxide obtained in these two reactions. If the coefficients of reaction (1) are doubled, it is readily seen that the summation of the reactions

$$2SiO_2 + 6C = 2SiC + 4CO$$

$$3Al_2O_3 + 2SiC + 4CO = 3Al_2O_3 \cdot 2SiO_2 + 6C$$

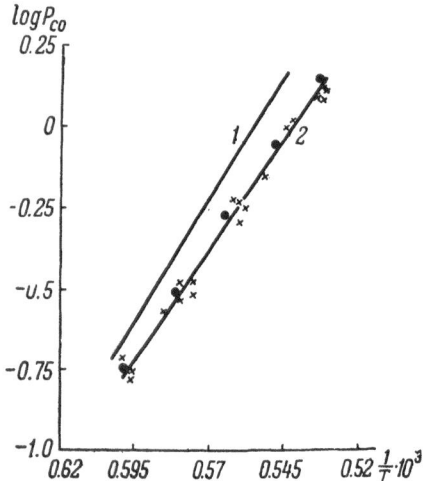

Fig. 58. Logarithmic relation of the partial pressure of CO to the reciprocal temperature for the reactions 1) $SiO_2 + 3C$ = $SiC + 2CO$; 2) $3Al_2O_3 \cdot 2SiO_2 + 6C$ = $3Al_2O_3 + 2SiC + 4CO$.

gives

$$3Al_2O_3 + 2SiO_2 = 3Al_2O_3 \cdot 2SiO_2.$$

The method used by Cooper and his co-workers was described in an article by Baird and Taylor (1958).

The same result, i.e., the determination of the free energy of formation of mullite, may be achieved if we consider the activity of silica in two equilibrium processes.

Gaseous silicon monoxide SiO is known to be formed in the reduction of silica and silicates and in the construction of the graphs given, appropriate corrections were introduced for the vapor pressure of silicon monoxide.

From the data in Fig. 58, the following expression was obtained for the free energy of formation of mullite (for the temperatures 1673-1873°K): $\Delta F = 26{,}500 - 18\,T$.

Cooper, Kay, and Taylor referred critically to the equation derived and considered that both ΔH and the entropy term differed considerably from the true values. It may be stated definitely that both of these terms are positive and large in absolute magnitude. The difference in these values, i.e., the free energy of formation of mullite, may be regarded as quite reliable, especially if restricted to the middle of the temperature region.

The entropy of mullite per mole of oxide present in this compound will equal $^{18}\!/_5$ = ~3.5 en. units. This generally high value is, however, commensurate with the corresponding values for some other silicates and aluminates. Thus, $\Delta S°$ per mole of oxide in the compound is as follows:

for $CaO \cdot SiO_2$ . . . . 0.4 en. unit
for $2CaO \cdot SiO_2$. . . . . 1.9 "   "
for $CaO \cdot Al_2O_3$. . . . . 2.1 "   "
for $12CaO \cdot 7Al_2O_3$ . . . . . . 2.6 "   "
for $3CaO \cdot Al_2O_3$. . . . . 1.6 "   "

From the equation given for the free energy of formation of mullite, it follows that with a fall in temperature mullite becomes less stable, and at a temperature of about 1200°C, the free energy of formation of mullite equals zero. Below this temperature mullite is thermodynamically unstable with respect to quartz and corundum. Taking into account the indefiniteness of the entropy term, we can only talk of some region close to 1200°C (within a range of 50-100°), where ΔF^0 of mullite equals zero.

In a discussion, Richardson (1961) also noted that a mixture of aluminum and silicon hydroxides should be heated to 1700°C for the x-ray diffraction diagram to have only lines characteristic of mullite. Taylor (1961) believes that the formation of mullite on heating of clays to 900°C is not contradictory to his thermodynamic conclusions. Mullite may be obtained as an intermediate metastable phase between the unstable product of the dehydrated argillaceous mineral and the stable mixture of silica and alumina. As a check on the conclusions of Cooper, Kay, and Taylor (1961), it would be interesting to determine the conditions of formation of mullite from the oxides in the complete absence of mineralizers.

Cooper, Kay, and Taylor give a new variant of the phase diagram of the system $Al_2O_3-SiO_2$, which provides for the decomposition of mullite below 1200°C. As the basis they took the phase diagram of this system proposed by Toropov and Galakhov (1958) with some changes introduced by Aramaki and Roy (1962). The phase diagram of the system $Al_2O_3-SiO_2$ according to Cooper, Kay, and Taylor is given in Fig. 59. Assuming that mullite decomposes in the solid state at 1200°C, the authors drew somewhat differently the boundary of the region of existence of solid solutions of mullite. Below 1200°C naturally there cannot be stable solid solutions of mullite with either silica or alumina.

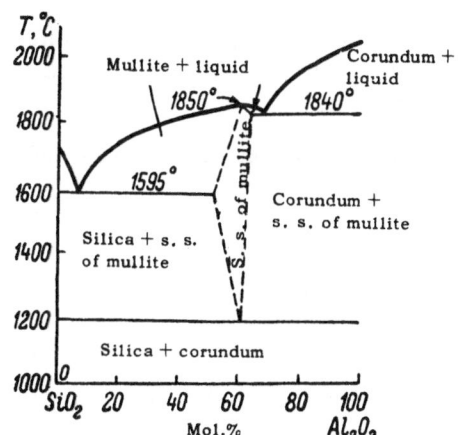

Fig. 59. Variant of the system Al_2O_3- $-SiO_2$, according to Cooper, Kay, and Taylor. (s.s. = solid solution.)

Cooper, Kay, and Taylor give "free energy of formation — temperature" diagrams for andalusite, sillimanite, and mullite (Fig. 60). Kyanite is not shown on this figure as Clark, Robertson, and Birch established that this modification is unstable at normal pressure, changing into andalusite and sillimanite. The relative positions of andalusite and sillimanite have not been established accurately up to now and the change in ΔF for them is shown approximately by broken lines. Andalusite is probably more stable, but taking into account the fact that sillimanite is formed from andalusite at high pressures and the molar volumes of these compounds are very similar (49.0 and 51.6 cm^3), it may be assumed that the free energies of formation of andalusite and sillimanite are similar.

Cooper and his co-workers consider that the free energies of formation of sillimanite and andalusite must be close to zero. It is arbitrarily assumed that the ΔF lines for sillimanite and andalusite intersect the line for mullite at a temperature of 1200°C.

The thermodynamic characteristics of the formation of the compound $12CaO \cdot 7Al_2O_3$ were obtained by studying the aluminum-thermal reduction of MgO in the presence of CaO. According to Grjotheim, Herstad, and Toguri (1961), this process may be represented by the equation

$$12CaO \text{ (solid)} + 21MgO \text{ (solid)} + 14Al \text{ (liq)} =$$
$$= 12CaO \cdot 7Al_2O_3 \text{ (solid)} + 21Mg \text{ (gas)}.$$

The authors discussed the possibility of the formation of other calcium aluminates by the reduction of magnesium. Taits (1957) considered that the reaction may proceed by the scheme

$$5CaO + 9MgO + 6Al = 5CaO \cdot 3Al_2O_3 + 9Mg.$$

There is also the possibility of the formation of tricalcium hexaaluminate

$$3CaO + 9MgO + 6Al = 3[CaO \cdot Al_2O_3] + 9Mg.$$

Grjotheim and his co-workers consider that at temperatures in the range of 886-1035°C, $12CaO \cdot 7Al_2O_3$ will be obtained. The compound $3CaO \cdot Al_2O_3$ must be formed at higher temperatures and below 1200°C it dissociates according to the reaction

$$7[3CaO \cdot Al_2O_3] \text{ (solid)} = 12CaO \cdot 7Al_2O_3 \text{ (solid)} + 9CaO \text{ (solid)}.$$

X-ray diffraction investigations also confirmed the formation of the compound $12CaO \cdot 7Al_2O_3$. Determination of the vapor pressure of magnesium by a method described in previous work of Grjotheim, and also the use of tabular data on the thermodynamic properties of the individual compounds, made it possible to find the change in free energy and entropy of this reaction, and then the formation of $12CaO \cdot 7Al_2O_3$

$$12CaO + 7Al_2O_3 = 12CaO \cdot 7Al_2O_3.$$

The changes in the standard enthalpy and entropy of the latter reaction are $\Delta H_{298}^0 = -18,500$ cal and $\Delta S_{298}^0 = 46.31$ en. units. According to King (1957), the standard entropy of the compound $12CaO \cdot 7Al_2O_3$, found from measurements of the low-temperature heat capacity, is 50 en. units, which is very close to the value found by studying the aluminum-thermal reduction. The free energy of formation of calcium aluminate $12CaO \cdot 7Al_2O_3$ from the oxides may be represented by the equation

$$\Delta F_T^\circ = -18\,500 - 50.6T \pm 1.5 \text{ kcal.}$$

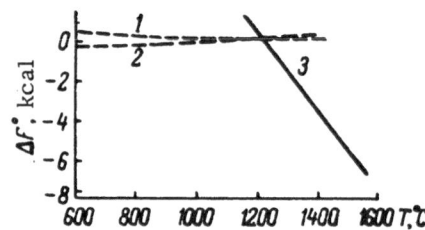

Fig. 60. Free energies of formation in relation to temperature. 1) Andalusite; 2) sillimanite; 3) mullite.

The work of Grjotheim, Herstad, and Toguri shows the reliability of obtaining thermodynamic values for compounds formed as a result of pyrometallurgical reactions.

Lebedev and Levitskii (1961) determined the free energy of the reactions

$$2Ni + SiO_2 + O_2 = Ni_2SiO_4, \tag{1}$$

$$2NiO + SiO_2 = Ni_2SiO_4 \tag{2}$$

by studying the two oxidation—reduction equilibria

$$0.5Ni_2SiO_4 + CO \rightleftarrows Ni + {}^1\!/_2SiO_2 + CO_2, \tag{3}$$

$$NiO + CO \rightleftarrows Ni + CO_2, \tag{4}$$

using a circulation method with an automatic gas analyzer.

For the free energy of the reaction

$$CO_2 \rightleftarrows CO + 0.5O_2$$

it was assumed that $\Delta F^0 = 67,500 - 20.75$ T, according to Peters and Mobius (1958). The free energies of the silicate formation reactions had the following values: for reaction (1), $\Delta F^0 = -119,800 + 39.35$ T (800-1100°C); for reaction (2), $\Delta F^0 = -4932 - 0.267$ T (800-1100°C).

The heat capacities of SiO_2, O_2, NiO, and Ni required for calculating the standard values of the enthalpy and entropy of formation of Ni_2SiO_4 were taken from the book of Kubaschewski and Evans (1954). The heat capacity of Ni_2SiO_4 was assumed to equal that of Fe_2SiO_4. Then, for reaction (1), we have $\Delta H^0_{298} = -124$ kcal/mole and $\Delta F^0_{298} = -110$ kcal/mole; and for reaction (2), $\Delta H^0_{298} = -7.0$ kcal/mole and $\Delta F^0_{298} = -6.4$ kcal/mole.

Taking for the enthalpy and free energy of the reaction

$$Si + O_2 = SiO_2$$

the values given in Circular No. 500 of the National Bureau of Standards, $\Delta H^0_{298} = -205.4$ kcal/mole and $\Delta F^0_{298} = -192.4$ kcal/mole, we find for the reaction

$$2Ni + Si + 2O_2 = Ni_2SiO_4$$

$\Delta H^0_{298} = -329.4$ kcal/mole and $\Delta F^0_{298} = -302.4$ kcal/mole.

3. Thermodynamic Activity of Oxides in Solid Solutions of Oxide Systems

The activity of oxides has been determined almost exclusively in oxide systems in the liquid (molten) state. These investigations have been particularly extensive with slags and a different method, namely an electrochemical method, has been used. The study of the activity of the components in solid oxide systems has only just begun.

Hahn and Muan (1961) determined the activity of nickel oxide in the solid solutions NiO—MgO and NiO—MnO. In the system NiO—MgO there is complete miscibility at 900°C. According to the data of Passerini (1929), there is limited solubility in the system NiO—MnO at 350°C; for the same system, Hahn and Muan observed complete solubility in the solid state above 1100°C.

For determination of the activity of nickel oxide in a solid solution, the following equilibrium was studied:

$$2NiO \, (\text{solid solution}) \rightleftarrows Ni \, (\text{metal}) + O_2 \, (\text{gas}).$$

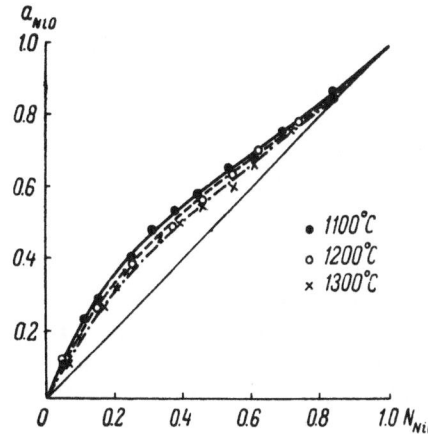

Fig. 61. Activity of NiO as a function of composition in the system NiO—MnO at 1100, 1200, and 1300°C.

The activity of nickel oxide in a solid solution is determined through the partial pressure of oxygen in equilibrium: 1) with metallic nickel and the solid solution, $P_{O_2}^*$, and 2) with metallic nickel and pure NiO, P_{O_2}.

$$a_{\text{NiO}} = \left(\frac{P_{O_2}}{P_{O_2}^*} \right)^{1/2}.$$

The composition of the solid solutions in samples after the equilibrium experiments was determined by x-ray diffraction. The interplanar distances $d(200)$ were determined beforehand for solid solutions of known compositions.

The equilibrium in the system NiO—Ni—O was determined over the range of 1100-1400°C at oxygen pressures between $10^{-8.920}$ and $10^{-5.758}$ atm. As a result of the experiments, the following relation was obtained for the partial pressure of oxygen in this system:

$$\log P_{O_2}^* = 9.043 - 24.730 \left[\frac{1}{T} \right].$$

For solid solutions of various compositions, equilibrium was established in the system solid solution — metallic nickel — oxygen of known partial pressure.

The equilibrium in the systems studied was established "from both directions." In the study of oxide solid solutions, into the reaction space were placed two samples, one beside the other: one sample contained more than the equilibrium amount of nickel oxide, while the second sample with metallic nickel and the oxide contained less than the equilibrium amount of NiO. The samples were kept at a given temperature and partial pressure of oxygen for 22 h. The oxygen pressure was controlled by the method used by Darken and Gurry (1945) and then Muan (1955). A mixture of carbon dioxide and hydrogen in a definite ratio was taken and the partial pressure of oxygen found from tables, for example those of Coughlin (1954).

The activity of nickel oxide in the system NiO—MgO was determined at 1100 and 1300°C and in the system NiO—MnO at 1100, 1200, and 1300°C. Figures 61 and 62 give the activities of NiO in relation to the concentration of solid solutions. The activities were obtained with quite high accuracy. The difference between two values of the activity for the same partial pressure of oxygen, but with different starting compositions, was ±0.02 for melts rich in nickel oxide and ±0.01 for melts poor in nickel oxide.

Figures 61 and 62 show that the system NiO—MgO may be regarded as ideal within the limits of experimental error. For solid solutions of the system NiO—MnO there was (in the temperature range of 1100-1300°C) a considerable deviation from an ideal straight line on the positive side. In its properties the system NiO—MnO approaches so-called regular solutions for which the excess partial molar free energy should be approximately proportional to $(1 - N_{\text{NiO}})^2$ and depend very little on temperature. Figure 63 shows that the rules characteristic of regular solutions hold.

To find the activities of nickel ferrite and magnetite in the system of solid solutions $NiFe_2O_4$—Fe_3O_4, Popov and Chufarov (1961) studied the reduction of nickel ferrite by hydrogen under equilibrium conditions. A sample of $NiFe_2O_4$ was placed in a sealed vacuum apparatus in which was circulated a mixture of $H_2 + H_2O$. The vapor pressure of water corresponded to the saturated vapor pressure at 0°C. Equilibrium was reached from both the oxidation and reduction sides and the mean values of the hydrogen pressure P_{H_2} were taken as the equilibrium values.

The reduction of $NiFe_2O_4$ formed the solid solution $NiFe_2O_4$—Fe_3O_4 and metallic nickel (with some iron). The composition of the solid solution formed was determined by x-ray diffraction. Knowing the equilibrium constant of the reduction reaction $K_p = P_{H_2O}/P_{H_2}$ and the molar composition of the solid solution obtained as a

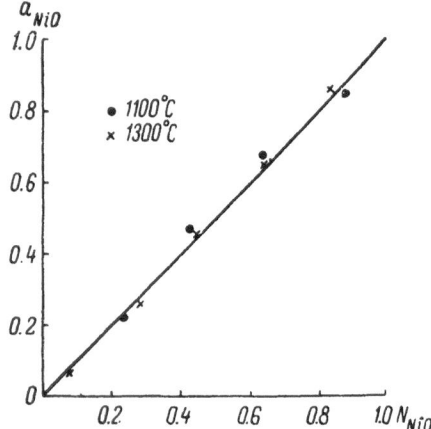

Fig. 62. Activity of NiO as a function of composition in the system NiO—MgO at 1100 and 1300°C.

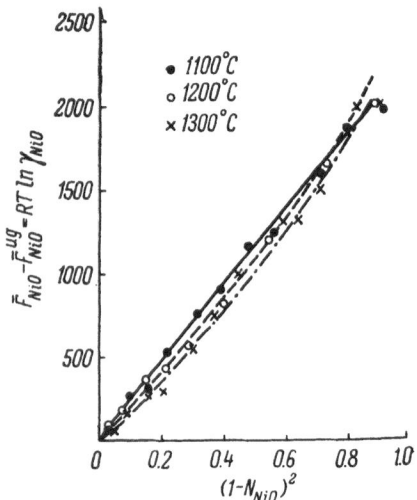

Fig. 63. Excess partial molar free energy of NiO dissolved in MnO at 1100, 1200, and 1300°C.

result of the reduction, Popov and Chufarov were able to calculate the activity of nickel ferrite a f and of magnetite a m in solid solutions of various compositions. The standard state adopted was ferrite 0.1% reduced. Figure 64 shows how the activity of ferrite and magnetite in the system $NiFe_2O_4 - Fe_3O_4$ varies with composition for different temperatures.

Popov and Chufarov consider that it is possible to apply the theory of regular solutions to the solid solutions studied so that it is possible to calculate the partial free energies, enthalpies, and entropies ($\Delta \bar{F}_i$, $\Delta \bar{H}_i$, $\Delta \bar{S}_i$), and then the total free energies of mixing, total enthalpies of mixing, and total entropies from additivity formulas, for example,

$$Pt \mid MeO + Me_2O_3 \mid Me_2O_3 \mid Pt \text{ etc.}$$

As a measure of the deviation of the structure of the solid solution from the ideal, they determined the excess entropies of mixing, which were the differences between the change in total entropy found and the change in entropy for an ideal solid solution. The latter value was determined from the formula

$$Pt \mid FeO - Al_2O_3 \mid Al_2O_3 \mid Pt .$$

Then the excess entropy of mixing will be

$$Pt \mid MgO - Al_2O_3 \mid Al_2O_3 \mid Pt .$$

Figure 65 gives the changes in total entropies of mixing and the excess entropies for nickel ferrite, magnetite, and the solid solution. In the first approximation these excess entropies characterize the degree of ordering of the solid solution. The degree of ordering changes differently for $NiFe_2O_4$, Fe_3O_4, and the solid solution as a whole.

Fischer (1958) made a series of investigations of solid solutions of the type $MeO - Me_2O_3$ using galvanic cells of the general form

$$\Delta F_{t_i} = \Delta \bar{F}_{f_i} \cdot N_{f_i} + \Delta \bar{F}_{m_i} \cdot N_{m_i}$$

in the temperature range of 1200-1800°C. Fischer and Hoffmann (1954, 1955a) studied solid solutions in the systems $FeO - Al_2O_3$ and $MgO - Al_2O_3$ and for this they constructed the cells

$$\Delta S_i^{id} = - R \ln N_i$$

and

$$\Delta S_i^{ex} = \Delta S_i - \Delta S_i^{id} ,$$

with various ratios of FeO and Al_2O_3 (and MgO and Al_2O_3, respectively).

The emf of the galvanic cells studied reflects well the physicochemical state of the systems and therefore its measurement gives some additional indications on the phase diagrams. Figures 66 and 67 give the relation of the emf to the FeO and MgO contents, respectively. Fischer found an explanation for the quite complex form of this relation.

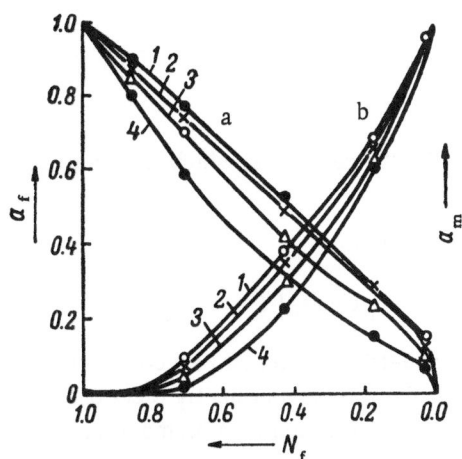

Fig. 64. Change in activity of ferrite (a) and magnetite (b) in the system $NiFe_2O_4-Fe_3O_4$ in relation to composition: 1) 900°; 2) 800°; 3) 700°; 4) 600°C.

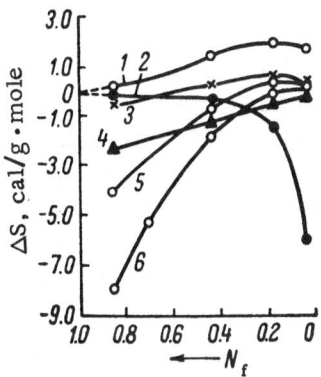

Fig. 65. Change in entropy of mixing in solid solutions of the system $NiFe_2O_4-Fe_2O_3$ in relation to the composition. 1) ΔS_f; 2) ΔS_f^{ex}; 3) ΔS_t; 4) ΔS_t^{ex}; 5) ΔS_m; 6) ΔS_m^{ex}.

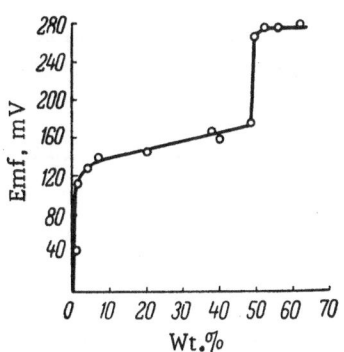

Fig. 66. Emf of the galvanic cell $Pt/FeO + Al_2O_3/Al_2O_3/Pt$ at 1500°C.

The sharp increase in the emf on the left of Fig. 66 corresponds to the formation of a solid solution of FeO in Al_2O_3. Saturation of the solid solution occurs at approximately 1.5% and this solid solution is in equilibrium with the spinel $FeO \cdot Al_2O_3$. It seemed that within the heterogeneous region of coexistence of these two phases the emf would remain constant, but there is some increase in it. This small increase in the emf is explained by the fact that trivalent iron, which may be present in the form of magnetite, forms a solid solution with spinel $FeO \cdot Al_2O_3$, so that with an increase in the amount of FeO in the system there is an increase in the magnetite content of the spinel solid solution. The appearance of the first portions of liquid wüstite above 48.8% FeO leads to a sharp jump in the emf from 170 to 270 mV. Subsequently, when there is the heterogeneous equilibrium of molten wüstite—spinel, the emf of the cell remains constant. On the basis of emf measurements, Fischer and Hoffmann (1955b) gave a new variant of the phase diagram of the system $FeO-Al_2O_3$ (Fig. 68).

Figure 67, which shows the results of measurements with the cell

$$Pt/MgO + Al_2O_3/Al_2O_3/Pt,$$

indicates the considerable solubility of aluminum oxide in spinel. The composition of this solid solution varies from 71.7% Al_2O_3 (the composition of spinel) to 80.5% Al_2O_3. In contrast to the cell with iron oxide, here the emf does not vary with an increase in the MgO content from 28.3% and above. This indicates that magnesium oxide is not soluble in spinel.

Fischer considered that the results of the emf measurements could be explained by the diffusion of di- and trivalent cations. The electrode in pure aluminum oxide is the negative pole of a galvanic cell, i.e., more positive charges depart from this electrode than arrive at it. Hence, it may be concluded that the rate of diffusion of aluminum ions is greater than that of iron ions. This conclusion of Fischer and Hoffmann was confirmed by the observation of the so-called Kirkendahl effect.

Knowing the transport numbers of aluminum and iron ions from the emf measurements, it is possible to determine the free energy of formation of ferrous spinel from the equation

$$\Delta F = -\left(n_{Al} - n_{Fe}\right) z \, \text{Far} \cdot E,$$

where z is the valence and n_{Al} and n_{Fe} are the transport numbers of the corresponding ions. According to Fischer's measurements, $n_{Al} - n_{Fe} = 0.2\text{-}0.4$. In accordance with the measurements of the emf, for ΔF we obtain a value from —7500 to —15,000 cal/mole (Fischer and Hoffmann, 1955a).

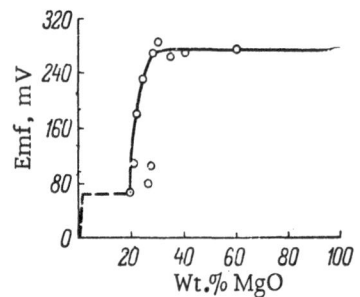

Fig. 67. Emf of the galvanic cell
Pt/ MgO + Al$_2$O$_3$/ Al$_2$O$_3$/ Pt at 1500°C.

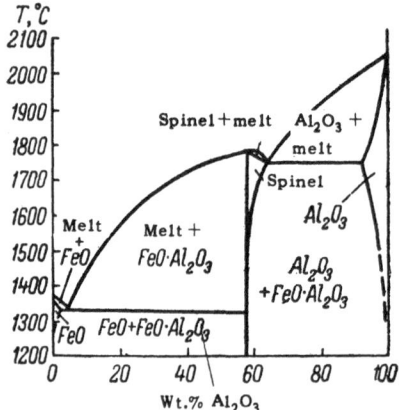

Fig. 68. Phase diagram of FeO— Al$_2$O$_3$
corrected on the basis of Fischer's
electrochemical work.

Thermochemical measurements (Fischer and Lorenz, 1957)
gave for the reaction

$$Al_2O_3 + FeO = FeAl_2O_4$$

a change in enthalpy of 14,600 ± 25 cal/mole, which is in good agreement with the emf measurements.

Investigation of the galvanic cell

$$Pt \mid FeO + Cr_2O_3 \mid Cr_2O_3 \mid Pt,$$

in the temperature range of 1200-1800°C with an FeO content from 10
to 50% showed that in this cell there is predominantly electronic conductivity. This is confirmed by the diffusion measurements of Lindner
and Åkerström (1956).

To elucidate the character of the semiconductor properties of
chromium oxide, Fischer and Lorenz (1957) investigated the thermo-
electromotive force of the cell Pt/ Cr$_2$O$_3$/ Pt over a wide temperature
range (from 800 to 1500°C). The results obtained also showed that
the electrical conductivity of Cr$_2$O$_3$ up to 1500°C is through defect
electrons.

The work of Fischer shows that measurements on galvanic cells
with solid electrolytes make it possible to obtain very diverse information on the systems studied. In addition to thermodynamic data, such
measurements make it possible to draw conclusions on the nature of
the electrical conductivity. Thus, in particular, in the systems FeO—
—Al$_2$O$_3$ and MgO—Al$_2$O$_3$ according to the measurements there is predominantly cationic conductivity. As we saw, electrochemical investigations even make it possible to find some characteristics of
phase diagrams which were not observed previously.

Aronson and Clayton (1960) used the electrochemical method to obtain the thermodynamic characteristics
of solid solutions in the system uranium dioxide—thorium dioxide.

As was shown by Trzebiatowski and Selwood (1950), uranium dioxide and thorium dioxide form a continuous series of solid solutions with a fluorite structure. According to Anderson et al. (Anderson, Edgington, Roberts,
and Wait, 1954), these solid solutions absorb oxygen, which enters interstitial positions, whereupon the uranium
may change from the tetra- to the pentavalent state

$$O_2 \text{ (gas)} + 4U^{4+} + 2 \; i.\; s. = 2O^i + 4U^{5+}$$

Here the symbol i.s. denotes a vacant interstitial position, while O^i is an oxygen ion in an interstitial position.
The solid solutions formed are represented by the general formula U$_y$Th$_{1-y}$O$_{2+x}$, in which the amount of excess
oxygen may be 0.32-0.34 without breakdown of the fluorite structure.

The solid solutions studied by Aronson and Clayton had compositions for which y varied from 0.9 to 0.3
and x from 0.02 to 0.16. The emf of the following galvanic cell was determined:

$$Fe, \; FeO \mid 0.85ZrO_2 + 0.15CaO \mid U_yTh_{1-y}O_{2+x}, \; Pt.$$

The solid solutions were prepared from the nitrates by precipitation of the hydroxides with ammonia and
subsequent calcination at 600°C. Then the product was reduced with hydrogen at 1300°C and oxidized with a
controlled amount of oxygen at 200-400°C. After oxidation the samples were annealed for a week at 800°C in
vacuum for homogenization. Measurements of the emf were made at 1150-1350°K and a linear temperature
dependence of the emf was found.

The partial molar free energy, entropy, and enthalpy of solution of oxygen in $U_y Th_{1-x} O_{2+x}$ were calculated from the emf data obtained. The partial molar free energy increased negatively with an increase in the oxygen content or an increase in the thorium content.

The above assumption that on oxidation of uranium—thorium solid solutions the uranium is converted into the pentavalent state is not the only possible one, though Aronson and Clayton considered this conversion most probable. From elementary statistical considerations these authors derived an equation giving the partial molar entropy of solution of oxygen in the solid solution

$$\bar{S}_{O_2} = -2R \ln[x/(1-x)] - 4R \ln[2x/(y-2x)] + Q.$$

This equation requires a negative increase in $\bar{S}_{O_2}$ with an increase in x or y, and this is in agreement with experimental data. The value Q appearing in the equation includes a term characterizing the decrease in entropy on removal of one mole of oxygen from the gas phase with a vapor pressure equal to 1 atm (this term equals 60 en. units at 1250°K) and a term characterizing the contribution to the vibrational entropy on introduction of oxygen ions into interstitial positions. The value of Q must remain practically constant with a change in x or y.

The check on the equation given consisted of the determination of Q for given values of x and y and the experimentally found values of $\bar{S}_{O_2}$. The value of Q did not remain constant at all with a change in the composition of the solid solutions studied. Only in approximate calculations is it possible to assume that Q has a constant value, which equals −40 en. units. If this value is adopted, then for the entropy contribution as a result of the solution of oxygen we obtain the value 20 en. units.

A comparison of the results obtained with the data from vapor pressure measurements of Robertson and his co-workers shows that the electrochemical method has a high accuracy.

BIBLIOGRAPHY

Anderson, Edgington, Roberts, and Wait. J. Chem. Soc. (1954), p. 3324.

Aramaki, S., and R. Roy. J. Am. Ceram. Soc. 45:5 (1962).

Aronson, S., and J. Belle. J. Chem. Phys. 29(1):151-158 (1958).

Aronson, S., and J. C. Clayton. J. Chem. Phys. 32(3):749-754 (1960).

Baird, J. D., and J. Taylor. Trans. Faraday Soc. 54:526 (1958).

Benz, R., and H. Schmalzried. Z. Phys. Chem., Neue Folge 29(1-2):77-82 (1961).

Benz, R., and C. Wagner. J. Phys. Chem. 65:1308 (1961).

Biltz, W., and H. Müller. Z. Anorg. Allgem. Chem. 163:257 (1927).

Blumenthal, R. N., and D. H. Whitmore. J. Am. Ceram. Soc. 44(10):508 (1961).

Boricke, F., and W. M. Bangert. Bureau of Mines Report of Investigation, 3813 (1945).

Brauer, G., K. A. Gingirich, and U. Holtschmidt. J. Inorg. Nucl. Chem. 16:77 (1960).

Busby, T. S. Trans. Brit. Ceram. Soc. 60(2):134 (1961).

Chen, H. M., and J. Chipman. Trans. Am. Soc. Metals 38:70 (1947).

Circular of the National Bureau of Standards, U. S. Printing Office, Washington, D. C., No. 500 (1952), p. 629.

Clark, S. P., E. C. Robertson, and F. Birch. Am. J. Sci. 255:628 (1957).

Cooper, A. C., D. A. R. Kay, and J. Taylor. Trans. Brit. Ceram. Soc. 60(2):124-134 (1961).

Coughlin, J. P. U. S. Bureau of Mines, Bulletin No. 542 (1954).

Croatto, U., and C. Bruno. Ric. Sci. 17:1998 (1947).

Darken, L. S., and R. W. Gurry. J. Am. Chem. Soc. 67:1398 (1945).

Eberius, E. Dissertation, Leipzig (1931).

Emmett, P. H., and J. F. Schultz. J. Am. Chem. Soc. 52:1782 (1930).

Fischer, W. A. In: Physical Chemistry of Steelmaking, J. Elliott (ed.), (1958), p. 79.

Fischer, W. A., and A. Hoffmann. Naturwissenschaften 41:162 (1954).

Fischer, W. A., and A. Hoffmann. Arch. Eisenhüttenw. 26:43 (1955).

Fischer, W. A., and A. Hoffmann. Arch. Eisenhüttenw. 26:63 (1955).

Fischer, W. A., and G. Lorenz. Arch. Eisenhüttenw. 28:497 (1957).

Flood, H., and W. J. Knapp. J. Am. Ceram. Soc. 40: 206 (1957).

Fricke, R., and G. Weitbrecht. Z. Elektrochem. 48: 87 (1942).

Gerasimov, Ya. I., and I. A. Vasil'eva. J. Chim. Phys. 56: 639 (1959).

Gerasimov, Ya. I., I. A. Vasil'eva, T. P. Chusova, V. A. Geiderikh, and M. A. Timofeeva. Dokl. Akad. Nauk SSSR 134: 1350-1352 (1960).

Glushkova, V. B. Zh. Neorgan. Khim. 2: 2438 (1957).

Glushkova, V. B., and É. K. Keler. Zh. Neorgan. Khim. 1: 2283 (1956).

Grjotheim, K., O. Herstad, and J.M. Toguri. Can. J. Chem. 39(11): 2290 (1961).

Hahn, W. C., and A. Muan. J. Phys. Chem. Solids 19(3-4): 338 (1961).

Hund, F. Z. Phys. Chem. 199: 142 (1952).

Karpachev, S. V., and S. F. Pal'guev. Trans. Inst. Electrochem. 1 (1960) (English Transl.).

Kelley, K. K. Contributions to the Data on Theoretical Metallurgy XIII, U. S. Bureau of Mines, Bulletin No. 584 (1959).

King, E. J. Am. Chem. Soc. 79: 5437 (1957).

Kiukkola, K. Acta Chem. Scand. 16(2): 326 (1962).

Kiukkola, K., and C. Wagner. J. Electrochem. Soc. 104(6): 379-387 (1957).

Kubaschewski, O., and E. Evans. Thermochemistry in Metallurgy [Russian translation], IL, Moscow (1954).

Kusenko, F. G., and P. V. Gel'd. Izv. Sibirsk. Otd. Akad. Nauk SSSR, No. 2 (1960).

Kuznetsov, F. A. Thermodynamic Investigation of Cerium Oxides, Author's abstract of dissertation, Moscow (1961).

Lavrent'ev, V. I. Thermodynamic Investigation of Niobium Oxides, Author's abstract of dissertation, IONKh Akad. Nauk SSSR, Moscow (1961).

Lebedev, B. G., and V. A. Levitskii. Zh. Fiz. Khim. 35(12): 2788 (1961).

Lindner, R., and A. Åkerström. Z. Phys. Chem., Neue Folge 6: 162 (1956).

Mchedlov-Petrosyan, O. P. Zh. Fiz. Khim. 24(11): 1299 (1950).

Mchedlov-Petrosyan, O. P. Zh. Fiz. Khim. 26(12): 1785 (1952).

Mchedlov-Petrosyan, O. P. In collection: Physicochemical Bases of Ceramics, Promstroiizdat, Moscow (1956), pp. 499-503.

Mchedlov-Petrosyan, O. P., and W. I. Babuschkin. Silikat. Tech. 9(5): 209 (1958).

Morozova, M. P., and L. L. Getskina. Zh. Otd. Khim. 29: 1049 (1959).

Morozova, M. P., and T. A. Stolyarova. Zh. Obshch. Khim. 30: 3848 (1960).

Muan, A. Trans. AIME 204: 965-976 (1955).

Ol'shanskii, Ya. I. In collection: Experimental Techniques and Investigation Methods at High Temperatures, Izd. Akad. Nauk SSSR (1959), pp. 402-410.

Pal'guev, S. F., S. V. Karpachev, A. D. Neuimin, and Z. S. Volchenkova. Dokl. Akad. Nauk SSSR 134: 1138-1141 (1960).

Passerini, L. Gazz. Chim. Ital. 59: 144 (1929).

Peters, H., and H. H. Mobius. Z. Phys. Chem. 209: 298 (1958).

Popov, G. P., and G. I. Chufarov. Dokl. Akad. Nauk SSSR 141: 877-879 (1961).

Popov, M. M., G. L. Gal'chenko, and M. D. Senin. Zh. Neorgan. Khim. 3: 1734 (1958).

Reinhold, H. Z. Anorg. Allgem. Chem. 171: 181 (1928).

Reinhold, H. Z. Elektrochem. 40: 361 (1934).

Rezukhina, T. N. Zh. Neorgan. Khim. 5: 1016-1021 (1960).

Rezukhina, T. N., Ya. I. Gerasimov, and Yu. P. Simanov. Vestn. Mosk. Univ., No. 6: 103 (1949).

Rezukhina, T. N., V. I. Lavrent'ev, V. A. Levitskii, and F. A. Kuznetsov. Zh. Fiz. Khim. 35(6): 1367-1369 (1961).

Rezukhina, T. N., V. A. Levitskii, and N. M. Kazimirova. Zh. Fiz. Khim. 35(11): 2639-2642 (1961).

Richardson, H. M. Trans. Brit. Ceram. Soc. 60(2): 133 (1961).

Richardson, F. D., and J. H. E. Jeffes. J. Iron Steel Inst. (London) 160(3): 261 (1948).

Richardson, F. D., and L. E. Webb. Trans. Inst. Mining Met. 64: 529 (1955).

Rose, B. A., G. J. Davis, and H. J. T. Ellingham. Discussions Faraday Soc. 4: 154 (1948).

Roth, W. A., and F. Wienert. Arch. Eisenhüttenw. 7: 455 (1934).

Sator, A. Comptes Rend. 234:2283 (1952).

Schenck, R., and H. Wesselkock. Z. Anorg. Allgem. Chem. 184:39 (1929).

Schmahl, G. N. Z. Elektrochem. 47:826 (1941).

Schmalzried, H. Z. Phys. Chem., Neue Folge 25(3-4):178-192 (1961).

Taits, A. Yu. Tsvetn. Metal. 30:56 (1957).

Taylor, J. Trans. Brit. Ceram. Soc. 60(2):133 (1961).

Torgeson, D. B., and T. G. Sahama. J. Am. Chem. Soc. 70:2156 (1948).

Toropov, N. A., and F. Ya. Galakhov. Izv. Akad. Nauk SSSR, Otd. Khim. Nauk, No. 1:8-11 (1958).

Treadwell, W. D., H. Amman, and T. Zürrer. Helv. Chim. Acta 19:1255 (1936).

Trzebiatowski, W., and P. W. Selwood. J. Am. Chem. Soc. 72:4504 (1950).

Tubandt, C. Handbuch der Experimentalphysik, Vol. 12 (1932).

Tubandt, C., S. Eggert, and Schibbe. Z. Anorg. Chem. 117:1 (1921).

Ure, R. W. J. Phys. Chem. 26:1363 (1957).

Wagner, C. Z. Phys. Chem., Abt. B 21:25 (1933).

Watanabe, M. Sci. Rept. Tohoku Imp. Univ. I, 22:893 (1933).

Westrum, E. F., and G. Grønvold. J. Am. Chem. Soc. 81:1777 (1959).

Yakovleva, R. Ya., and T. N. Rezukhina. Zh. Fiz. Khim. 34:819 (1961).

CHAPTER VI

DIFFUSION PROCESSES AND KINETICS OF REACTIONS
IN THE SOLID STATE

In the well-known book of P. P. Budnikov and A. M. Ginstling, Reactions in Mixtures of Solids (1961), much space is devoted to the description of the mechanism and kinetics of reactions in systems of solids where original views are developed largely on the basis of their own experimental data. After the publication of this book, the work of Ginstling became widely known in the Soviet Union and there is no need to repeat it.

The aim of the present chapter is to report some work of mainly foreign schools on the mechanism and kinetics of solid-phase reactions. In these schools it is mainly investigations of diffusion processes that have been developed and in the examination of chemical interactions in crystalline substances, insufficient attention has been paid to the accompanying phenomena, whose importance was clearly demonstrated by A. M. Ginstling.

In articles on the kinetics of solid-phase reactions, foreign authors usually attempt to find a single formula which describes the overall rate of the process without separating the phenomenon into separate stages, which require special formulas for their kinetic description, as Ginstling showed.

A small section in this chapter reviews work on the effect of the gaseous medium on the rate and character of processes in solids.

1. Investigation of Reactions in the Solid State

The determination of the nature of the diffusion particles is of great importance in the understanding of the mechanism of reactions in the solid state. The first reaction for which the diffusing particles were established was the formation of the double silver—mercury iodide salt Ag_2HgI_4 from AgI and HgI_2. Koch and Wagner (1936) showed that in the formation of this double iodide in the solid state there is counterdiffusion of the cations, i.e., Ag^+ and Hg^{2+}, through the anionic lattice of the reaction product. The anionic lattice formed by iodide ions is rigid and immobile. These authors put forward the hypothesis that spinel formation proceeds by the same mechanism.

The determination of the mechanism of formation of compounds of the general formula AB_2O_4 (spinels, ferrites, and chromites) has been attracting great attention in recent years. It seemed that the mechanism of solid-phase reactions proposed by Wagner (1936) applied in the formation of these compounds. However, the situation is more complex here and not all spinels, ferrites, etc., can apparently be covered by one mechanism.

At the present time many data have been accumulated on the diffusion of ions A and B in compounds of the general formula AB_2O_4. Table 17 gives the results of investigations of self-diffusion in spinels obtained by Lindner et al. (Lindner, 1956a).

In the chromites and zinc ferrite given, the activation energies of self-diffusion of di- and trivalent ions are practically identical and this makes it probable that there is diffusion of both these ions in the compounds AB_2O_4. Lindner and Åkerström (1956) give temperature dependences of the self-diffusion constants of Ni and Cr in $NiCr_2O_4$ and the rate constants of the reaction

$$NiO + Cr_2O_3 = NiCr_2O_4.$$

Table 17. Self-Diffusion Coefficient in Spinels $D = D_0 \exp(-Q/RT)$

Spinel	Diffusing ion	D_0, cm^2/sec	Q, kcal/mole
$MgAl_2O_4$	Mg^{2+}	$2 \cdot 10^2$	86
$ZnAl_2O_4$	Zn^{2+}	$2.5 \cdot 10^2$	78
$NiAl_2O_4$	Ni^{2+}	$3 \cdot 10^{-4}$	55
$ZnFe_2O_4$	Zn^{2+}	$8.8 \cdot 10^2$	86
$ZnFe_2O_4$	Fe^{3+}	$8.5 \cdot 10^2$	82
$ZnCr_2O_4$	Zn^{2+}	60	85
$ZnCr_2O_4$ . . . . ·	Cr^{3+}	9	81
$NiCr_2O_4$	Ni^{2+}	0.85	75
$NiCr_2O_4$	Cr^{3+}	0.75	73
$SnZn_2O_4$	Sn^{4+}	$2.3 \cdot 10^5$	109
$SnZn_2O_4$	Zn^{2+}	37	76

As Fig. 69 shows, a line drawn through the circles will be parallel to the broken lines characterizing the activation energies of the Ni^{2+} ion (75 kcal) and the Cr^{3+} ion (73 kcal). This is a convincing demonstration that Wagner's mechanism of counterion diffusion is justified in this case. The use of the "inert label" method is not possible in the system examined, as platinum labels are strongly attacked by oxides, mainly chromium oxide.

The activation energy of the reaction

$$ZnO + Cr_2O_3 = ZnCr_2O_4$$

according to the determinations of Lindner and Åkerström, equals approximately 80 kcal/mole, which is close to the activation energies of self-diffusion of Zn^{2+} (85 kcal) and Cr^{3+} (81 kcal) in zinc chromite given in the table. This again indicates that the formation mechanism proposed by Wagner applies to zinc chromium spinel.

The formation of zinc iron spinel $ZnFe_2O_4$ also proceeds through the counterdiffusion of Zn^{2+} and Fe^{3+} in a stationary lattice of oxygen ions (Wagner's mechanism). Even a comparison of the diffusion constants of zinc and iron in zinc iron spinel and the rate constant of formation of zinc iron spinel makes Wagner's mechanism probable. As Fig. 70 shows, the constant of formation of $ZnFe_2O_4$ has a value between the self-diffusion constants of zinc and iron (Lindner, 1952a, 1952b, 1955b).

However, the work of Szabo, Batta, and Solymosi (1961) showed that the activation energy of solid-phase reactions is strongly dependent on small impurities and, therefore, conclusions cannot always be drawn on the mechanism of the reaction from a comparison of the activation energies of reaction diffusion and self-diffusion. Szabo, Batta, and Solymosi studied the effect of some additives on the formation of nickel ferrite. Here the activation energy of the reaction changes in accordance with the effect of the additive on the defect structure of the reacting substances. When pure NiO and Fe_2O_3 were used, the activation energy of the reaction was 95 kcal, when 1% Cr_2O_3 was added to the NiO, it fell to 59 kcal, and when 1% TiO_2 was added to the Fe_2O_3, it rose to 132 kcal. The authors observed an effect of the oxygen concentration in the surrounding atmosphere on the reaction rate, and this may be explained if the oxides are regarded as semiconductors. It is readily seen that the activation energy of the reaction is changed by additives which affect the mobility of the cations (see p. 100).

The Wagner mechanism for the formation of zinc iron spinel was demonstrated by experiments with inert labels. Before the reaction, the surface of a tablet of iron oxide was covered with small pieces (~2 mg) of thin platinum foil ("inert labels"). After the reaction, the pieces of platinum were found to lie inside the spinel disk obtained. The distance of the "labels" from the outer surfaces (i.e., the surfaces of contact) of the disks was 127 and 453 mμ (the thickness of the starting disks was 580 mμ), i.e., the ratio of the thicknesses of the layers to the left and right of the "labels" was 1 : 3.56. This ratio differs slightly from the required value of 1 : 3, and this may be explained by some deviation from stoichiometry (deficit in zinc) in the spinel. Figure 71 shows that the amount of zinc transported is less than that which corresponds to the ratio of the thicknesses of the layers. For the reaction

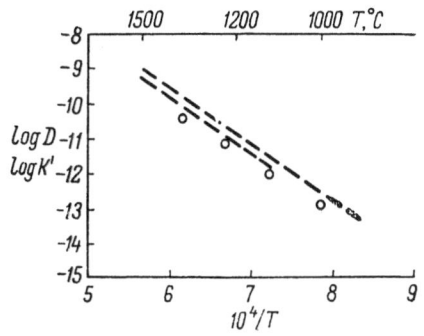

Fig. 69. Temperature dependence of the rate constant for spinel formation (circles) and the self-diffusion coefficient (broken line) for $NiCr_2O_4$.

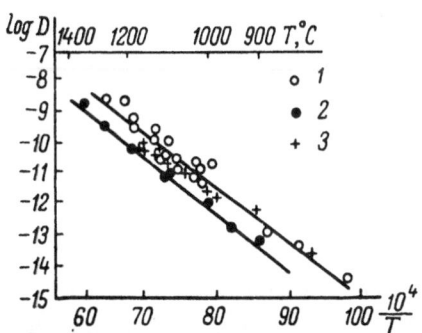

Fig. 70. Temperature dependence of the self-diffusion coefficients of Fe^{3+} (1) and Zn^{2+} (2), and formation constants of zinc ferrite (3).

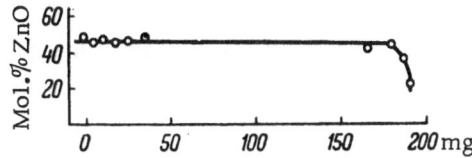

Fig. 71. Distribution of radioactive zinc in a layer of zinc ferrite $ZnFe_2O_4$.

$$ZnO + Fe_2O_3 = ZnFe_2O_4$$

we may thus regard it as established that there is counterdiffusion of the cations and no mobility of the oxygen.

The study of diffusion in aluminates is more complex, as there is no radioactive isotope of aluminum suitable for practical measurements. Table 17 gives only the self-diffusion coefficients of Mg^{2+}, Zn^{2+}, and Ni^{2+} in the corresponding aluminates.

Lindner and Åkerström, who studied the formation of nickel spinel from NiO and Al_2O_3, gave the following expression for the rate constant of the reaction:

$$K' = 3 \cdot 10^6 \exp (-127\,000/RT) \text{ cm}^3/\text{sec.}$$

For determining the rate of the reaction

$$NiO + Al_2O_3 = NiAl_2O_4$$

in the high-temperature region, the given authors used a gravimetric method. It was observed that the newly formed spinel was wholly on the alumina tablet. At lower temperatures there was transfer of radioactive Ni^{65}. The "inert labels" method was difficult to use because of the thinness of the layer of spinel formed. The activation energy obtained, namely 127 kcal, is considerably higher than the self-diffusion coefficient of nickel and in the opinion of Lindner this indicates the migration of oxygen in the system examined.

The formation of zinc spinel $ZnAl_2O_4$ was studied by Bengtson and Jagitsch (1947). They used the method of weighing plates and the "inert labels" method. For the reaction

$$ZnO + Al_2O_3 = ZnAl_2O_4$$

it is possible to postulate four different mechanisms.

I. The cations Zn^{2+} and Al^{3+} migrate in opposite directions.

II. ZnO migrates as a molecule or as a zinc cation and an oxygen anion in one direction.

III. The "acid oxide" Al_2O_3 migrates in one direction.

IV. The two oxides migrate in opposite directions.

If at the beginning we take plates or pressed tablets of zinc oxide and aluminum oxide whose weights correspond to a gram-molecule of each of these compounds, then after the experiment (i.e., after the formation of the spinel), there will be different changes in the weights of the plates, depending on the mechanism.

Mechanism I.

$$\frac{4ZnO}{-3Zn^{2+}} \qquad \frac{4Al_2O_3}{+3Zn^{2+}}$$

$$\frac{+2Al^{3+}}{ZnAl_2O_4} \qquad \frac{-2Al^{3+}}{3ZnAl_2O_3}$$

Ratio of weights of plates 1 : 3.

Mechanism II.

$$\frac{\begin{array}{c|c} \text{ZnO} & \text{Al}_2\text{O}_3 \\ -\text{ZnO} & +\text{ZnO} \end{array}}{\begin{array}{c|c} 0 & \text{ZnAl}_2\text{O}_4 \end{array}}$$

Ratio of weights of plates $0:1$.

Mechanism III.

$$\frac{\begin{array}{c|c} \text{ZnO} & \text{Al}_2\text{O}_3 \\ +\text{Al}_2\text{O}_3 & -\text{Al}_2\text{O}_3 \end{array}}{\begin{array}{c|c} \text{ZnAl}_2\text{O}_4 & 0 \end{array}}$$

Ratio of weights of plates $1:0$.

Mechanism IV (with the condition that the diffusion rates of the two oxides are the same).

$$\frac{\begin{array}{c|c} 2\text{ZnO} & 2\text{Al}_2\text{O}_3 \\ -\text{ZnO} & +\text{ZnO} \end{array}}{\begin{array}{c|c} +\text{Al}_2\text{O}_3 & -\text{Al}_2\text{O}_3 \\ \hline \text{ZnAl}_2\text{O}_4 & \text{ZnAl}_2\text{O}_4 \end{array}}$$

Ratio of weights of plates $1:1$.

As a result of weighing the plates, Bengtson and Jagitsch came to the conclusion that the formation of zinc spinel proceeds by the second mechanism, i.e., there is migration of zinc and oxygen. The increase in weight of the aluminum oxide plate was equal to the decrease in weight of the zinc oxide plate (the volatility of zinc oxide was taken into account).

This conclusion was confirmed by experiments with "inert labels" (platinum grains). Bengtson and Jagitsch placed corundum crystals in zinc oxide powder and observed that the corundum was covered with reaction products after heating to 1380°C, while "inert labels" (which were initially on the surface of the corundum) were found on the surface of the spinel layer formed.

The reaction rate constant could be determined from the weights of the plates. The temperature dependence of the constant is represented by the formula

$$K'' = 3.2 \cdot 10^8 \exp\left(\frac{98\,000}{RT}\right) \text{g}^2 \cdot \text{cm}^{-4} \cdot \text{hr}^{-1}.$$

The conclusions of Jagitsch and Bengtson were confirmed by Lindner, who did not observe penetration of "inert labels" into the spinel layer in the system $\text{ZnO}-\text{Al}_2\text{O}_3$. Determination of the radioactivity of the zinc in the spinel formed (by stepwise grinding away) also led to the conclusion that zinc oxide diffuses. The concentration of radioactive zinc in the whole layer of spinel (thickness 120 mμ) was constant and corresponded to the stoichiometric composition. At the spinel—zinc oxide boundary there was a sharp jump in the radioactivity from that characteristic of the spinel to that corresponding to pure zinc oxide. The whole of the spinel was attached to the aluminum oxide plate, while a constant concentration of radioactive zinc was found through the whole of the remaining zinc oxide plate.

From the brief report of Lindner (Lindner and Åkerström, 1956), it follows that in the formation of zinc tin spinel SnZn_2O_4 there is considerable diffusion of tin ion.

Hopkins (1949) demonstrated that Fe_2O_3 diffuses through ZnFe_2O_4. It is true that this conclusion was based on indirect evidence. Thirsk and Whitmore (1940) studied the formation of spinel by depositing nickel oxide powder on a corundum surface and using the electron diffraction method. They considered that there was diffusion of only one sort of ion.

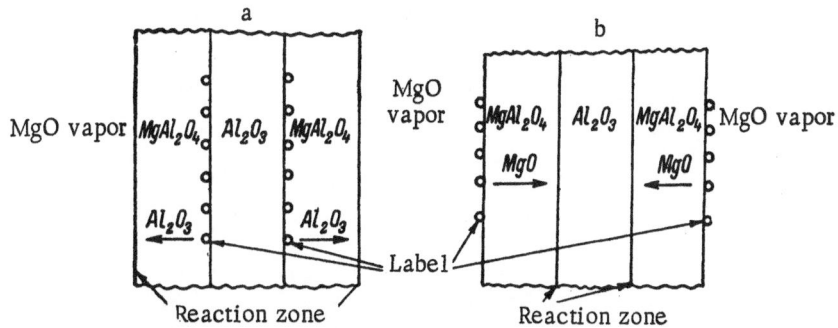

Fig. 72. Scheme of experiment on the formation of spinel using "inert labels." a) Only Al_2O_3 in the form of ions or "molecules" moves through the layer of $MgAl_2O_4$; b) only MgO in the form of ions or "molecules" moves through the layer of $MgAl_2O_4$.

In discussing the results of investigations of reactions of the type

$$AO + B_2O_3 = AB_2O_4,$$

Lindner and Åkerström had difficulty in finding a reason for the difference in the mechanisms of formation of aluminates and the formation of ferrites, chromites, and stannates (Wagner's mechanism was established for the last three groups of compounds). All the compounds listed above have a generally similar structure and similar parameters and, therefore, it is impossible to find an explanation in structural differences.

In the recent work of Carter (1961a), the "inert labels" method was used to elucidate the mechanism of formation of $MgAl_2O_4$. A disk cut from a single crystal of aluminum oxide or from a polycrystalline product obtained by sintering was wound with fine molybdenum wire, which served as the "inert label." For holding the wire better, the disk was fixed on a ring of alumina. A similar ring was fixed on top of the disk. The disk with the rings was placed into a closed molybdenum boat, which contained magnesium oxide in excess as compared with the stoichiometry of the reaction. There was no direct contact between the MgO and Al_2O_3. The system was kept in a hydrogen atmosphere at 1900°C for 48 h, i.e., sufficient time for completion of the reaction according to all the data. The magnesium oxide entered the reaction sphere through the gas phase, i.e., as vapor.

The idea of the use of "inert labels" is clear from the schemes given (Fig. 72). If only Al_2O_3 diffuses (in the form of Al^{3+} and O^{2-} ions or the whole group Al_2O_3), then the molybdenum wires will be at the Al_2O_3- $-MgAl_2O_4$ interface, i.e., at the surface of the Al_2O_3 disk (Fig. 72a). When the whole of the alumina participates in the reaction, the upper and lower wires must get nearer. If only MgO diffuses through the $MgAl_2O_4$ layer, the wires must be at the $MgO-MgAl_2O_4$ boundary, i.e., at the surface of the newly formed spinel (Fig. 72b). Finally, if there is simultaneous diffusion of Al^{3+} and Mg^{2+} ions, the molybdenum wires will lie inside the disk after the experiment.

In Carter's experiments the molybdenum wires were found inside the spinel disk formed and this indicated diffusion of Al^{3+} and Mg^{2+} ions through a rigid oxygen lattice and the occurrence of reactions at both the $Al_2O_3-MgAl_2O_4$ and the $MgO-MgAl_2O_4$ boundaries. The position of the "inert labels" (molybdenum wires) was exactly as required on the condition that two Al^{3+} ions moved through the oxygen lattice toward the magnesium oxide (i.e., toward the outside of the disk) and three Mg^{2+} ions moved to the aluminum oxide disk. At the $MgO-MgAl_2O_4$ and $Al_2O_3-MgAl_2O_4$ boundaries there occurred the processes

$$2Al^{3+} + 3O^{2-} + MgO = MgAl_2O_4,$$
$$3Mg^{2+} + 3O^{2-} + 3Al_2O_3 = 3MgAl_2O_4.$$

The ratio of the thickness of the $MgAl_2O_4$ layer above the molybdenum wires to the thickness of the layer below these wires was within the limits of experimental error equal to 1 : 3, and this confirmed the reaction mechanism presented above.

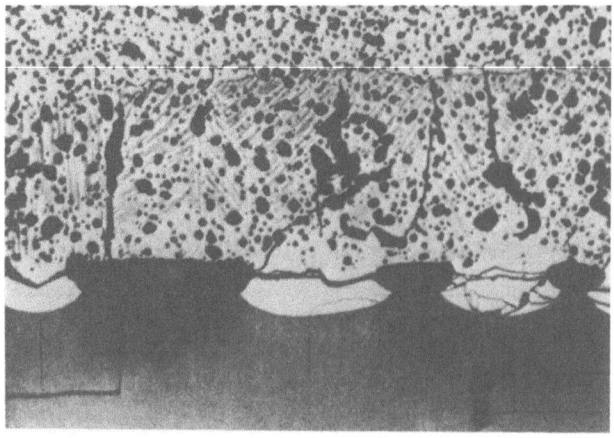

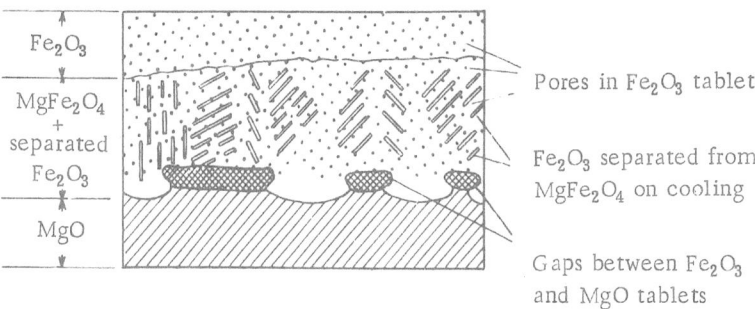

Fig. 73. Microsection of the boundary of MgO and Fe_2O_3 disks
and a schematic representation of it.

Carter observed a difference in the microstructures of the spinels formed at the Al_2O_3—$MgAl_2O_4$ and MgO·
—$MgAl_2O_4$ boundaries. The oxygen lattice of the newly formed spinel at the boundary with Al_2O_3 was formed
as a result of very slight rearrangements of the original oxygen lattice of the aluminum oxide. As a result of
this, the pores present in the alumina disk remained in the spinel disk formed. The spinel was formed quite differently at the MgO—$MgAl_2O_4$ boundary. Here the oxygen lattice of the $MgAl_2O_4$ was formed from oxygen
which previously belonged to gaseous magnesium oxide and, therefore, the spinel formed without any pores. By
taking in one case a single crystal of aluminum oxide free from pores and, in the other, a polycrystalline
sample obtained by sintering, Carter obtained spinels of different structures, namely nonporous and porous.

In the samples of spinel obtained, the molybdenum wires were not only in the thickness of the samples,
but were also observed in the form of pieces at the very surface. Carter was unable to explain the appearance
of these pieces. He suggested that during the heating the wires underwent strong extension with the result that
they split and broke. It is possible that in using the method of "inert labels," previous authors (for example,
Lindner), who considered that the appearance of the wires on the surface was the result of diffusion processes,
came to incorrect conclusions on the migration of only NiO (in the system NiO—Al_2O_3) and ZnO (in the system
ZnO—Al_2O_3).

In the study of the reaction in the system MgO—Fe_2O_3, it was impossible to use metallic wires because of
the action of the iron oxide on them and, therefore, Carter used as the "inert labels" the pores which were in
the original disk of iron oxide. Such disks, which contained 17% of pores, were obtained by heating at 1380°C
in an oxygen atmosphere for 20 h. These disks were placed between two smooth plates of single-crystal magnesium oxide and heated at 1370°C in oxygen for 39 h. Figure 73 shows a microsection and a schematic representation of it. The observed boundary of the pores indicates that there was counterdiffusion of the cations in
this system. As in the case of the system MgO—Al_2O_3, the ratio of the thickness of the compact ferrite to the
thickness of the porous ferrite was close to 1 : 3. This indicates that here the process had a mechanism analogous to that examined above for the formation of spinels, namely, counterdiffusion of the cations through a
rigid oxygen lattice.

Table 18. Quantitative Characteristics of Self-Diffusion of Cations
in Silicates

System	Diffusing ion	D_0, cm^2/sec	Q, kcal/ mole
α-CaSiO$_3$	Ca^{2+}	$7 \cdot 10^4$	112
β-CaSiO$_3$	Ca^{2+}	0.2	78
Ca$_3$Si$_2$O$_7$	Ca^{2+}	10^{-2}	73
α-Ca$_2$SiO$_4$	Ca^{2+}	$2 \cdot 10^{-2}$	55
α'-Ca$_2$SiO$_4$	Ca^{2+}	$3.6 \cdot 10^{-2}$	65
BaSiO$_3$	Ba^{2+}	$5 \cdot 10^2$	80
PbSiO$_4$	Pb^{2+}	85	59.5
Pb$_2$SiO$_4$	Pb^{2+}	8.2	47
BaTiO$_3$	Ba^{2+}	0.8	89

The diffusion of cations in silicates has been studied mainly by Lindner (1955b). A review of his work on diffusion was given by Lindner in 1955. Table 18 gives the activation energies and pre-exponential factors in the equation $D = D_0 \exp(-Q/RT)$ for some silicates and barium titanate.

In 1956, Lindner (1956b) made a detailed investigation of the mechanism of solid-phase formation of lead silicates and calcium silicates. Still earlier, Jagitsch and his co-workers (Jagitsch, Bengtson, 1946; Jagitsch, Perlström, 1946) had studied the mechanism of formation of Pb$_2$SiO$_4$ and Pb$_4$SiO$_6$ by using plates of PbO/PbSiO$_3$ and PbO/Pb$_2$SiO$_4$ on top of each other. Platinum black ("inert label") was deposited on the boundary surface before heating. After the experiment the platinum black was at the surface of contact in an unchanged form. The silicate plate had increased in weight, while there was no detectable formation of lead silicate on the lead oxide plate. The reaction layer grew only on the lead silicate plate. Hence, it was concluded that there was the transfer of lead and oxygen in some form through the reaction layer and no transfer of silicon in the opposite direction.

To understand the mechanism of solid-phase reactions, it is very important to know the quantitative characteristics of the diffusion of the possible participants in the reaction. The self-diffusion of lead in metallic lead and in its compounds has been studied repeatedly. Lindner (1950) determined the self-diffusion of lead in lead meta- and orthosilicates by using the isotope Pb[212] (thorium B with a half-life of 10.6 h). The temperature dependence of the self-diffusion coefficients in these two salts is given by the equations

$$PbSiO_3, \quad D = 85 \exp(-59{,}500/RT) \text{ cm}^2/\text{sec}$$

$$Pb_2SiO_4, \quad D = 0.2 \exp(-47{,}000/RT) \text{ cm}^2/\text{sec}.$$

No allowance was made here for the relative diffusion maxima which are connected with polymorphic conversions at 585°C for PbSiO$_3$ and 620°C for Pb$_2$SiO$_4$.

Lindner (1956b) carried out special experiments to observe the diffusion of silicon using irradiated and unirradiated plates of lead silicate. Irradiation in an atomic reactor gave radioactive Si[31] with a half-life of 157 min. These experiments showed that no migration of Si[31] occurs, and this confirmed Jagitsch's observation that there was no diffusion of silicon.

Lindner studied the formation of Pb$_2$SiO$_4$ from PbO and PbSiO$_3$ and PbSiO$_3$ from Pb$_2$SiO$_4$ and SiO$_2$ over a wide temperature range. For the reaction

$$PbO + PbSiO_3 = Pb_2SiO_4$$

he obtained the following temperature dependence of the reaction rate constant:

$$K'' = 6 \cdot 10^5 \exp(-61{,}300/RT) \text{ g}^2 \cdot \text{cm}^{-4} \cdot \text{sec}^{-1}.$$

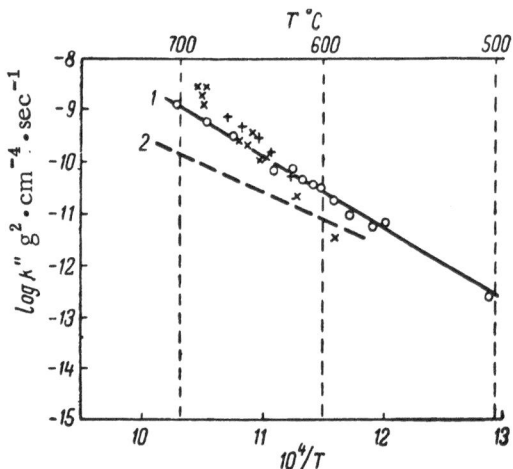

Fig. 74. Temperature dependence of the reaction rate constant for Pb₂SiO₄ formation (1) and the self-diffusion coefficient of Pb²⁺ in Pb₂SiO₄ (2).

As Fig. 74 shows, the lines of the temperature dependence of the reaction rate constant in the formation of Pb_2SiO_4 and the self-diffusion coefficient of Pb^{2+} in Pb_2SiO_4 are not parallel, i.e., the activation energies of self-diffusion of lead and reaction diffusion of lead are different. Hence, it follows that the process determining the reaction kinetics requires a higher activation energy than the activation energy of self-diffusion of lead. It is natural to assume that, in addition to the migration of lead there must migrate some other component part of the system. It is most likely that there is transfer of oxygen in the system examined. Direct experimental measurements of the mobility of oxygen using, for example, the isotope O^{18}, are required to confirm this hypothesis.

In the study of the reaction

$$Pb_2SiO_4 + SiO_2 = 2PbSiO_3,$$

Lindner obtained less clear results, and this is probably connected with the complicating effect of the polymorphic conversions of quartz and lead metasilicate.

In the investigation of the mechanism of formation of calcium silicates, Lindner first determined the self-diffusion coefficients of calcium in its silicates, using the radioactive isotope Ca^{45} with a half-life of 180 days.

The diffusion of calcium in the high-temperature modification of wollastonite $CaSiO_3$ was determined in the range of 1130-1140°C (Lindner, 1956b), and the following temperature dependence was obtained for the self-diffusion coefficient:

$$D = 7 \cdot 10^4 \exp\left(-112{,}000/RT\right) \text{ cm}^2/\text{sec.}$$

Because of the slowness of diffusion, only approximate data could be obtained for the low-temperature modification of $CaSiO_3$. Lindner gave an equation for the self-diffusion coefficient, which must be regarded as preliminary:

$$D = 0.2 \exp\left(-78{,}000/RT\right) \text{ cm}^2/\text{sec.}$$

The diffusion of radioactive calcium in tricalcium bisilicate $Ca_3Si_2O_7$ was studied by Lindner and Obermayer (1955) in the temperature range of 1000-1400°C. At 1260°C, there was a break in the temperature dependence of the diffusion coefficient, which was connected with the polymorphic conversion of $Ca_3Si_2O_7$. Below 1260°C (for the β-modification of $Ca_3Si_2O_7$), the relation of D and T is expressed by the formula

$$D = 10^{-2} \exp\left(-73{,}000/RT\right) \text{ cm}^2/\text{sec.}$$

The activation energy of the diffusion of calcium in $\alpha\text{-}Ca_3Si_2O_7$, i.e., above 1260°C, is approximately 58 kcal.

In work with Ca_2SiO_4, because of the α'-β conversion, the radioactivity of the calcium was measured during the actual heating with a specially constructed furnace (Lindner and Spicar, 1954). The α'-α conversion, which was observed as a break on the temperature dependence of the diffusion coefficient, occurred at 1370°C, which differs from the data given in the literature. The following values of the self-diffusion coefficient of calcium were obtained for the α'- and α-modifications of calcium orthosilicate:

$$\alpha'\text{-}Ca_2SiO_4 \quad D = 3.6 \cdot 10^{-2} \exp\left(-65{,}000/RT\right) \text{ cm}^2/\text{sec}$$

$$\alpha\text{-}Ca_2SiO_4 \quad D = 2.0 \cdot 10^{-2} \exp\left(-55{,}000/RT\right) \text{ cm}^2/\text{sec.}$$

Table 19. Activation Energies of Self-Diffusion of Ca^{2+} in Various Calcium Silicates, kcal

Calcium silicates	α-Modification	$\beta(\alpha')$-Modification
Ca_2SiO_4	55	65
$Ca_3Si_2O_7$	(58)	73
$CaSiO_3$	112	(78)

In studying the formation of calcium silicate in powders, Brune and Lindner (1953) obtained for the reaction diffusion constant at 1300°C approximately the same value as that given for α'-Ca_2SiO_4. This supports the hypothesis that the process determining the rate of the reaction between CaO and SiO_2 is the diffusion of Ca^{2+} in Ca_2SiO_4.

Table 19 gives the activation energies of self-diffusion of Ca^{2+} in various calcium silicates.

The activation energy of calcium diffusion increases with an increase in the silica content of the silicate, i.e., with a change to compounds with a high degree of binding of the SiO_4 tetrahedra. This makes it understandable why Ca_2SiO_4 is the primary product of the reaction between CaO and SiO_2.

To elucidate the possibility of the diffusion of silicon in the solid-phase formation of calcium silicates, Lindner (1956b) investigated the transfer of activity from calcium silicate containing radioactive Si^{31} to an unirradiated quartz plate or tablet of sintered calcium metasilicate. As in the case of lead silicates, the results obtained indicated the absence of diffusion of silicon. Thus, for the formation of calcium silicates we cannot adopt the mechanism proposed by Wagner (counterdiffusion of Ca^{2+} and Si^{4+} ions). The second diffusing component may be oxygen. Special experiments are required for a final solution of the problem of oxygen diffusion.

To study the mechanism of the reaction of calcium carbonate in the solid state with some silicates and aluminosilicates (albite, sillimanite, nepheline, etc.), Jagitsch (1947) used plates superposed on each other. The experiments were carried out at 700-750°C. On the basis of the change in weight of the plates (the Na_2CO_3 plate decreased in weight and the silicate plate increased in weight), and also the result of experiments on the deposition of an "inert label" (platinum black) on an albite plate, Jagitsch concluded that sodium oxide diffused from soda into the silicate in his experiments.

Convincing demonstrations of the migration of the oxide in a solid-state reaction were obtained by Jagitsch in an investigation of the formation of $Mg_3(PO_4)_2$ from $Mg_2P_2O_7$ and MgO (Jagitsch, Perlström, 1946). The reaction

$$Mg_2P_2O_7 + MgO = Mg_3(PO_4)_2$$

may proceed in three ways: 1) counterdiffusion of Mg^{2+} and P^{5+}; 2) one-way diffusion of MgO; 3) one-way diffusion (in the direction opposite to the diffusion of MgO) of P_2O_5 aggregates. To determine which of the possible mechanisms occurred in actual fact, Jagitsch made a detailed, meticulous investigation, which led to the conclusion that the last of the mechanisms holds.

The experiments were carried out with disks of magnesium pyrophosphate and oxide on top of each other. Before heating, platinum black was deposited as an "inert label" on the surface of the tablets. X-ray investigation showed that the reaction product was formed on both sides of the "inert label" with the layer on the side of the MgO tablet thin and readily crumbling, and that on the side of the pyrophosphate tablet, strong and adhering to the base layer. Weighing the tablets after the experiments showed that the pyrophosphate plate decreased in weight, while the magnesium oxide tablet increased. As an average of 22 experiments, the ratio of the decrease in weight of the $Mg_2P_2O_7$ tablet to the increase in weight of the MgO tablet was 1:0.95, which is within the limits of experimental error. These results provided a strong demonstration of the migration of P_2O_5.

Additional experiments with radium confirmed the proposed reaction mechanism. A radium preparation was deposited on one or another of the tablets. In neither case was there transfer of radioactivity to the neighboring tablet. This indicates that MgO, with which the radium should migrate, did not diffuse from the magnesium oxide tablet to the pyrophosphate tablet. Jagitsch was able to weigh the layers of newly formed magnesium orthophosphate. The ratio of the weights of these two layers was close to 2:1, which is in good agreement with the mechanism proposed:

$$3Mg_2P_2O_7 - P_2O_5 = 2Mg_3(PO_4)_2,$$
$$3MgO + P_2O_7 = Mg_3(PO_4)_2.$$

The ratio of the weight of the orthophosphate film (on the MgO tablet) to the decrease in weight of the pyrophosphate tablet was 1.4, which may be regarded (taking into account the difficulty of separating the layers) as sufficiently close to the required value:

$$\frac{Mg_3(PO_4)_2}{P_2O_5} = \frac{263}{142} = 1.85.$$

The large group of exchange reactions studied by Hedvall (1952), for example

$$BaO + CaWO_4 = BaWO_4 + CaO,$$

also probably proceeds with the diffusion of oxides, but as yet there have been no direct investigations of the mechanism of exchange reactions.

Some data on exchange reactions were obtained in an investigation of them after the preliminary introduction into the crystal lattice of the reacting solids of small amounts of impurities — ions of different valence. This method is similar to that used in semiconductor technology, where this introduction of impurities is called "doping." These additives cannot be called mineralizers in the sense that is usually understood for the latter, but at the same time "doping" appreciably affects the mobility of the diffusing particles and hence the rate of reactions in solid substances.

Schwab, Kohler-Rau, and Ehrenstorfer (1961) showed, for example, that the exchange reaction

$$ZnO + CuSO_4 = ZnSO_4 + CuO$$

is facilitated (the reaction rate is increased) on addition of monovalent lithium to the zinc oxide and is hampered by the addition of trivalent gallium. Thus, the activation energy of this reaction is 25 kcal/mole in the case of pure ZnO, 8 kcal/mole with the addition of 1% Li_2O, and 36 kcal/mole with the addition of 1% Ga_2O_3. On the other hand, the addition of trivalent chromium to nickel oxide (which characteristically had p-conductivity) facilitates the reaction

$$NiO + MoO_3 = NiMoO_4,$$

The activation energy is reduced from 82 kcal/mole for pure NiO to 54 kcal/mole on addition of 1% Cr_2O_3.

The temperatures at which exchange reactions begin to proceed at sufficient rates in solid phases also change considerably when the participants in the reactions are "doped." Thus, the reaction

$$2CuCl + BaO = Cu_2O + BaCl_2$$

begins at 241°C with pure cuprous chloride, while the reaction temperature is reduced to 201°C on addition of 1% $AlCl_3$ to the CuCl and to 116.5°C on addition of 1% Cu_2S. The action of the additive introduced is determined by the amount of it. Thus, the "doping" of $PbCl_2$ with KCl leads to such results. The reaction

$$PbCl_2 + BaO = PbO + BaCl_2$$

begins at 272°C with pure $PbCl_2$, at 250°C with the addition of 1% KCl, at 233°C with the addition of 2% KCl, and at 216°C with 3% KCl.

In the case of cation-conducting substances (CuCl), the reaction temperature is reduced by the addition of highly charged cations (the number of "cationic holes" is increased) and highly charged anions (the number of cations in interstitial positions is increased). In the case of anion-conducting substances ($PbCl_2$), the reaction is facilitated by the introduction of highly charged anions (the number of anionic "holes" is increased) and cations of low charge (for the same reason).

Table 20. Diffusion of Oxygen in Oxides

Oxide	Temperature range, °C	D_0, cm^2/sec	E_a, kcal/mole	S, cal/mole·deg	Investigators
CdO	640—820	$8 \cdot 10^6$	93	+43	Haul, Just, Dümbgen, 1961.
Cu$_2$O	1030—1120	$6.5 \cdot 10^{-3}$	39.3	—3.4	Moore, Ebisuzaki, Sluss, 1958.
TiO$_2$	860—1030	1.1	73	+12	Haul, Just, Dümbgen, 1961.
UO$_{2.002}$	550—800	$1.2 \cdot 10^3$	65.3	—	—
UO$_{2.004}$	450—600	$7.0 \cdot 10^{-6}$	29.7	—	Belle, Auskern, Bostrom, Susko, 1961.
UO$_{2.063}$	320—500	$2.06 \cdot 10^{-3}$	29.7	—	—
ZnO	1100—1300	$6.5 \cdot 10^{11}$	165	+59	Moore, Williams, 1959.
Zr$_{0.85}$Ca$_{0.15}$O$_{1.85}$	680—900	$5.1 \cdot 10^{-3}$	29.8	+1.4	Kingery, 1959.
BeO	1600—1900	$5.2 \cdot 10^{-7}$	4.29	—	Austerman, Meyer, Swarthout, 1961.

On the basis of his experiments, Schwab came to the conclusion that the process determining the rate of the reaction in the powder mixtures studied is self-diffusion in the starting substance. After the work of Schwab and his co-workers, it is impossible to talk of the constancy of the temperature at which an exchange "reaction" begins, as was stated by Hedvall.

We have shown that until now there have been no direct data on the diffusion of oxygen in the reactions of solid oxides, but work in recent years has established without doubt the self-diffusion of oxygen in many oxides.

The usual method of studying the self-diffusion of oxygen in oxides consists of observing isotopic exchange between gaseous oxygen containing O^{18} and crystals of the oxide. The decrease in the O^{18} concentration in the surrounding gas, which is mixed vigorously, is measured. A knowledge of the surface of the oxide powder is required for calculating the diffusion coefficient. The procedure was described, for example, by Kingery (1959b). The diffusion of oxygen in uranium oxides was studied in detail by Belle et al. (Belle and Auskern, 1959). The authors considered that in these oxides of variable composition there migrates oxygen in interstitial positions.

Kingery studied the diffusion of the oxygen ion in refractory oxides, which are most important from a practical point of view. He studied the self-diffusion of O^{2-} in Al$_2$O$_3$ (Oishi, Kingery, 1960a), in MgO (Oishi, Kingery, 1960b), etc. Austerman, Meyer, and Swarthout (1961) studied the diffusion of O^{18} from a surface layer of BeO^{18} into polycrystalline beryllium oxide by the method of removal of layers.

Boreskov and Popovskii (1961) give the diffusion coefficients and activation energies of oxygen in Co$_3$O$_4$, CuO, Fe$_2$O$_3$, V$_2$O$_5$. They note the high activation energies for CuO (130 kcal/mole) and Fe$_2$O$_3$ (100 kcal/mole). It is possible that this activation energy refers not only to the movement of oxygen vacancies, but also to the temperature dependence of the equilibrium concentration of vacancies. Of the given oxides, vanadium pentoxide has the highest oxygen diffusion coefficient ($2 \cdot 10^{-13}$ m/h at 523°C), and this is probably connected with the peculiarities of the crystal structure of this oxide.

A thorough investigation of the diffusion of oxygen in oxides was made by Haul, Just, and Dümbgen (1961). The study of the diffusion of oxygen in silica and silicate compounds is of considerable interest for establishing the mechanism of the formation of silicates. Haul and his co-workers showed that in "Pyrex" glass at 540°C the oxygen has no mobility. Even at a temperature only 60° below the softening point of the glass there is no exchange with oxygen O^{18}. In a very fine powder of silica, the diffusion of oxygen is appreciable at 500-600°C. Diffusion of oxygen is observed in quartz only at 1070°C. The diffusion coefficient of oxygen in fine quartz fibers equals $4 \cdot 10^{-14}$ cm^2/sec and in quartz crystals, $5 \cdot 10^{-17}$ cm^2/sec.

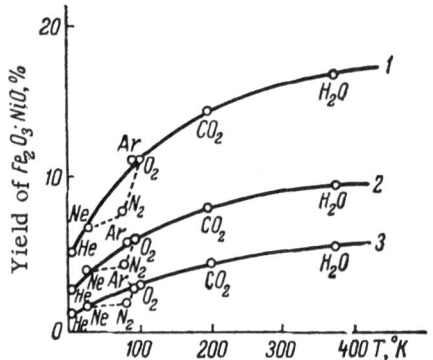

Fig. 75. Relation of the yield of nickel ferrite in atmospheres of various gases. 1) 700; 2) 650; 3) 600°C. Heating time 15 min.

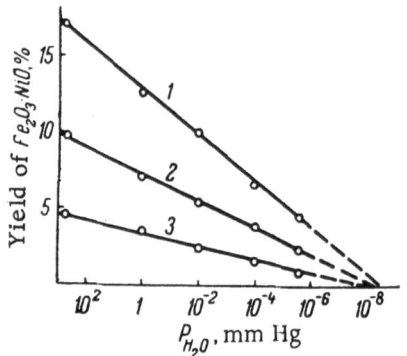

Fig. 76. Relation of the yield of ferrite to the water vapor pressure. 1) 700; 2) 650; 3) 600°C.

In Table 20 we give some of the results of the investigation of the diffusion of oxygen in oxides. Here we give D_0 and E_a, i.e., the constants in the equation $D = D_0 \exp(-E_a/RT)$, and the entropy of activation, determined from the formula $S = R \ln D_0/d^2 \nu$, where d is the distance between neighboring oxygen atoms and ν is the frequency ($3 \cdot 10^{12}$ sec^{-1}).

2. Effect of the Gaseous Atmosphere on Reactions in Solids

A long series of studies of the effect of the gaseous atmosphere on the rate of reactions between solids, the temperature of polymorphic conversions, and sintering was begun in 1946 by Forestier and his co-workers (Forestier, Perlat, 1946; Forestier, Kiehl, Maurer, Stahl, 1952). Forestier explains the action of gases by their adsorption and the change in the character of the surface of the interacting particles.

In work with Kiehl (Forestier, Kiehl, 1949, 1950), Forestier investigated the effect of helium, neon, nitrogen, argon, oxygen, carbon dioxide, and water vapor on the formation of nickel ferrite at 600-700°C when diffusion in the crystal lattice of the iron and nickel oxides is still quite insignificant. The authors observed a clearly expressed dependence of the rate of formation of the spinel on the liquefaction point of the gas (Fig. 75). For the same gas, the kinetics of the reaction depend on the gas pressure. The relation of the reaction rate to the water vapor pressure from 760 to $6 \cdot 10^{-6}$ mm Hg is shown in Fig. 76.

The data presented in Figs. 75 and 76 made it possible for Forestier (1956) to formulate the two following principles: 1) "The rate of the reaction increases symbatically with the liquefaction point of the gas used and consequently depends on the amount of adsorbed gas"; 2) "for the same gas, the reaction rate is a linear function of the logarithm of the gas pressure and the reaction rate tends to zero as the pressure falls to zero." Consequently, there should be no reaction between solids in a complete vacuum on condition that the observations are made below the temperature at which appreciable intracrystalline diffusion begins.

The gas has an effect on a reaction in the solid state, according to Forestier, only up to a certain pressure, after which an increase in the pressure has no further effect. This is evident from Fig. 77, where the relation of the nickel ferrite yield to the argon pressure is given. The effect of the argon is appreciable only at low pressures. A pressure above 1 atm has the same effect, even up to 100 atm.

The identical effect of inert gases and also gases such as oxygen and water vapor indicates the physical nature of the adsorption (van der Waals forces). This is also indicated by the relation between the rate of nickel ferrite formation and the heat of evaporation of the gases studied (Fig. 78).

This relation obviously fits quite well on a straight line.

Forestier and Stahl (Stahl, 1952) also established the effect of gases on the temperature of polymorphic conversions of a series of salts (KNO$_3$, AgI, AgNO$_3$, and Na$_2$MoO$_4$) and also tin dioxide and quartz. Figure 79 shows how the temperature of the α-β conversion of quartz changes in relation to the surrounding atmosphere. The polymorphic conversion of a crystalline solid is a process which occurs inside the solid and the effect of a gas may be explained if we assume that it penetrates inside the crystal.

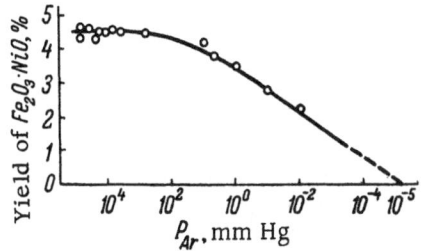

Fig. 77. Relation of the nickel ferrite yield to the argon pressure. Temperature 350°C. Heating time 30 min.

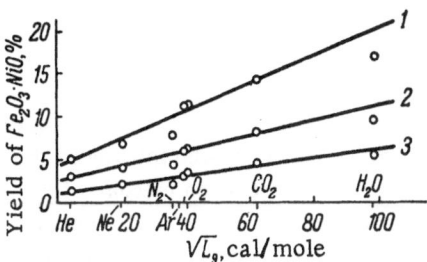

Fig. 78. Relation of the nickel ferrite yield to the surrounding atmosphere. 1) 700°; 2) 650°; 3) 500°C. Along the abscissa axis is plotted the square root of the heat of evaporation of the given substances. Heating time 15 min.

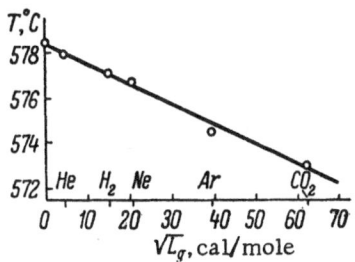

Fig. 79. Relation of the α-β quartz conversion temperature to the surrounding atmosphere. Along the abscissa axis is plotted the square root of the heat of evaporation of the given substances.

Weyl (1956) considers that gases adsorbed on the surface of a crystal, even chemically inactive gases such as nitrogen or the inert gases, modify the electronic structure of the surface atoms and thus have a considerable effect on the chemical bonds in the solids. Trambouze and Silvent (1961) studied the reaction

$$NiO + MoO_3 = NiMoO_4$$

in atmospheres of various gases. A stoichiometric mixture of the given oxides was outgassed in vacuum at 200°C and then kept at a higher temperature (up to 500°C) in various gaseous atmospheres or vacuum. The gases could be divided into two groups with respect to their effect: inert gases, which included helium, nitrogen, argon, and krypton, and reactive gases which included oxygen, carbon dioxide, and water vapor. The former had much less effect than the reactive gases, but still the reaction proceeded more rapidly in them than in vacuum. Of the second group, water vapor had the greatest effect and oxygen the least.

The experiments of Trambouze and Silvent did not confirm the relation between the reaction rate and the boiling point of the gases noted by Forestier and his co-workers (in the study of the formation of ferrites). Trambouze and Silvent explained the effect of reactive gases (O_2, CO_2, H_2O) by their chemisorption on the nickel oxide and consider that the main process of the reaction is surface diffusion.

A considerable effect of the external atmosphere on solid-phase processes was observed by Leonov. The partial pressure of oxygen in the surrounding atmosphere is of great importance in sintering and reactions between oxides. In an atmosphere of moist hydrogen with a partial pressure of oxygen at 1500-1700°C of about 10^{-12} atm, recrystallization of periclase occurs at a considerable rate. Thus, when finely disperse MgO was sintered at 1750°C for 2 h, the mean size of the periclase crystals was 24 × 24 μ in air, 160 × 160 μ in hydrogen, and 40 × 40 μ in argon (Leonov, 1959).

The considerable effect of the partial pressure of oxygen on the rate of formation of magnesial spinel from MgO and Al_2O_3 is shown by the following figures: the amount of $MgAl_2O_4$ formed at 1150°C in 30 min in air was 11.8%, and in hydrogen 41.0% (Leonov, 1961a). The reaction

$$ZnO + Al_2O_3 = ZnAl_2O_4$$

proceeded much more rapidly in purified argon (at 900°C, the spinel yield after 60 min was 12%) than in argon containing 0.4% O_2 (the spinel yield was 5.5%) (Leonov, 1960).

With his experiments Leonov demonstrated that processes involving solid oxides proceed more rapidly when the partial pressure of oxygen in the gas phase is commensurate with the dissociation pressure of the oxides. Leonov calls this increase in the activity of the processes "dissociative activation," and the substance, in his terminology, is in a "transition activated state" under these conditions.

Table 21. Effect of Water Vapor on the Formation of Silicates and Aluminates

Composition of starting mixture	T, °C	vacu-um (1 mm Hg)	Amount of substance formed, %			
			Oxygen atmosphere containing water vapor, mm Hg			
			0.1	17.5	92.5	355
CaO + SiO$_2$ (quartz)	1200	39.6	48.2	56.7	58.6	62.4
CaO + SiO$_2$ (amorph)	1000	34.7	46.5	59.2	67.5	75.5
MgO + Al$_2$O$_3$	1250	16.8	18.0	26.0	32.8	38.0
CaO + Al$_2$O$_3$	1250	52.5	68.5	85.0	89.0	96.0
BaO + Al$_2$O$_3$	1250	40.0	64.0	80.0	84.0	92.0
ZnO + Al$_2$O$_3$	1000	13.0	16.5	28.4	39.4	54.2

Note. Heating time 60 min.

In the light of the work examined it becomes clear that the factor which determines the chemical stability of oxides at high temperatures is not their melting point, as believed by Tamman, who introduced the concept of "reaction temperature," which depends on the melting point, but the dissociation pressure of the oxides. If we are considering kinetic processes, of the three oxides examined below the most stable one in a reducing medium will be corundum, despite the fact that its melting point (2050°C) is considerably lower than that of chromium oxide (2300°C) and magnesium oxide (2800°C). Corundum differs from the given oxides in having the lowest dissociation pressure.

It is much more difficult to explain the strong effect of water vapor on high-temperature chemical reactions between oxides observed by Leonov (1961b). Water vapor not only increases the rate of formation of silicates and aluminates appreciably, but also causes the formation of different chemical compounds. Thus, in the system CaO—Al$_2$O$_3$ in vacuum and dry oxygen there are predominantly formed 3CaO · Al$_2$O$_3$ and CaO · Al$_2$O$_3$ (in the range of 1000-1300°C). In very wet oxygen (P$_{H_2O}$ = 355 mm Hg), together with CaO · Al$_2$O$_3$ the predominating phase is 5CaO · 3Al$_2$O$_3$. Modificational changes in the presence and in the absence of water vapor proceed differently. The effect of water vapor on chemical reactions between oxides in solid phases is shown by Table 21.

The explanation of Leonov of the effect of water vapor by its chemical adsorption with the formation of a "transition activated state" requires further experimental checking.

3. Kinetics of Reactions in Crystalline Oxide Systems

In 1941, Serin and Ellickson (1941) proposed an equation for the rate of a reaction between crystalline substances, which represents an extension of the thermal conductivity equation (Carslaw, Jaeger, 1959) to chemical reactions. , The Serin—Ellickson equation has the form

$$(1 - x) = \frac{6}{\pi^2} \sum_{n=1}^{\infty} \frac{1}{n^2} \exp\left(-\frac{n^2 \pi^2 D t}{r_0^2}\right) = \frac{6}{\pi^2} \sum_{n=1}^{\infty} \frac{1}{n^2} \exp\left(-n^2 k t\right),$$

where $k = \pi^2 D / r^2$, D is the diffusion coefficient, x is the degree of conversion, t is the time, and r_0 is the radius of a particle.

These authors checked their equation on the data of Jander (1927) for the reaction of CaCO$_3$ with SiO$_2$ and MoO$_3$, and obtained quite good agreement with the data of the author. The Serin—Ellickson equation was checked by Carter (1961b) on the results of the oxidation of spherical nickel particles in oxygen. Carter plotted the relation of k to x (the fraction of nickel oxidized). The value of k only approximately remained constant; k changed particularly strongly when x approached unity. Hlavač (1961), who studied the kinetics of the formation of spinel from MgO and Al$_2$O$_3$, demonstrated that his data agree with the Serin—Ellickson equation.

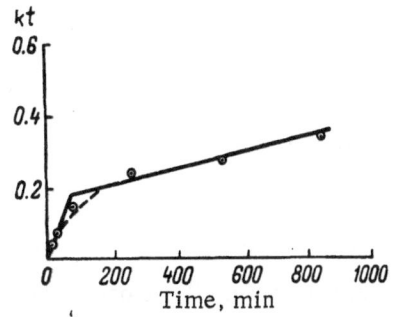

Fig. 80. Relation of kt−t for the reaction of CaO with Ag_2SO_4. Temperature 530°C.

Mason (1957) and Riemen (Riemen and Daniels, 1957) showed that the Serin−Ellickson equation describes the kinetics of several reactions in the solid state: 1) exchange reactions in the systems NaBr−KCl, CsCl−KBr, and NaI−KBr; 2) the reaction of Ag_2SO_4 with CaO and SrO; 3) the formation of silver sulfide Ag_2S from silver and sulfur (Riemen, 1957).

To facilitate the use of the Serin−Ellickson equation, Mason calculated the values of $(1-x)$ for various values of the exponential term using a Card Programmed Calculator IBM Model II. He showed that the Serin−Ellickson equation, which was derived for spherical particles, may be applied to particles of any form.

The reactions between halides at 400-500°C were followed by x rays. The kinetics of these reactions corresponded completely to the Serin−Ellickson equation, as was shown by the linear relation of kt − t obtained. The reaction of silver sulfate with calcium oxide (or strontium oxide) convenient to study as the silver oxide formed by the double exchange

$$Ag_2SO_4 + CaO \rightarrow Ag_2O + CaSO_4$$

decomposes with the liberation of oxygen. The amount of the substances reacting may be determined both from the loss in weight and by determination of the residual silver sulfate.

The relation of kt − t in the temperature region of 400-600°C for the reactions examined was not linear for the whole time of the reactions, but could be represented by two intersecting straight lines (Fig. 80), since the reaction proceeded in two stages. The diffusion mechanism in the crystalline lattice described by the Serin−Ellickson equation applied 40-60 min after the beginning of the reaction. Riemen and Daniels were unable to interpret the first stage of the reaction and put forward two possible explanations: 1) in the first moments there is a surface reaction, and only when the surface is completely covered with the new compound does the diffusion process begin in the volume; 2) as a result of the exothermic nature of the reaction, in the first moment the reaction mixture heats up and its temperature is above that of the furnace.

Riemen carried out special experiments in which coarse (0.5 cm) particles of silver sulfate (or calcium oxide) were coated with fine (about 50 μ) particles of calcium oxide (or silver sulfate) and kept at 625°C. These experiments showed that in the first period there is a rapid surface reaction. However, the authors still did not exclude the possibility of self-heating of the powder in the initial stage.

The kinetic model for reactions between solid particles proposed by Jander (1927) has been examined and criticized repeatedly (Budnikov and Ginstling, 1961; Avgustinik and Kurdevanidze, 1946). In a recent paper, Carter (1961b) again pointed out the weak aspects of Jander's theses and proposed a new equation.

Jander considered a particle (grain) in the form of a sphere of radius r on which, as a result of the reaction with another substance, there forms a layer of reaction product of thickness y. Jander considered that the rate of growth of the layer of reaction product was inversely proportional to its thickness, i.e.,

$$\frac{dy}{dt} = \frac{k}{y}, \tag{1}$$

and, after integration,

$$y^2 = 2kt. \tag{2}$$

The volume of the substance unreacted up to time t will equal

$$V = {}^4/_3\pi (r-y)^3 \tag{3}$$

or

$$V = {}^4/_3\pi r^3 (1-x), \tag{4}$$

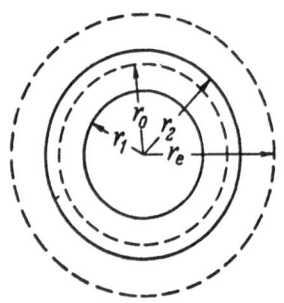

Fig. 81. Plan of reacting particle of
spherical form.

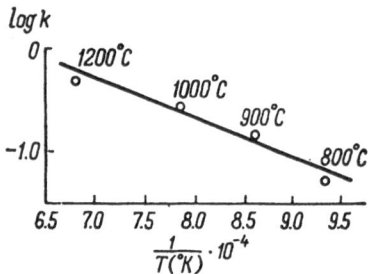

Fig. 82. Relation of the rate constant
of the formation of a solid solution in
the system ZnO—NiO to temperature.

where x is the part of the initial sphere that has reacted.

From the equation

$$^4/_3\pi\,(r-y)^3 = {}^4/_3\pi r^3\,(1-x)$$

we have

$$y = r\,[1-(1-x)^{1/3}].\qquad(5)$$

By replacing y by the value for it from Eq. (2), we obtain the well-known equation of Jander,

$$[1-(1-x)^{1/3}]^2 = \frac{2kt}{r^2} = Kt.\qquad(6)$$

If this equation holds, the relation of $[1-(1-x)^{1/3}]^2$ to t should be represented by a straight line.

Carter pointed out the two following vulnerable assumptions in Jander's derivation: 1) strictly speaking, Eq. (1) applies to a planar surface and in the case of spherical particles it is justified only at small values of x; 2) Eqs. (3) and (4) are justified only when x = 0, i.e., at the beginning of the reaction.

As was to be expected, Jander's equation does not hold at high values of x. Carter's kinetic equation has the following form:

$$[1+(z-1)\,x]^{2/3}+(z-1)\,(1-x)^{2/3} = z+2\,(1-z)\,\frac{kt}{r_0^2}\,.$$

This equation was derived from the following considerations (Fig. 81): let us assume that a sphere of radius r_0 reacts over the whole of its surface with a fine powder or gas. As in Jander's derivation, it is assumed that the process determining the kinetics is diffusion. Let us assume that r_0 is the initial radius of the particle; r_e is the final radius of the reaction product when x = 1; r_1 is the instantaneous radius of the particle of A after some reaction time, which varies from r_0 (initial moment of the reaction, x = 0) to 0 (final moment of reaction, x = 1); r_2 is the instantaneous radius of the sphere of the unreacted component A plus the reaction product, which varies from r_0 (initial moment) to r_e (final moment). We denote by z the volume of reaction product formed per unit volume of component A consumed, i.e., the ratio of the equivalent volumes.

The amount of substance A present at moment t will be

$$Q_A = {}^4/_3\pi r_1^3.\qquad(7)$$

The rate of change of Q_A, i.e., dQ_A/dt, equals the amount of substance diffusing through a spherical shell of thickness $r_2 - r_1$. According to Barrer (1948),

$$\frac{dQ_A}{dt} = -\frac{4\pi k r_1 r_2}{r_2 - r_1}\,,\qquad(8)$$

where k is the reaction rate constant.

As the total volume of the sphere r_2 consists of the volume of the unreacted component A and a layer of the reaction product, we may write

$$r_2^3 = z r_0^3 + r_1^3\,(1-z).\qquad(9)$$

The radius r_1 is expressed in terms of the initial radius of the particle

$$r_1 = (1-x)^{1/3}\,r_0.\qquad(10)$$

By differentiating Eq. (7) with respect to time and equating the expression obtained to the right-hand side of Eq. (8), we obtain

$$\frac{r_1 dr_1}{dt} = -k \frac{r_2}{r_2 - r_1}.$$

(11)

By substituting in this equation the value of r_2 from Eq. (9), we obtain

$$\left\{ r_1 - \frac{r_1^2}{\left[zr_0^3 + r_1^3 (1-z) \right]^{1/3}} \right\} dr_1 = -kdt.$$

(12)

By integrating Eq. (12) from r_0 to r_1, we obtain

$$\left[(1-z) r_1^3 + zr_0^3 \right]^{2/3} - (1-z) r_1^2 = zr_0^2 + 2(1-z) kt.$$

(13)

By substituting in this equation the value of r_1 from Eq. (10), we arrive at the final form of Carter's equation

$$\left[1 + (z-1) x \right]^{2/3} + (z-1)(1-x)^{2/3} = z + 2(1-z) \frac{kt}{r_0^2}.$$

(14)

To check Eq. (14), Carter studied the oxidation of small spherical nickel particles, 74 and 149 μ in size, at 1040 and 1130°C, respectively. By plotting along the ordinate axis the expression

$$\left[1 + (z-1) x \right]^{2/3} + (z-1)(1-x)^{2/3}$$

and time along the abscissa, Carter obtained points which lay well on straight lines. Carter did not check his equation for other reactions.

The kinetics of the formation of a solid solution in the system $NiO-ZnO$ was studied with powder mixtures by Kedesdy and Drukalski (1954). The amount of solid solution formed by the reaction

$$(1-x) NiO + x ZnO = (Ni_{1-x} Zn_x) O$$

was determined with x rays. The limiting solid solution contained 40 mol.% ZnO. The data of the authors corresponded to an equation obtained by combining Jander's equation

$$y = (1 - \sqrt[3]{1-z}) r$$

and a parabolic equation

$$y^2 = 2Dt.$$

Thus, the following equation was obtained:

$$\frac{2D}{r^2} = (1 - \sqrt[3]{1-z})^2 / t = k,$$

where k is the reaction rate constant.

The authors considered the relation of log k to $1/T$ (Fig. 82) as linear even though the experimental points can evidently be described by a different relation. The activation energy E_a obtained from the normal equation

$$k = c \cdot \exp \left(\frac{-E_a}{RT} \right),$$

equals 17,200 cal/mole or 0.75 eV. This activation energy corresponds to the activation energy for the diffusion of cations in ionic lattices and agrees well with the activation energies for the diffusion of ZnO found from measurements of the electrical conductivity of this substance.

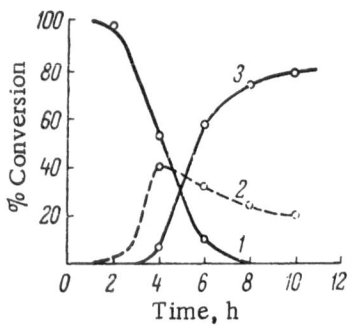

Fig. 83. Kinetics of conversion of quartz into cristobalite. 1) Quartz; 2) transition phase; 3) cristobalite.

The study of the kinetics of the formation of solid solutions is of great interest, as here the only possible process leading to the formation of the final product is diffusion and there is no fundamental rearrangement of the crystal lattice. Kedesdy and Drukalski consider that the formation of the solid solution occurs as a result of the diffusion of ZnO (in the form of the cation and anion) into the NiO lattice with this lattice remaining cubic after the formation of the solid solution. At elevated temperatures there arise defect structures in both the starting lattices.

Roberts and his co-workers (Roberts, 1959) made a detailed investigation of the kinetics of conversion of quartz into cristobalite. In the opinion of these authors, such investigations should help to elucidate the complex process of the interconversions of silica. The kinetic investigations were carried out by quantitative thermal analysis after the sample had been in the furnace for a long time. The furnace constructed made it possible to maintain for 260 h a constant high temperature (up to 1450°C).

The kinetics of quartz degeneration were studied by Roberts with highly pure rock crystal from Madagascar (this material contains 99.96% SiO_2). Brazilian quartz was also investigated. Samples of powdered Madagascar quartz with two grain sizes, the fine with a mean diameter of 0.058 mm and coarser material, were kept for long periods at 1270, 1320, 1370, and 1450°C. It was found that the rate of degeneration of quartz depends strongly on the particle size of the starting material. For the fine fraction the relation of the logarithm of the concentration of unreacted quartz to time was a straight line (for 1320 and 1370°C). This relation was more complex for coarser grains of the starting quartz.

Roberts did not observe tridymite in degenerate quartz at higher temperatures (1500-1650°C), but stated that the conversion of quartz into cristobalite proceeds through an intermediate noncrystalline phase with a density of 2.30 g/cm³ (Chaklader, Roberts, 1961). Here it is necessary to examine the kinetics of two processes: 1) the conversion of quartz into the intermediate phase, and, 2) the conversion of the intermediate phase into cristobalite. As the experimental data show, both of these processes may be regarded as reactions of the first order. Figures 83 and 84 give data showing the kinetics of conversion of quartz into cristobalite.

Starting from the assumption that the process is first order, we may represent the concentration of the intermediate substance in the following way:

$$c_B = a \frac{k_1}{k_2 - k_1} (e^{-k_1 t} - e^{-k_2 t}).$$

We assume that the conversion of quartz into cristobalite proceeds by the scheme $A \xrightarrow{k_1} B \xrightarrow{k_2} C$, where A, B, and C are quartz, the intermediate phase, and cristobalite, respectively. The concentration of substance A after a time t will be

$$c_A = a e^{-k_1 t},$$

where a is the initial concentration of the starting substance. The rate with which C is formed from B will be

$$\frac{dc_C}{dt} = k_2 c_B.$$

The rate of accumulation of B in the system will be the difference between the rate of formation of B from A and the rate of decomposition of B with the formation of C, i.e.,

$$\frac{dc_B}{dt} = -\frac{dc_A}{dt} - \frac{dc_C}{dt} = k_1 c_A - k_2 c_B.$$

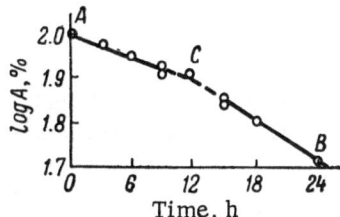

Fig. 84. Logarithmic relation of the amount of quartz converted into cristobalite to time. Diameter of quartz grains 0.137 mm. Temperature 1530°C.

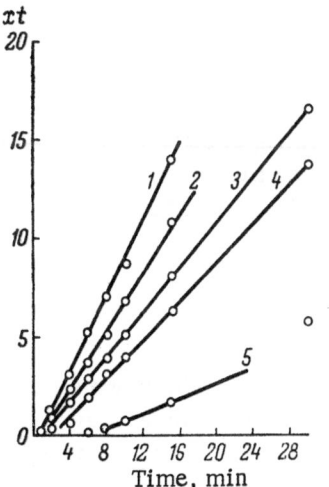

Fig. 85. Relation xt—t at various temperatures for the mixture NiO + Fe$_2$O$_3$. 1) 1200°C, a = 0.975; 2) 1100°C; 3) 1000°C, a = 0.57; 4) 900°C, a = 0.50; 5) 800 °C, a = 0.20.

Consequently, the concentration of the intermediate phase at any moment t will be

$$c_B = a \frac{k_1}{k_2 - k_1} (e^{-k_1 t} - e^{-k_2 t}).$$

In general, the calculated concentration of the intermediate phase was found to be quite close to that found experimentally, and this confirms the type of reaction chosen (first-order reactions). Subsequently, Chaklader (1961) investigated the conversion of quartz into cristobalite in the presence of Al$_2$O$_3$. The character of the process remained unchanged, but the quantitative effect of the alumina on the kinetics was very marked.

The kinetics of the formation of ferrites have been studied recently. Economos and Clevenger (1960) consider that the kinetics of the formation of nickel ferrite may be described by the equation

$$\frac{dy}{dy} = \frac{k}{y},$$

where k is constant. Blum and Li (1961) demonstrated that the best results of investigating the kinetics of the formation of nickel ferrite may be represented by the equation

$$\frac{dx}{dt} = \frac{a - x}{t},$$

where a is constant. This equation is considerably simpler than Jander's equation.

Blum and Li give the relation of xt to the heating time t (Fig. 85) for various temperatures at which the synthesis of nickel ferrite was carried out. As the figure shows, the experimental points lie very well on a straight line, which confirms the validity of the kinetic equation

$$\frac{dx}{dt} = \frac{a - x}{t}.$$

Blum and Li showed that the data of Economos and Clevenger, and also those of Fresh (1957), fit the equation proposed well, but not Jander's equation. If the time dependence of $(1 - \sqrt[3]{1 - x})^2$ is plotted, the experimental points do not lie on one line.

Blum and Li's equation must be regarded as purely empirical. These authors point out that the constant a is determined by the ratio of the surfaces of the Fe$_2$O$_3$ and NiO powders. The proposed equation does not give any indication of the character of the process determining the kinetics. The authors consider that in the initial stage there is predominantly a surface reaction and then in the main reaction there will be diffusion in the crystal lattice.

In connection with the study of the mechanism of formation of zinc spinel, Bengtson and Jagitsch (1947) determined the rate of growth of a layer of spinel in the reaction between ZnO and Al$_2$O$_3$ in the solid state. As in the case of oxide films, the kinetics of growth of the spinel layer may be expressed by the parabolic formula

$$k = \left(\frac{\Delta m}{s}\right)^2 \frac{1}{t},$$

where Δm is the change in the layer thickness in g; s is the cross section of the layer in cm^2, and t is the time in h. For the constant k these authors obtained the following temperature dependence:

$$k = 3.2 \cdot 10^8 \exp\left(-\frac{98\,000}{RT}\right) \text{ g}^2 \cdot \text{cm}^{-4} \cdot \text{h}^{-1}.$$

In the study of reactions between tablets of oxides it is convenient to determine the change in weight of the tablets. For this case the kinetic formula may be written in the form $(\Delta m)^2 = kt$. Then Δm will characterize the increase in weight of one of the tablets as a result of the reaction.

In the study of the reaction of soda with albite and other silicates, Jagitsch (1949) determined the increase in weight of a plate of the silicate on which was placed a plate of soda. The author found that his data fitted the parabolic relation well. For the temperature dependence of the constant k, Jagitsch gives the following expressions: 1) for the reaction of sodium carbonate with sillimanite:

$$k \cong 2 \cdot 10^{13} \exp\left(-\frac{90\,000}{RT}\right) \text{ g}^2 \cdot \text{cm}^{-4} \cdot \text{h}^{-1};$$

2) for the reaction of sodium carbonate with metakaolin:

$$k \cong 8 \cdot 10^{14} \exp\left(-\frac{95\,000}{RT}\right) \text{ g}^2 \cdot \text{cm}^{-4} \cdot \text{h}^{-1}.$$

An extensive investigation of the kinetics of formation of mullite in the heating of certain kaolin clays was carried out by Lundin (1956). He noted that the first work on the kinetics of mullite formation was carried out by Budnikov and his co-workers (Budnikov and Shmukler, 1946; Budnikov and Gevorkyan, 1951), who proposed an empirical expression for the kinetics:

$$m = a \log t + b.$$

Lundin studied pure kaolins and those containing mica. Observing that there is no crystalline mullite at 950°C, he began to determine mullite only at 1050°C and proceeded to 1200°C, using quantitative x-ray analysis. Lundin started from the hypothesis that the kinetic equation must reflect a diffusion mechanism. He proposed two equations of which only the first has a theoretical basis. The first equation has the form

$$\frac{m_\infty - m}{m_\infty} = f\left(\frac{t}{\theta}\right),$$

where m_∞ is the mullite content of the system (in %) when the mullite formation reaction is complete ($t = \infty$), m is the mullite content (in %) at a moment of time t, and θ is the characteristic constant with the dimensions of time.

Lundin arbitrarily assumed that θ is the time required for the formation of 50% mullite. The change in the constant with temperature is well expressed by the Arrhenius equation

$$\theta = ae^{\frac{E_a}{RT}}.$$

The activation energy E_a was found to equal 192 kcal/mole, which is four times as great as that of Budnikov. To determine the amount of mullite, Budnikov used treatment with hydrofluoric acid, assuming that the residue after this treatment was pure mullite.

In the derivation of the second formula for the kinetics of mullite formation, Lundin assumed that the process is a reaction of the first order. However, he considers that in the kinetic equation there should appear not the whole mullite content of the system at a given moment, but some other value which may be regarded as the mullite content of the reaction part of the system. This value is determined by the formula

$$c_m = \frac{m_\infty - m}{(100 - m - b)} \cdot 100,$$

where b is the content of inactive components (in %).

In accordance with an equation of the first order,

$$-\frac{dc_m}{t} = kc_{m'}$$

where k is the rate constant with the dimensions of reciprocal time. Taking into account the fact that when t = 0, $c_m = m_\infty/(100 - b)$, after integration we obtain

$$\ln\left(\frac{c_m(100 - b)}{m_\infty}\right) = -kt.$$

This equation corresponded well to the data from the measurements. The relation of log k to time was plotted to obtain the activation energy. The activation energy obtained in this way agreed well with that found by the first formula.

Although the kinetics of formation of mullite may be described by both formulas, Lundin preferred the first, which was derived assuming a diffusion mechanism. No definite mechanism for the process formed a basis for the second formula, but this formula may be valuable for representing practical results.

BIBLIOGRAPHY

Auskern, A. B., and J. Belle. J. Nucl. Mater. 3(3): 267, 311 (1961).

Austerman, S. B., R. A. Meyer, and D. G. Swarthout. U. S. Atom. Energy Comm. NAA-SR-6427 (1961), 20 pp.

Austerman, S. B., R. A. Meyer, and D. G. Swarthout. Chem. Abstr. 56: 4113 (1962).

Avgustinik, A.I., and O. K. Kurdevanidze. Zh. Prikl. Khim. 19: 1189 (1946).

Barrer, R. Diffusion in Solids [Russian translation], IL, Moscow (1948).

Belle, J., and A. B. Auskern. Kinetics of High-Temperature Processes (1959), pp. 44-49.

Belle, J., A. B. Auskern, W. A. Bostrom, and F. S. Susko. Reactivity of Solids (1961), p. 452.

Bengtson, B., and R. Jagitsch. Arkiv. Kemi 24: 18 (1947).

Blum, S. L., and P. C. Li. J. Am. Ceram. Soc. 44(12): 611-617 (1961).

Boreskov, G. K., and V. V. Popovskii. Kinetika i Kataliz 2(5): 657 (1961).

Brune, U., and R. Lindner. Arkiv. Kemi 5: 26 (1953).

Budnikov, P. P., and K. O. Gevorkyan. Zh. Prikl. Khim. 24: 141 (1951).

Budnikov, P. P., and A. M. Ginstling. Reactions in Mixtures of Solids, Promstroiizdat, Moscow (1961), pp. 175-227.

Budnikov, P. P., and S. G. Tresvyatskii. Dokl. Akad. Nauk SSSR 95(5): 1041 (1954).

Budnikov, P. P., and K. M. Shmukler. Zh. Prikl. Khim. 19: 1029 (1946).

Carslaw, H. S., and C. Jaeger. Conduction of Heat in Solids, 2nd ed., Oxford, England (1959).

Carter, R. E. J. Am. Ceram. Soc. 44(3): 116-120 (1961).

Carter, R. E. J. Chem. Phys. 34(6): 2010-2015 (1961).

Carter, R. E. J. Chem. Phys., Vol. 35 (1961).

Castell, H. S., S. Dilont, and M. Warrington. Nature 153: 653 (1944).

Chaklader, A. C. D. J. Am. Ceram. Soc. 44(4): 175-180 (1961).

Chaklader, A. C. D., and A. L. Roberts. J. Am. Ceram. Soc. 44(1): 35-41 (1961).

Charap, S. H., and E. A. Giess. J. Am. Ceram. Soc. 45(4): 200 (1962).

Economos, G., and T. R. Clevenger. J. Am. Ceram. Soc. 43: 48 (1960).

Forestier, H. Quelques Problèmes de Chimie Minerale, Dixieme Conseil de Chimie, Brussels (1956), p. 505.

Forestier, H., and C. Haaser. Compt. Rend. 227: 123 (1948).

Forestier, H., and J. P. Kiehl. Compt. Rend. 229: 197 (1949).

Forestier, H., and J. P. Kiehl. J. Chim. Phys. 47: 165 (1950).

Forestier, H., Kiehl, Maurer, and Stahl. Proceedings of the International Symposium on the Reactivity of Solids, Göthenburg (1952), p. 41.

Forestier, H., and N. Perlat. Compt. Rend. 223: 575 (1946).

Fresh, D. L. Proceedings of the Symposium of the Microwave Research Institute, Polytechnic Institute of Brooklyn 7: 233-243 (1957).

Haul, R., D. Just, and G. Dümbgen. Reactivity of Solids, Am. Elsevier Publishing Co., Inc., New York (1961), p. 65.

Hedvall, J. A. Enführung in die Festkörperchemie (1952).

Hedvall, J. A. Trans. Brit. Ceram. Soc. 55(1): 1-12 (1956).

Hedvall, J. A. J. Soc. Glas. Technol. 40: 196, 405, 556 (1956).

Hedvall, J. A. Bull. Soc. Chim. France 1: 12-15 (1961).

Hedvall, J. A. Kinetika i Kataliz 3(1): 3-12 (1962).

Hlaváč, J. Reactivity of Solids, Am. Elsevier Publishing Co., Inc., New York (1961), p. 129.

Hopkins, D. W. J. Electrochem. Soc. 96(3): 195 (1949).

Jagitsch, R. Nature 159: 166 (1947).

Jagitsch, R. Naturforsch. 4a(2): 97-101 (1949).

Jagitsch, R., and B. Bengtson. Arkiv Kemi 22A: 6 (1946).

Jagitsch, R., and G. Perlström. Arkiv Kemi 22A: 5 (1946).

Jander, W. Z. Anorg. Allgem. Chem. 163: 1 (1927).

Jorgensen, P. J., M. E. Wadsworth, and I. B. Cutler. J. Am. Ceram. Soc. 42: 613 (1959).

Kedesdy, U., and A. Drukalski. J. Am. Chem. Soc. 76(23): 5941-5946 (1954).

Kiehl, J. P. Compt. Rend., 232: 1666 (1951).

Kingery, W. D. (ed.). Kinetics of High Temperature Processes, The M.I.T. Press, Cambridge, Mass. (1959), pp. 37-44.

Kingery, W. D. J. Am. Ceram. Soc. 42(8): 293 (1959).

Kingery, W. D. J. Am. Ceram. Soc. 43(9): 473 (1960).

Koch, E., and C. Wagner. Z. Physik. Chem. B34(3-4): 317-320 (1936).

Leonov, A. I. Izv. Akad. Nauk SSSR, Otd. Khim. Nauk, No. 12: 2073 (1959).

Leonov, A. I. Izv. Akad. Nauk SSSR, Otd. Khim. Nauk, No. 9: 1529 (1960).

Leonov, A. I. Zh. Fiz. Khim. 35(10): 2328 (1961).

Leonov, A. I. Izv. Akad. Nauk SSSR, Otd. Khim. Nauk, No. 8: 1411 (1961).

Lindner, R. Acta Chem. Scand. 5: 735 (1950).

Lindner, R. K. Vet. Akad. 4: 26 (1952).

Lindner, R. Acta Chem. Scand. 6: 457 (1952).

Lindner, R. Z. Naturforsch. 10a(12): 1027 (1955).

Lindner, R. J. Chem. Phys. 23(2): 410 (1955).

Lindner, R. Z. Elektrochem. 59(10): 967-970 (1955).

Lindner, R. Quelques Problèmes de Chimie Minerale, Inst. Intern. Chim. Solvay, Dixieme Conseil de Chimie, Brussels (1956), pp. 459, 478.

Lindner, R. Z. Physik. Chem., Neue Folge 6(3-4): 129-142 (1956).

Lindner, R. Trans. Chalmers Univ. of Technol. Göthenburg (1956).

Lindner, R., and A. Åkerström. Z. Physik. Chem., Neue Folge 6(3-4): 162-177 (1956).

Lindner, R., and A. S. Obermayer. J. Chem. Phys. 23: 988 (1955).

Lindner, R., and E. Spicar. Arkiv Kemi 7: 62 (1954).

Lundin, S. T. Geol. Foren. Stockholm Förh. 80(4): 458-480 (1956).

Mason, H. F. J. Phys. Chem. 61(6): 796-802 (1957).

Moore, W. J., Y. Ebisuzaki, and J. S. Sluss. J. Phys. Chem. 62: 1038 (1958).

Moore, W. J., and E. L. Williams. Discussions Faraday Soc. 28: 86 (1959).

Oishi, Y., and W. D. Kingery. J. Chem. Phys. 33: 480 (1960).

Oishi, Y., and W. D. Kingery. J. Chem. Phys. 33: 905 (1960).

Pavlyuchenko, M. M. (ed.). Heterogeneous Chemical Reactions, Minsk (1961).

Riemen, W. P. J. Phys. Chem. 61(6): 813-814 (1957).

Riemen, W. P., and F. Daniels. J. Phys. Chem. 61(6): 802-805 (1957).

Roberts, A. L. In: Kinetics of High Temperature Processes, The M.I.T. Press, Cambridge, Mass. (1959), p. 392.

Schwab, G. M., M. Kohler-Rau, and S. Ehrenstorfer. Reactivity of Solids (1961), p. 392.

Secco, E. A., and W. J. Moore. J. Chem. Phys. 26: 942 (1957).

Serin, B., and. R. T. Ellickson. J. Chem. Phys. 9(10): 742 (1941).

Stahl, P. Compt. Rend.. 232: 1669 (1952).

Szabo, S. G., I. Batta, and F. Solymosi. Reactivity of Solids Am. Elsevier Publishing Co., Inc., New York (1961), p. 409.

Thirsk, H. R., and E. J. Whitmore. Transactions Faraday Soc. 36: 565-574 (1940).

Trambouze, J. Y., and A. Silvent. Reactivity of Solids, Am. Elsevier Publishing Co., Inc., New York (1961), pp. 549-555.

Wagner, C. Z. Physik. Chem. B34: 309 (1936).

Weyl, W. A. Quelques Problèmes de Chimie Minerale, Dixieme Conseil de Chimie, Brussels (1956), pp. 401, 447.

CHAPTER VII

LOWER COMPOUNDS OF SILICON WITH OXYGEN

At the present time it is assumed that the following oxides of silicon exist: SiO_2, SiO, and Si_2O_3. The last two compounds are stable only at high temperatures. It is usually assumed that the composition of silica corresponds exactly to the formula SiO_2. However, in view of the fact that silicon monoxide is readily formed at high temperatures, there is the possibility that silica and silicon monoxide will give a compound of variable composition SiO_{2-x}, where x is probably a low value.

An indication of the formation of compounds with an oxygen deficit is found in the article of Ewles and Youell (1951). These authors heated silica with reducing agents, namely silicon, carbon, or hydrogen, or in vacuum. The oxygen deficit in the silica, which was determined with 0.1 N $K_2Cr_2O_7$ solution, was characterized by the following data: silica heated in hydrogen at 750°C for 2 h had an oxygen deficit of 0.05%, and silica heated in vacuum at 1250°C for 4 h had an oxygen deficit of 0.01%. These figures were obtained after the silica that had been treated thermally had been stored for six months.

The strong changes in the properties of quartz, tridymite, cristobalite, and vitreous silica after irradiation with neutrons (Primak, Fuchs, Day, 1955; Sosman, 1955) may also be connected with the formation of defect structures.

The state of oxygen dissolved in silicon was studied in connection with the investigation of the semiconductor properties of silicon. At its melting point, silicon dissolves about $2 \cdot 10^{18}$ cm^{-3} of oxygen. Silicon crystals grown in a silica crucible usually contain a somewhat smaller amount of oxygen. (up to $1.5 \cdot 10^{18}$ cm^{-3}).

Measurements of the infrared absorption of silicon crystals obtained after fast cooling from the melting point show that the oxygen and silicon form various structures from $Si-O-Si$ to SiO_4. The ionization of the SiO_4 group

$$SiO_4 \rightleftarrows SiO_4^+ + e^-$$

is responsible for the donor properties of silicon containing oxygen (Fuller, 1960).

Papers continue to appear in which attempts are made to prove the individuality of the compound Si_2O_3. By condensation from the gas phase containing monoxide vapor and oxygen, Cremer, Faessler, and Krämer (1959) obtained films with a composition between SiO and SiO_2. In a high-vacuum spectrograph the authors observed the K_α-doublet characteristic of silicon, which they attempted to explain by the presence of the compound Si_2O_3 in the film and not a mixture of SiO and SiO_2.

Dadape and Margrave (1961) definitely stated that they obtained a compound corresponding to the formula Si_2O_3. The authors obtained this compound of silicon by evaporating tablets of very pure materials in vacuum (p < 0.5 mm Hg) at 2200°C, which was attained with a solar mirror. They studied the following mixtures: Si + SiO_2 (molar ratios of 1:1, 1:2, 4.32:1, and 2.14:1), Si + Al_2O_3 (2:1), Si + B_2O_3 (1:1), Al + SiO_2 (1:1), and B + SiO_2 (1:1).

In all condensates they observed a cubic phase (other phases were sometimes present). The condensate obtained from various mixtures of Si + SiO_2 gave weak x-ray lines which indicated very fine crystals. These crystals could be made larger by pressing (pressure 70,000 atm) at 800°C. The lattice constant of cubic Si_2O_3, a = 5.77 ± 0.03 A.

The composition of the compound obtained was established by determination of its increase in weight on oxidation with oxygen to SiO_2. The authors were unable to measure the density of Si_2O_3, though with a microscope they observed very fine isotropic crystals of this compound. The difference from silicon monoxide appeared in the infrared spectra. For the substance obtained, the most intense absorption band was observed at 1061 cm^{-1}, which lies between the band characteristic of amorphous SiO (1000 cm^{-1}) and the band observed with single crystals of silicon contaminated with oxygen (1105 cm^{-1}).

1. Solid Silicon Monoxide

The disagreement on the conditions of existence of solid silicon monoxide still continues. Is the sublimate obtained by heating a mixture of Si + SiO_2 in vacuum (deposited on the cold parts of the reaction chamber) solid silicon monoxide or a fine mixture of silicon and silica formed as a result of disproportionation of the monoxide? Can the solid monoxide be thermodynamically stable in some temperature range, or is it stable only as a gas, while the solid may be obtained only in a quenched, frozen state? In the articles of Gel'd (1951) and also the book of Gel'd and Esin (1957), we find much information on actually existing or only supposed solid silicon monoxide.

In the literature there are reports of phenomena, which supposedly indicate the existence of solid silicon monoxide in large masses, but a more detailed investigation often leads to the conclusion that a mixture of Si + SiO_2 is present in some peculiar state. Thus, it is difficult to explain Potter's observation that a substance assumed to be SiO does not become liquid at 1700°C. Had SiO been unstable at this temperature and disproportionated into Si and SiO_2, fusion should have occurred earlier, as SiO_2 softens at this temperature and Si melts much lower. Brewer and Edwards (1954) repeated Potter's experiments and also observed that silicon monoxide (possibly a mixture of Si + SiO_2) has a higher melting point than any of its components.

Gel'd made an electron microscope study of silicon monoxide, which was obtained as thin films by condensation of SiO vapor in high vacuum. Only one phase was observed on photographs with a magnification of 21,000, and this indicates that there was not a mixture of silicon and silica formed as a result of disproportionation. To be sure that the observed homogeneous substance is silicon monoxide, it is necessary to prevent completely the oxidation of the monoxide to silica.

To determine the possibility of the reaction

$$Si \ (solid) + SiO_2 \ (solid.) = 2 \ SiO \ (solid.)$$

it is necessary to find methods of observing this reaction at a high temperature.

Brewer and Edwards passed an electric current through a mixture of Si and SiO_2 in a tube. It was observed that at some moment a narrow luminous zone appeared with a temperature of about 1450°C. They believed that at this temperature there was the rapidly developing reaction

$$^1/_2 Si \ (solid) + {}^1/_2 SiO_2 \ (solid) = SiO \ (solid).$$

The formation of a narrow luminous zone and not general heating of the mixture is explained by Brewer and Edwards by a strong increase in electrical resistance at sites of formation of the first nuclei of solid SiO and local superheating which, in its turn, produces rapid formation of fresh amounts of silicon monoxide.

On the basis of these observations, Brewer and Edwards believe that silicon monoxide is stable above 1450°K and thermodynamically unstable at all temperatures below 1450°K. The adoption of this temperature as the limit of stability of SiO is also supported by the anomaly in the vapor pressure of a mixture of Si (solid) + SiO_2 (tridymite) at 1460°K observed by Schäfer and Hörnle (1950). The change in the character of the temperature dependence of the vapor pressure may be explained by the formation of SiO at 1450-1460°K. Brewer and Edwards consider that silicon monoxide must melt above 1975°K.

As yet there is no reliable evidence based on high-temperature x-ray studies. As regards x-ray investigations of quenched samples of silicon monoxide, according to Gel'd's (1951) careful experiments, it is only possible to talk of amorphous SiO. In preparations containing 50 atom.% oxygen according to analysis, Gel'd observed only lines corresponding to silicon. Literature reports of the existence of crystalline silicon monoxide have not been strictly proved. Some authors (Grube, Speidel, 1949; Hass, 1950) state that on x-ray diffraction patterns there is one ring characteristic of this compound with d = 3.6 A ("characteristic ring").

In his determinations of the enthalpy of silicon, Olett (1958) placed silicon (2-4 g) in a silica glass capsule and sealed it off. After the experiments, on the inner wall of the capsule was a brown powder. X-ray diffraction investigations of this powder showed only lines characteristic of silicon and α-cristobalite, though the formation of this powder was undoubtedly concerned with the condensation of the monoxide.

A high-temperature x-ray study of a stoichiometric mixture of Si + SiO$_2$ was carried out by Brewer and Edwards and then by others. The given authors were able to raise the temperature in the x-ray camera to only 900°C, and then did not observe any lines apart from those belonging to Si and SiO$_2$. No new lines were observed either from samples of a mixture of Si and SiO$_2$ heated at 1300°C and then quenched and investigated in a normal x-ray camera.

Hoch and Johnston (1953) made an x-ray investigation of a stoichiometric mixture of Si and amorphous SiO$_2$ in a high-temperature camera up to higher temperatures. In parallel in the same camera with heating to the same temperatures, x-ray diffraction patterns were obtained of the silicon and silica samples used. Up to 1250°C the x-ray diffraction pattern corresponded to the mixture taken and no new lines were observed; at this temperature there appeared weak lines which the authors ascribed to SiO. Judging by the x-ray diffraction patterns, at 1300°C the formation of SiO was complete in 9 h. Hoch and Johnston stated that the monoxide observed at 1300°C had a cubic lattice with the constant a = 7.135 A. They considered that by rapid cooling (from 1300 to 850°C in 2 sec) they were able to obtain the monoxide at 25°C with Si and SiO$_2$, which formed as a result of disproportionation of a large part of the monoxide. At 25°C the lattice constant of the crystals obtained corresponded to 7.09 A.

Hoch and Johnston attempted to obtain the monoxide in larger amounts and for this they heated a mixture of Si and SiO$_2$ in a tantalum container at 1300°C for 9 h and then quenched it. However, only two weak lines ascribed to SiO were observed together with sharp lines of silicon. The authors consider that the inadequate rate of quenching (from 1300 to 850°C in 10 sec) was insufficient to retain the monoxide, which disproportionated. Hoch and Johnston observed that silica remained amorphous up to 1300°C and they did not attempt to make certain that the lines which they ascribed to the monoxide actually could not belong to SiO$_2$.

Geller and Thurmond (1955) had doubts on the data of Hoch and Johnston and carried out some experiments, though they did not repeat exactly the experiments of the latter. A mixture of high-purity silicon and vitreous silica was heated for 4-19 h in a sealed evacuated tube of quartz glass at 1300°C and then quenched rapidly (2-5 sec) by immersion in water. The x-ray diffraction pattern of the product obtained showed lines corresponding to silicon and diffuse bands from vitreous silica.

When Geller and Thurmond, like Hoch and Johnston, used polystyrene or cellulose acetate as a binder for preparing the samples, on the x-ray patterns in the high-temperature camera there appeared lines characteristic of β-SiC (cubic form). With an x-ray camera with which observations could be made up to 300°C, they obtained x-ray patterns of β-cristobalite and β-SiC. Table 22 shows that most of the values of d found by Hoch and Johnston are found with β-cristobalite or β-carborundum. Only two values, namely 1.67 and 2.05 A, are not in the interplanar distances of these two compounds.

On the basis of their work, Geller and Thurmond came to the conclusion that the lines on the x-ray diffraction patterns obtained by Hoch and Johnston should be ascribed not to the monoxide SiO, but a mixture of β-cristobalite (high-temperature form) and β-SiC (cubic form). However, it should be remembered that not all the lines observed by Hoch and Johnston could be explained, and that β-cristobalite and β-SiC have lines which were not on the x-ray diffraction patterns of Hoch and Johnston.

In 1959, Brady (1959) made a detailed Fourier analysis of the radial distribution curve obtained for the amorphous product in which silicon monoxide was believed to be present. As Fig. 86 (curve 1) shows, on this curve there are peaks at 1.63 and 4.1 A and two weakly expressed peaks lying between 2 and 3 A. The figure also gives the radial distribution curve for vitreous silica according to the data of Warren, Krutter, and Morningstar (1936). The two peaks on curve 3 at 2.35 and 3.8 A correspond quite well to peaks characteristic of

Table 22.   Comparison of Values of d for the Product Obtained by Hoch and
Johnston with the Corresponding Values for β-Cristobalite and β-SiC Obtained by Geller and Thurmond at 300°C

Data of Hoch and Johnston		β-Cristobalite		β-SiC	
d	I/I_0	d	I/I_0	d	I/I_0
4.13	Strong	4.125	100	—	—
2.53	Strong	2.523	26	2.515	69
2.18	Weak	—	—	2.177	17
2.05	Weak	2.059	9	—	—
1.67	Very weak	—	—	—	—
1.64	Very weak	1.636	13	—	—
1.54	Average	—	—	1.542	46
—	—	1.456	10	—	—
—	—	1.373	3	—	—
1.32	Average	—	—	1.315	32
—	—	1.260	2	1.259	6
—	—	1.206	5	—	—
—	—	1.128	3	—	—
—	—	—	—	1.090	8
0.99	Very weak	—	—	1.000	16
0.98	Very weak	—	—	0.9753	8
0.89	Weak	—	—	0.8899	24
0.84	Weak	—	—	0.8394	24

elementary silicon. The peak at 2.35 A corresponds to the nearest Si—Si and the peak at 3.8 A, to the next distance between silicon atoms. On the basis of these data, Brady concluded that the amorphous product corresponding to the formula SiO was a fine mixture of amorphous silicon and silica.

The most intense peaks for amorphous SiO_2 and Si are observed for the Bragg angles of 10.0 and 13.0°, respectively. The "characteristic ring" of silicon monoxide corresponds to a Bragg angle of 11.8°. In view of the diffuseness of the "characteristic ring," Brady considers that it covers both of the given peaks.

Even after Brady's work, the question of the existence of solid silicon monoxide was not conclusively settled. The sample taken as silicon monoxide which was used by this author was a factory product, whose preparation method was not described. As stated above, the x-ray characteristics obtained for this product did not correspond exactly to a mixture of Si and SiO_2. Brady himself notes that the second peak for silicon corresponds to approximately eight nearest atoms and not six, as should be.

Because of the impossibility of obtaining unequivocal data on the existence of solid silicon monoxide by x-ray investigations, optical investigation methods, especially in the infrared region, assume great importance. In the work of Hass and Salzberg (1954), infrared measurements with a film of silicon monoxide covered the spectra region of 0.24-14 μ. These authors state that the film that they investigated, which was obtained by sublimation and subsequent condensation of commercial "monox," was actually an individual compound with the formula SiO. The increase in thickness observed on oxidation actually corresponded to that required for the conversion SiO → SiO_2.

Howarth and Spitzer (1961) extended the region of investigation up to 30 μ with the result that additional data were obtained confirming that the film obtained in this way, as for Hass and Salzberg, actually is silicon monoxide and not a mixture of Si + SiO_2. As the infrared measurements showed, the starting "monox" was not the monoxide, but rather a stoichiometric mixture of Si and SiO_2. In Howarth and Spitzer's opinion, the most convincing evidence in favor of the existence of a

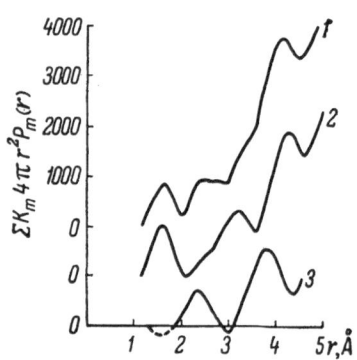

Fig. 86. Radial distribution functions. 1) SiO; 2) $\frac{1}{2}SiO_2$; 3) result of subtracting curve 2 from curve 1.

Table 23. Thermochemical Properties of Liquid Silica at High Temperatures

Temperature, °K	$H_T^\circ - H_{298}^\circ$, cal/mole	$H_T^\circ - H_{2000}^\circ$, cal/mole	C_p, cal/mole·deg	Heat of formation ΔH of liquid SiO_2, cal/mole
2000	29010 ± 580	0	21.66 ± 0.43	−215200 ± 1000
2500	39840 ± 1100	10830 ± 500	21.66 ± 1.0	−212453 ± 1615
3000	50670 ± 1600	21660 ± 1000	21.66 ± 1.0	−210467 ± 2230

film of SiO is the fact that this film does not show a strong absorption band at $\lambda > 20\ \mu$ (about 22.5 μ), which is characteristic of vitreous silica. There is apparently no doubt that solid silicon monoxide can exist as a thin amorphous film.

2. Thermodynamic Properties of Silicon Dioxide at High Temperatures

For the high-temperature chemistry of silicates it is very important to know the properties of silica and other oxygen-containing compounds of silicon at high temperatures. Handbooks of thermal properties usually give data for SiO_2 up to 2000°K. At the same time, contemporary technology requires data on the state of substances at higher temperatures, at least up to 3000°K.

The lack of experimental data makes it necessary to look for methods of finding the thermal properties of substances at high temperatures by extrapolation and with various assumptions. Schick (1960) gives a thermodynamic treatment of high-temperature reactions in the silicon—oxygen system. The material of this article is of great interest for understanding high-temperature processes in which silicon compounds participate.

Enthalpy and Heat of Formation of SiO_2 above 2000°K. To find the difference in heat content $H_T^0 - H_{2000}^\circ$ of liquid (vitreous) silica and its heat of formation for 2500 and 3000°K, Schick makes the assumption that the molecular heat capacity of silica at the given temperatures remains the same as for 2000°K, namely 21.66 cal/mole·deg. Table 23 gives the difference in heat content and heat of formation of silica at high temperatures.

The heat of formation of silica given was calculated from the equation

$$\Delta H_T^\circ = \Delta H_{2000}^\circ + (H_T^\circ - H_{2000}^\circ)_{SiO_2\,(liq)} - (H_T^\circ - H_{2000}^\circ)_{Si\,(liq)} - (H_T^\circ - H_{2000}^\circ)_{O_2}.$$

The corresponding data for silicon and oxygen are given by the following figures: for silicon, $H_{2500} - H_{2000} = 3500 \pm 70$ cal/mole, $H_{3000} - H_{2000} = 7000 \pm 140$ cal/mole; for oxygen, $H_{2500} - H_{2000} = 4583 \pm 45$ cal per mole, $H_{3000} - H_{2000} = 9297 \pm 90$ cal/mole. These data are taken from Stull and Sinke's handbook (1956). The data given will undoubtedly be corrected to some extent by future experimental determinations.

For the heat of atomization of SiO_2 (gas), i.e., ΔH_{298} of the reaction

$$SiO\ (gas) = Si\ (gas) + 2O\ (gas),$$

Brewer and Rosenblatt (1961) give the value 300 kcal/mole.

Evaporation of SiO_2. The evaporation of silica is best examined in connection with the thermodynamic analysis of the reaction

$$SiO_2\ (liq) = SiO\ (gas) + \tfrac{1}{2}O_2\ (gas),$$

as the evaporation of silica under neutral conditions is accompanied by this reaction.

Experimental studies of the evaporation of SiO_2 were made by Brewer and Mastick (1951), using the effusion method, and by Porter, Chupka, and Inghram (1955) with a mass spectrometer. The evaporation of silica was recently studied by Nesmeyanov and Firsova (1960). Still earlier, Ruff et al. (Ruff, Schmidt, 1921; Ruff, Konschak, 1926) and the Japanese scientists Inuzuka and Ageha (1942) used Langmuir's method to determine the vapor pressure of silica deposited on a molybdenum or tungsten wire. As a result of the reduction

$$SiO_2 \text{ (solid)} + {}^1/_3 Mo \text{ (solid)} = SiO \text{ (gas)} + {}^1/_3 MoO_3 \text{ (gas)}$$

the vapor pressure of silica may have been higher than under neutral conditions.

As yet the only study in which mass spectrometry was used to determine the composition of the vapor over silica and a stoichiometric mixture of Si + SiO$_2$ is that of Porter, Chupka, and Inghram. These authors used a Knudsen cell of very pure aluminum oxide with a metallic housing and covered the temperature range of 1200-1950°K.

Over the mixture Si + SiO$_2$, SiO molecules were observed at low temperatures and Si$_2$O$_2$ at higher temperatures. Under neutral conditions, SiO$_2$ and O$_2$ molecules and atomic oxygen were present over silica. From the brief description of Porter, Chupka, and Inghram it is difficult to judge which modification of silica was used by these authors. Without sufficient proof, the authors themselves consider that they used cristobalite.

Table 24 gives the partial pressures of the above components of the vapor over pure SiO$_2$ and the mixture Si + SiO$_2$.

Schick (1960) pointed out that in the calibration of the mass spectrometer, Porter and his co-workers used old data on the vapor pressure of silver and gold. If we use the new values obtained by Honig (1957) for the vapor pressure of these metals, the partial pressures of SiO should be approximately doubled. Moreover, it should be remembered that there was still a reducing atmosphere in the measurements of Porter et al.

In 1960, Nesmeyanov and Firsova (1960) determined the pressure of the vapor in equilibrium with solid silicon dioxide in the temperature range of 1600-1759°K. These authors used the effusion method in the so-called integral method which they developed (Nesmeyanov and Belykh, 1960a, 1960b). The effusion cells of molybdenum, tantalum, or platinum were heated with high-frequency currents. The substance evaporating was condensed on a quartz dome cooled with water and the rate of effusion was found from the amount of silicon in the sublimate deposited.

Nesmeyanov and Firsova carried out calculations on the assumption that the vapor contained only SiO and oxygen molecules, though they did not deny the presence of SiO$_2$ molecules in a very small amount in the vapor (two orders less than P$_{SiO}$) and recognized that neglecting this led to some error. They considered that the solid silica in the effusion cell was tridymite, though the starting material was only examined by the x-ray method and no check was made on the low-temperature conversions of tridymite. It was assumed that in the evaporation there occurred the reaction

$$SiO_2 \text{ (tridymite)} = SiO \text{ (gas)} + x O \text{ (gas)} + {}^1/_2 (1 - x) O_2 \text{ (gas)}$$

with the equilibrium constant

$$K_p = P_{SiO} P_O^x P_{O_2}^{{}^1/_2(1-x)},$$

where x is the degree of dissociation of oxygen.

The partial pressures of the components of the vapor were found from the following equations:

$$P_{SiO} = 2.256 \cdot 10^{-2} \frac{n_{SiO} \cdot M_{SiO}}{st} \sqrt{\frac{T}{M_{SiO}}},$$

$$P_O = 2.256 \cdot 10^{-2} \frac{n_{SiO} \cdot x M_{SiO}}{st} \sqrt{\frac{T}{M_O}},$$

$$P_{O_2} = 2.256 \cdot 10^{-2} \frac{n_{SiO} {}^1/_2 (1-x) M_{O_2}}{st} \sqrt{\frac{T}{M_{O_2}}}.$$

Table 24. Partial Vapor Pressures of Silicon Compounds (According to Porter, Chupka, and Inghram), in atm

T, °K	SiO	SiO₂	Si₂O₂	O₂	O

Over SiO₂ (cristobalite)

T, °K	SiO	SiO₂	Si₂O₂	O₂	O
1800	$6.8 \cdot 10^{-6}$	$1.6 \cdot 10^{-8}$	—	$5.2 \cdot 10^{-8}$	—
1900	$3.4 \cdot 10^{-5}$	$1.0 \cdot 10^{-7}$	—	$8.2 \cdot 10^{-8}$	$2.2 \cdot 10^{-7}$

Over Si + SiO₂ (tridymite)

T, °K	SiO	SiO₂	Si₂O₂	O₂	O
1345	$5.0 \cdot 10^{-6}$	—	—	—	—
1463	$1.1 \cdot 10^{-4}$	—	$4.5 \cdot 10^{-8}$	—	—

Table 25. Thermodynamic Properties of Gaseous Silicon Dioxide

T, °K	C_p, cal/mole·deg (according to Schick)	$S°$, entropy units (according to Schick)	Φ_T, cal/mole·deg (according to Bergman and Medvedev)	$\frac{H° - H_0°}{T}$ cal/mole·deg (according to Schick)	$H° - H_0°$, cal/mole (according to Schick)	$H - H_{298}$ cal/mole (according to Schick)	$H° - H_0°$, cal/mole (according to Bergman and Medvedev)
298	10.80	54.65	—	8.53	2542	0	—
500	12.55	—	—	9.83	4915	2373	—
1000	14.10	—	—	11.66	11660	9118	—
1200	—	—	60.132	—	—	—	14339
1500	14.50	—	62.854	12.57	18855	16313	18662
1800	—	78.57	65.156	12.90	23220	20678	23031
1900	—	79.37	65.850	12.99	24681	22139	24495
2000	14.67	80.06	66.513	13.06	26120	23578	25962
2500	14.73	83.37	69.449	13.40	33500	30958	33323
3000	14.79	85.96	71.902	13.62	40860	38318	40710
3200	—	—	72.781	—	—	—	43670

Table 26. Heat of Sublimation of SiO₂ According to Data of Different Authors

L_{subl}, kcal/mole	Temperature range, °K	Determination method	Authors
102	2073—2503	Boiling	Ruff, Schmidt. 1921.
115	2333	Boiling	Ruff, Konschak, 1926.
103	1273—1473	Langmuir	Inuzuka, Aqeha, 1942.
122	1840—1951	Effusion	Brewer, Mastick, 1951.
140	1800—1900	Mass spectrometric	Porter, Chupka, Inghram, 1955.

Here, n_{SiO} is the number of moles of SiO equal to the number of moles of SiO₂ evaporating with subsequent dissociation. For the equilibrium constant of the reaction O₂ = 2O, we may write

$$K'_p = \frac{P_O^2}{P_{O_2}} = 6.38 \cdot 10^{-2} \frac{x^2}{1-x} n_{SiO} \sqrt{T}.$$

Taking the value of K'_p from appropriate handbooks and n_{SiO} from experimental data, it is possible to find values of x which, in their turn, make it possible to find P_O and P_{O_2}. Having found the equilibrium constant of the reaction K_p, by using the required additional data from existing thermodynamic tables, the authors calculated the heat of formation of gaseous SiO. From 24 experiments they found the mean value ΔH_{form} for SiO (gas, 0°K) = −20.525 kcal/mole.

Thermodynamic Properties of SiO_2 (Gas). There are no data in the literature on the structure and spectra of gaseous SiO_2. Evidently, the only experimental demonstration of the existence of SiO_2 molecules is the mass spectrometric investigation of Porter, Chupka, and Inghram (see p. 120), who observed a beam of SiO_2^+ ions, formed by the direct ionization of SiO_2 molecules.

The thermodynamic properties of gaseous molecules were calculated by Bergman and Medvedev (1959). These authors assumed that SiO_2 molecules have a symmetrical linear structure, i.e., assumed an analogy between SiO_2 and CO_2. In the calculation of the moment of inertia of SiO_2 (gas) it was assumed that r_{Si-O} = 1.554 A. The vibration frequencies of SiO_2 molecules were calculated from equations of a "unified field" model, derived by Herzberg (1949). The following force constants were used: $K_1 = 7.5 \cdot 10^5$ dyne/cm; $K_{12} = 0.7 \cdot 10^5$, $K_\gamma = 0.93 \cdot 10^{-11}$ dyne · radian/cm, which were obtained by comparing the corresponding force constants of the molecules SiO, CO, and CO_2. As a result of these calculations, the following vibration frequencies were adopted for the SiO_2 molecule: $\gamma_1 = 940$ cm^{-1}, $\gamma_2 = 420$ cm^{-1}, $\gamma_3 = 1240$ cm^{-1}.

The entropy of SiO_2 (gas) was calculated by Schick. He adopted somewhat different vibration frequencies for the gaseous SiO_2 molecule, but the thermodynamic functions of SiO_2 (gas) he calculated were quite close to those obtained by Bergman and Medvedev.

The entropy was calculated by Schick from the statistical mechanics equation for a linear polyatomic molecule

$$S = \frac{3}{2} R \ln M + \frac{7}{2} R \ln T - R \ln P + R \ln I - R \ln \sigma + \sum S \,(\text{Einstein}) + S_{el} + 175.385.$$

The moment of inertia of the SiO_2 molecule

$$I = 2M_0 r_{Si-O}^2,$$

where M_0 is the mass of an oxygen atom $16 \cdot 1.66 \cdot 10^{-24}$ g; r_{Si-O} is the interatomic distance, equal to 1.54 A ($1.54 \cdot 10^{-8}$ cm), whence $I = 1.26 \cdot 10^{-38}$ g · cm^2. The entropy contribution of S_{el}^0 in the relevant temperature region is insignificant and may be neglected; the symmetry number σ is assumed to equal two. The calculation of ΣS (Einstein) was carried out for four vibrational degrees of freedom of a linear triatomic molecule.

Schick also calculated the enthalpy function

$$\left(\frac{H^0 - H_0^0}{T} \right)_{\text{trans + rot + el}} = \frac{7}{2} R = 6.954.$$

Table 25 gives the various thermodynamic properties of the SiO_2 (gas) molecule according to the data of Schick and also Bergman and Medvedev.

Thermochemistry of the Reaction SiO_2 (cond.) = SiO_2 (gas). The heat of evaporation (sublimation) of silica was calculated from vapor pressure measurements. Table 26 gives the values of the heat of sublimation of silica obtained from work carried out by Ruff et al. (Ruff, Schmidt, 1921; Ruff, Konschak, 1926), Inuzuka and Ageha (1942), Brewer and Mastick (1951), and Porter, Chupka, and Inghram (1955).

It should be pointed out that under the conditions of the experiments of Ruff et al. there was partial reduction of the SiO_2 to SiO by the material of the vessels and therefore they obtained high values for the vapor pressure. It is probable that the vapor pressures in the work of Inuzuka and Ageha were also high. The best of these data are those obtained by Brewer and Mastick, but as there is dissociation of SiO_2 vapor to SiO and oxygen on evaporation of silica, the value they give for the heat of sublimation (122 kcal/mole), which was calculated without allowance for the dissociation, is only the lower limit of the possible values.

The data of Porter, Chupka, and Inghram should be regarded as most accurate, though these authors did not have the pure reaction

$$SiO_2 \text{ (cond)} \rightarrow SiO_2 \text{ (gas)}$$

under their experimental conditions.

Silicon monoxide and oxygen were obtained at the same time. Moreover, in using the data of Porter et al. it is usual to correct the vapor pressure of silica in accordance with more accurate values for the vapor pressure of the calibration material (see p. 120).

Taking the value $4.3 \cdot 10^{-5}$ atm as the vapor pressure of gold at 1750°K and making appropriate corrections to the data of Porter et al., Brewer and Rosenblatt obtained a heat of evaporation of silica L_{evap} = 135 kcal/mole (according to the second law of thermodynamics) and L_{evap} = 137 kcal/mole (according to the third law of thermodynamics).

The calculations of Porter, Chupka, and Inghram themselves lead to quite similar results. To find the heat of sublimation of silica under standard conditions, these authors adopted for the mean heat capacity of cristobalite over the range of 298-2000°K the value 16.5 cal/deg (Kelley, 1950) and for the heat capacity of SiO_2 (gas) the value 13 cal/deg. For the reaction

$$SiO_2 \text{ (cristobalite)} = SiO_2 \text{ (gas)}$$

the authors adopted L_{298} = 136 ± 8 kcal/mole. For 0°K, after appropriate calculation we obtain L_0^0 = 140 kcal per mole. At 2000°K silica will be in a liquid state. Taking for the heat of fusion of cristobalite 2500 cal per mole, we obtain for the heat of evaporation of liquid silica L_{2000} = 134,100, L_{2500} = 130,000, and L_{3000} = 127,000 cal/mole.

The change in free energy of the reaction

$$SiO_2 \text{ (cristobalite)} = SiO_2 \text{ (gas)}$$

may be obtained, for example, from data on the vapor pressure of silica from the measurements of Porter et al. For 1800°C,

$$\Delta F^\circ = RT \ln P_{SiO_2} = -RT \ln 1.6 \cdot 10^{-8} = 64,214 \text{ cal/mole,}$$

and for 1900°C, ΔF^0 = 60,860 cal/mole.

The free energy of the evaporation of liquid silica may be calculated knowing the entropies of the participants in the reaction and the corresponding thermal effects. Taking the value of the entropy of liquid silica from the tables of Kelley (1950), we have for the change in entropy of evaporation $\Delta S^0 = S_{SiO_2 \text{(gas)}} - S_{SiO_2 \text{(liq)}}$: 38.99 (2000°K, 37.24 (2500°K), 35.93 (3000°K) en. units. The change in free energy of evaporation $\Delta F^0 = \Delta H^0 - T\Delta S^0$: 56,147 (2000°K), 37,577 (2500°K), 19,416 (3000°K) cal/mole. The change in free energy of the process examined equals zero when T = 3540°K; the vapor pressure of liquid silica then equals 1 atm (boiling point) with the assumption that only gaseous SiO_2 is in the vapor. At 3000°K the vapor pressure of silica P_{SiO_2} = $e^{(-19.416/RT)}$ = 0.039 atm.

Boiling Point of SiO_2. It should be pointed out that the boiling point of silica, which is such an important constant for its high-temperature chemistry, has not yet been determined accurately. Present-day handbooks usually give the old data of Ruff, a careful examination of which leads to the conclusion that they are not accurate. Ruff and Schmidt (1921) observed the evaporation of silica in an atmosphere of carbon vapor and therefore the boiling point they give for silica, 2230°C, is somewhat low.

Later, Ruff and Konschak (1926) used an iridium container, but placed it in a graphite vessel and consequently, in this case, there was also a reducing atmosphere. The iridium container was so strongly attacked that it was possible to make only one measurement at 2060°C. On the basis of this measurement, by extrapolation the temperature 2590°C was taken as the boiling point of silica. Ruff himself acknowledged that this temperature was low, and in 1935 (Ruff, 1935), modified it to 2950°C.

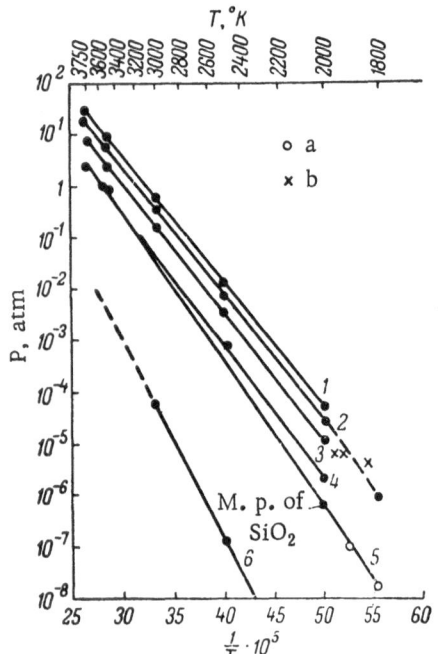

Fig. 87. Temperature dependence of the partial vapor pressures of all the substances which may be present above silica under neutral conditions (allowing for the dissociation of oxygen). a) Experimental points of Porter, Chupka, and Inghram for P_{SiO_2}(gas); b) experimental points of Brewer and Mastick for P_{SiO}(gas). 1) P_{tot}; 2) P_{SiO}; 3) P_{O_2}; 4) P_O; 5) P_{SiO_2}; 6) $P_{Si_2O_2}$.

In 1961, Schick calculated the boiling point of silica, taking into account the new data of Brewer and Mastick, and Porter, Chupka, and Inghram from the investigation of the evaporation of silica. Figure 87 gives the temperature dependence of the partial vapor pressure of all the substances which could be above silica under neutral conditions. Line 1 represents the total pressure. As Fig. 87 shows, this total pressure equals atmospheric pressure at 3070°K. The temperature of 3070°K should be considered as the normal boiling point of silica. At this temperature we have the following partial pressures of the components of the vapor: P_{SiO} = 0.62 atm, P_{O_2} = 0.26 atm, P_O = 0.074 atm, P_{SiO_2} = 0.058 atm, $P_{Si_2O_2}$ = 0.0001 atm.

Schick substantiates the accuracy of the proposed boiling point of silica by the unpublished data of Sherman from the observation of the surface temperature in the ablation of quartz. These observations led to a boiling point of 2680 ± 125°C, which agrees with that calculated by Schick within the limits of experimental error.

Thermodynamics of the Reaction SiO_2 (cond.) = SiO (gas) + $\frac{1}{2}O_2$ (gas). The heat effect of the reaction given is the difference in the heats of formation of silicon monoxide and silica. The heat of formation of silica at high temperatures was given at the beginning of this section. The heat of formation of silicon monoxide was determined in connection with the study of chemical equilibria involving SiO (see p. 128). Taking for the heat of formation of SiO (gas) the values −36,135 (2000 °K), −37,527 (2500°K), and −38,954 (3000°K), we find for the heat of the reaction given,

$$\Delta H_T^\circ = \Delta H_{form\ SiO\ (gas)}^\circ - \Delta H_{form\ SiO_2\ (liq)}^\circ$$

the following values:

179,065 (2000°K), 174,926 (2500°K), and 171,513 (3000°K) cal/mole.

In the determination of the free energy of the reaction examined, the greatest difficulties arise in finding the entropy of liquid silica above 2000°K. As yet it is necessary to take an approximate value as the heat capacity of SiO_2 has not been studied in the region of 2000-3000°K. If it is assumed that the heat capacity in the region of 2000-3000°K remains unchanged and equals 21.66 cal/mole (i.e., as at 2000°K), then for the entropy of liquid silica at 2000, 2500, and 3000°K we obtain the values of 41.07, 46.13, and 50.03 en. units, respectively. All these values make it possible to give the following temperature dependence of the free energy of the reaction examined:

$$\Delta F_T^\circ = 173,000 - 54.2T \text{ (cal/mole)}.$$

The accuracy of the values of the free energy calculated by this equation were estimated as ± 4000 cal per mole. The equation derived may be used in the temperature range of 2000-3000°K. For temperatures below 2000°K (to 1800°K), Schick (1960) recommends the following equation:

$$\Delta F^\circ = 180,800 - 58.11T \text{ (cal/mole)}.$$

By using this equation it is possible to calculate the vapor pressure of SiO and hence to draw conclusions on the accuracy of the experimental data of Brewer and Mastick and also Porter, Chupka, and Inghram.

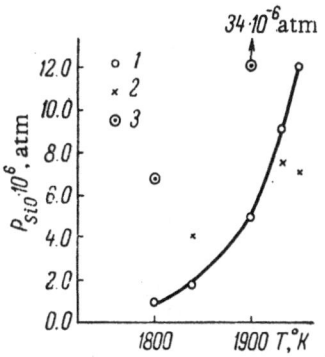

Fig. 88. Comparison of the calculated partial pressure of SiO (gas) (in equilibrium with cristobalite) with the experimental data for the temperature region 1800-2000°K. 1) Calculated from the relation $P_{SiO} = e^{19.72\left(1 - \frac{3075}{T}\right)}$; 2) experimental points of Brewer and Mastick (effusion cell) under neutral conditions; 3) experimental points of Porter, Chupka, and Inghram (mass spectrometer) probably obtained under reducing conditions.

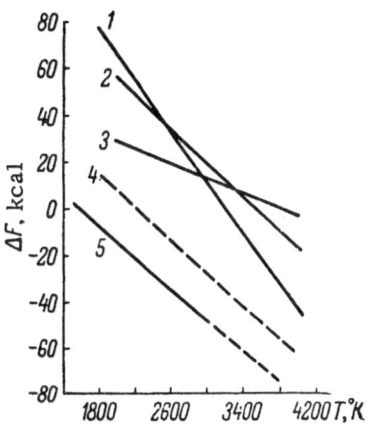

Fig. 89. Change in free energy of some reactions involving silica.

The free energy of the reaction examined is related as follows to the vapor pressure of SiO:

$$\Delta F^{\circ}_T = -RT \ln K = -RT \ln P_{SiO} \cdot P^{1/2}_{O_2}.$$

Under neutral conditions $P_{SiO} = 2P_{O_2}$ and, therefore, by substituting in the equation for ΔF the value $P_{O_2} = P_{SiO}/2$ and solving this equation with respect to P_{SiO}, we have $P_{SiO} = e^{19.72\left(1 - 3075/T\right)}$. The temperature dependence of the partial pressure of SiO calculated from this equation is given in Fig. 88.

As Fig. 88 shows, the measurements of Brewer and Mastick agree well with the calculated values of P_{SiO}. The mass spectrometric measurements give values which are approximately six times as great.

Under oxidizing conditions, for example in an air atmosphere, there will be completely different partial pressures of SiO; the relative amounts of the individual constituents will also be different. Under these conditions, there will be the reaction:

$$SiO \ (gas) + {}^1/_2 O_2 = SiO_2 \ (gas),$$

leading to an increase in the concentration of gaseous silica. In accordance with the observations of Porter, Chupka, and Inghram, the equilibrium constant of this reaction at 1800°K equals 10. Hence it follows that with an oxygen pressure equal to atmospheric the partial pressure of SiO_2 will be ten times as great as the partial pressure of SiO. At 1900°K the partial pressure of SiO_2 may be $10^{-4}-10^{-5}$ atm.

Summarized Data on Reactions Involving SiO_2. Figure 89 gives the temperature dependence of the change in free energy for some reactions involving silica. The free energy of the reaction Si_2O_2 (gas) $\rightleftharpoons 2SiO$ (gas) is also given. Porter, Chupka, and Inghram not only established the existence of the dimer Si_2O_2, but also determined its partial vapor pressure. For the reactions referred to in Fig. 89 we give the corresponding formulas for the temperature dependence of the free energy:

$$1 - \begin{cases} 1 - SiO_2 \ (cristobalite) = SiO \ (gas) + {}^1/_2 O_2 \ (gas) \ \Delta F^{\circ} = \\ = 180,800 - 58.11T, \\ SiO_2 \ (liq) = SiO \ (gas) + {}^1/_2 O_2 \ (gas) \Delta F = 173,000 - 54.2T, \end{cases}$$

$$2 - SiO_2 \ (liq) = SiO_2 \ (gas)$$
$$\Delta F^{\circ} = 127,200 - 36.0T,$$

$$3 - {}^1/_2 O_2 \ (gas) = O \ (gas),$$
$$4 - SiO \ (solid) = SiO \ (gas),$$
$$5 - Si_2O_2 \ (gas) = 2SiO \ (gas)$$
$$\Delta F^{\circ} = 52,500 - 33.4T.$$

3. Investigation of Oxidation – Reduction Equilibria Involving Silicon Monoxide

The thermodynamic properties of silicon monoxide may be obtained by a study of the reaction

$$SiO_2 \ (cond) \rightleftarrows SiO \ (gas) + {}^1/_2 O_2 \ (gas),$$

Table 27. Vapor Pressure Over "Solid" SiO (according to Günther)

Opening in crucible, $cm^2 \cdot 10^3$	T, °K	$P \cdot 10^4$, torr	Opening in crucible, $cm^2 \cdot 10^3$	T, °K	$P \cdot 10^4$, torr
4.1	1468	1300	58.5	1202	0.98
	1520	3690		1245	3.44
				1249	4.35
17.6	1307	22.4		1257	5.26
	1329	41.0		1303	23.2
	1339	50.7		1309	20.7
	1359	89.0		1335	44.1
	1366	110		1361	99.0
	1392	195			
	1419	377			
	1439	632			

Table 28. Values of Constants in Equation for Vapor Pressure of Silicon Monoxide $\log P = A - B/T$ Over the Mixture Si + SiO$_2$, According to Data of Various Authors

A	B	Authors
10.203	16660	Gel'd and Koechnev, 1948.
11.00	16800	Schäfer, Hornle, 1950.
8.42	12730	Tombs, Welch, 1952.
13.28	20800	Günther, 1958.

which has already been examined above. However, more reliable results are obtained by investigating the oxidation−reduction reactions

$$SiO_2 \text{ (cond)} + Si \text{ (solid)} \rightleftarrows 2SiO \text{ (gas)}, \qquad (1)$$

$$SiO_2 \text{ (cond)} + H_2 \text{ (gas)} \rightleftarrows SiO \text{ (gas)} + H_2O \text{ (gas)}. \qquad (2)$$

Finally, it is possible to use other reduction reactions of silica, for example

$$SiO_2 \text{ (cond)} + C \text{ (solid)} \rightleftarrows CO \text{ (gas)} + SiO \text{ (gas)}, \qquad (3)$$

$$2SiO_2 \text{ (cond)} + SiC \text{ (solid)} \rightleftarrows 3SiO \text{ (gas)} + CO \text{ (gas)}. \qquad (4)$$

The most reliable data on the thermodynamic properties of silicon monoxide were obtained by studying reactions (1) and (2).

The reaction

$$Si \text{ (solid)} + SiO_2 \text{ (solid)} \rightleftarrows 2SiO \text{ (gas)}$$

was investigated for the first time in 1948 by Gel'd and Kochnev (1948). These authors first synthesized the solid monoxide (this may not have been the monoxide, but a mixture of Si and SiO$_2$ obtained as a result of disproportionation), with which the investigation for determining the vapor pressure of SiO was carried out. The monoxide was synthesized from a prepared mixture of silicon and silica, which was heated in vacuum $(10^{-4}-10^{-5}$ mm Hg) at 1250-1350°C. The silicon monoxide vapor was condensed on cold parts of the protective tube of Armco iron.

Gel'd and Kochnev used the flow method and the effusion method to determine the vapor pressure of SiO. By the first of these methods it was only possible to establish that at 1000-1200°C silicon monoxide has a very

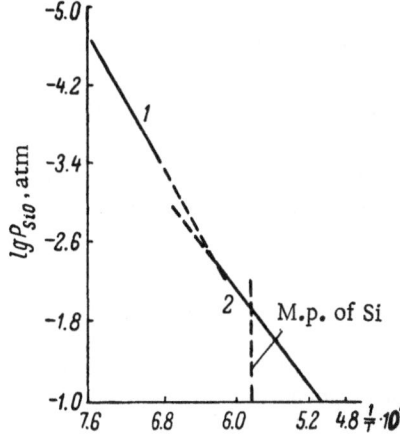

Fig. 90. Temperature dependence of the vapor pressure of SiO over a mixture of Si + SiO$_2$ (or solid SiO). 1) Data of Schäfer and Hörnle; 2) data of Tombs and Welch.

low vapor pressure (less than 1 mm Hg). The second method made it possible to find the absolute values of the vapor pressure, expressed by the equation

$$\lg P_{SiO} \, (\text{atm}) = -\frac{16\,660}{T} + 7.3218.$$

The data of Gel'd and Kochnev have remained valid up to the present time. Their statement that they obtained the monoxide was disputed by subsequent investigators. For the final results it is not important whether the sublimate obtained by Gel'd and Kochnev was silicon monoxide or an intimate mixture of Si + SiO$_2$.

Grube and Speidel (1949) were the first to study the reaction

$$SiO_2 \, (\text{solid}) + H_2 \, (\text{gas}) \rightleftarrows SiO \, (\text{gas}) + H_2O \, (\text{gas}),$$

and used the flow method over the temperature range of 1200-1500°C.

Schäfer and Hörnle (1950) used the effusion method to study the pressure of SiO formed by heating a stoichiometric mixture of Si (98.2% purity) and quartz. The effusion cell was made of fused silica. The authors expressed the vapor pressure obtained as a result of the experiments by the equation

$$\log P_{SiO} \, (\text{mm Hg}) = \frac{-W_P}{4.57T} + \frac{\Delta S}{4.57} + \log 760.$$

The heat of the reaction

$$SiO \, (\text{gas}) = \tfrac{1}{2}Si \, (\text{solid}) + \tfrac{1}{2}SiO_2 \, (\text{solid})$$

was found to equal ΔH = 77.0 kcal and the entropy ΔS = 37.1 cal/deg.

An extensive investigation of the thermodynamic properties of silicon monoxide was carried out by Tombs and Welch (1952). They studied reactions (1) and (2) of the reduction of silica to silicon monoxide. The reaction

$$SiO_2 \, (\text{solid}) + Si \, (\text{solid}) = 2\,SiO \, (\text{gas})$$

was studied by the flow method. Argon was passed over a mixture of SiO$_2$ + Si; the amount of silicon monoxide entrained was determined by two methods, namely by the loss in weight and by weighing the condensate. Tombs and Welch studied a wider temperature range than previous investigators, namely from 1200 to 1650°C.

In the study of the reaction

$$SiO_2 + H_2 \rightleftarrows SiO + H_2O \, ,$$

Tombs and Welch rapidly quenched the reaction mixture and then analyzed the mixture of hydrogen and water vapor with an analyzer, which measured the thermal conductivity of the gas mixture.

The vapor pressure of a mixture of Si + SiO$_2$ and also of previously prepared silicon monoxide was determined in 1957 by Günther (1958). He used Knudsen's effusion method in the experimental form of Herlet and Reich (1957). The effusion cell was a graphite beaker in whose lid there was an opening for the escape of vapor. The material investigated was contained in a pure alumina (sometimes silica) crucible, which was placed in the graphite beaker. The vapor was condensed in a trap fixed to the pointer of microbalances. This made it possible to follow the evaporation continuously.

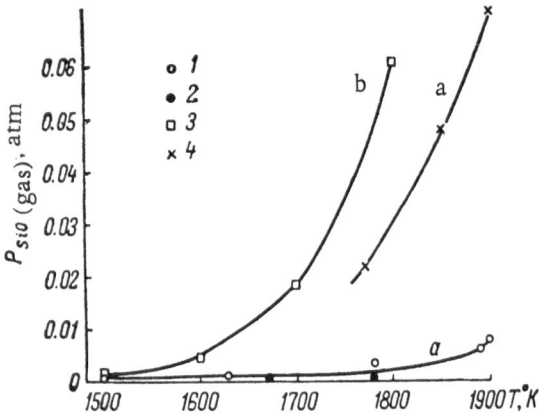

Fig. 91. Temperature dependence of the partial pressure of SiO (gas) in the decomposition of silica by hydrogen (a) or silicon (b) according to the data of various authors. 1,4) Tombs and Welch; 2) Grube and Speidel; 3) Humphrey et al.

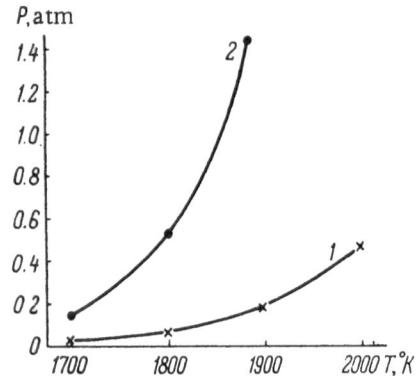

Fig. 92. Temperature dependence of the partial pressures of the most important components of the vapor in the decomposition of SiO_2 by C or Si (according to the data of Humphrey et al.). 1) P_{SiO} for the reaction $\frac{1}{2}SiO_2$(solid) $+ \frac{1}{2}Si$ (liq) $\rightarrow SiO$(gas); 2) P_{CO} for the reaction $SiO_2 + C$(graphite) $\rightarrow SiO$(gas) $+ CO$ (gas) with the condition that P_{SiO} is fixed by the reaction $\frac{1}{2}SiO_2$(solid) $+ \frac{1}{2}Si$ (liq) $\rightarrow SiO$(gas).

In the measurement of the vapor pressure of silicon monoxide formed by heating a mixture of silicon and silica, a different vapor pressure was observed depending on the diameter of the effusion opening and with a decrease in the opening, the pressure increased. This may be explained by the pressure inside the crucible not reaching the equilibrium value because of the low rate of the reaction

$$Si + SiO_2 \rightleftarrows 2SiO \ .$$

On using prepared monoxide (or, if it is assumed that it had already undergone disproportionation, a very fine, ideal mixture of SiO_2 and Si), Günther obtained the same vapor pressure, which was also equal to the equilibrium pressure, with all effusion openings. The vapor pressure of silicon monoxide obtained by Günther is given in Table 27.

The values obtained by different authors for the vapor pressure of SiO over a mixture of Si + SiO_2 are given in Table 28 in the form of the constants A and B of the formula $\log P = A - B/T$.

Figure 90 gives the temperature dependence of the vapor pressure of SiO over a mixture of Si + SiO_2 (or over solid SiO). These data may be regarded as quite accurate.

In 1961, Ramstad, Richardson, and Bowles (1961) studied the reactions

$$SiO_2 + H_2 \rightleftarrows SiO + H_2O,$$
$$SiO_2 + Si \rightleftarrows 2SiO,$$

using the flow method. The first reaction was investigated in the temperature range of 1425-1600°C by measurement of the amount of SiO_2 entrained by the stream of H_2 or $H_2 + H_2O$. In the study of the second reaction (in the range of 1310-1485°C), argon was used as the gas entraining the SiO. It is interesting to compare the vapor pressure of silicon monoxide in the two reduction processes examined (Fig. 91).

As Fig. 91 shows, the partial pressure of silicon monoxide obtained in a hydrogen atmosphere is less than in the reduction of SiO_2 with Si. This indicates that under conditions of reduction with hydrogen it is not possible to form solid SiO or a mixture of Si + SiO_2.

The vapor pressure of silicon monoxide may be studied by using other reducing agents for silica. Thus, Humphrey, Todd, Coughlin, and King (1952) studied the reaction

$$SiO_2 \text{ (solid)} + C \text{ (solid)} = SiO \text{ (gas)} + CO \text{ (gas)}.$$

Figure 92 gives the partial pressures of CO and SiO under conditions when these substances are obtained by decomposition of silica by carbon. The figure shows that a pressure of 1 atm (P_{CO} = 0.89 atm, P_{SiO} = 0.11 atm), is reached at a temperature of 1845°K.

4. Thermodynamics of Reactions Involving Silicon Monoxide and Thermodynamic Properties of SiO

Thermodynamics of the Reaction $\frac{1}{2}SiO_2$(solid) + $\frac{1}{2}$Si(solid) = SiO(gas) and SiO(solid) = SiO(gas). For determining the free energy of the reaction

$$^1/_2 SiO_2 \text{ (cristobalite)} + {}^1/_2 Si \text{ (solid)} = SiO \text{ (gas)}$$

the two following forms of equilibria were studied:

$$SiO_2 \text{ (solid)} + H_2 \text{ (gas)} \rightleftarrows SiO \text{ (gas)} + H_2O \text{ (gas)}, \qquad (1)$$

$$^1/_2 SiO_2 \text{ (solid)} + {}^1/_2 Si \text{ (solid)} \rightleftarrows SiO \text{ (gas)}. \qquad (2)$$

The following values were obtained for the free energy of reaction (1): Grube and Speidel, ΔF^0 = 112,000 −35.5 T (1200-1500°C); Tombs and Welch, ΔF^0 = 12,800 − 21.0 T (1228-1653°C).

The vapor pressure of the monoxide over a mixture of silicon and silica was studied by Schäfer and Hörnle (1950) and Tombs and Welch (1952). The latter authors, working in the region of high temperatures, had liquid silicon in the reaction mixture and therefore, to the free energy value they obtained ΔF^0 = 58,550 − 25.45 T there should be added the free energy of fusion of silicon, which for one-half gram-atom of silicon is ΔF^0 = 6050 − 3.6 T.

From the free energy of reaction (1) it is possible to obtain the free energy of reaction (2), if the free energy of formation of water vapor and cristobalite is known.

As a result of appropriate calculations, and also from direct determinations, we obtain the following values for the free energy of reaction (2):

Tombs and Welch (from the pressure of SiO over the mixture Si + SiO$_2$), ΔF^0 = 64,600 − 29.05 T.

Schäfer and Hörnle (from a pressure of SiO over a mixture of Si + SiO$_2$), ΔF^0 = 77,000 − 37.1 T.

Tombs and Welch (from a study of the reduction of SiO$_2$ with hydrogen), ΔF^0 = 32,550 − 13.2 T.

Grube and Speidel (from a study of the reduction of SiO$_2$ with hydrogen), ΔF^0 = 66,350 − 27.75 T.

As a result of studying reduction equilibria (see section 3, Ch. VII), Ramstad and Richardson (1961) arrived at the following equations for the free energy of reaction (2):

$$\Delta F^\circ = 162{,}930 - 77.25T \quad (1550-1685° \text{ K}),$$
$$\Delta F^\circ = 151{,}300 - 70.07T \quad (1685-1800° \text{ K}).$$

For reaction (1) these authors give the equation for the free energy,

$$\Delta F^\circ = 127{,}100 - 45.07T \quad \text{cal /mole.}$$

Figure 93 gives all data available up to the present time from the study of the equilibria of reactions (1) and (2). As the figure shows there is quite a considerable discrepancy in the results of different authors.

To assess the accuracy of the results of the investigation of oxidation−reduction equilibria, it is necessary to determine the entropy of the participants in the reaction. If the entropy obtained is close to the value adopted on the basis of other considerations, then it may be concluded that the other thermodynamic functions, found by investigating the equilibria, such as the heat of reaction, are accurate.

From the equations presented for ΔF^0 according to the data of Tombs and Welch, Schäfer and Hörnle, and Grube and Speidel, we have the following values for ΔS: 29.05, 37.1, 13.2, and 27.72. To find the standard values of the entropy of silicon monoxide, it is necessary to use the following equation:

$$S_{SiO \text{ (gas)}} = {}^1/_2 S_{SiO_2 \text{ (solid)}} + {}^1/_2 S_{Si \text{ (solid)}} + \Delta S_{298}.$$

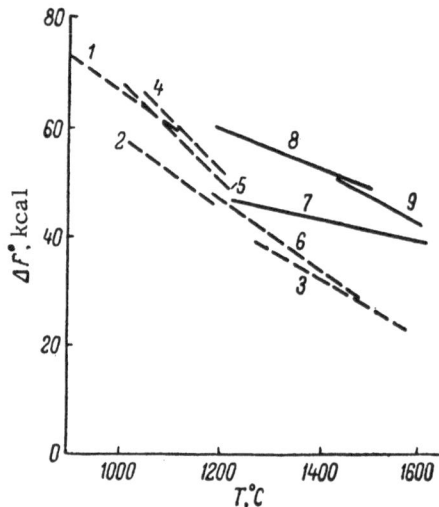

Fig. 93. Temperature dependence of the
free energy of the reactions SiO$_2$ (solid)
+ H$_2$(gas) = SiO(gas) + H$_2$O(gas) (solid
lines) and $\frac{1}{2}$ Si(solid) + SiO$_2$(solid) = SiO
(gas) (broken lines) according to the data
of different authors. 1) Gel'd and Kochnev; 2) Schäfer and Hörnle; 3,7) Tombs
and Welch; 4) Porter, Chupka, and
Inghram; 5) Günther; 6,9) Ramstad and
Richardson; 8) Grube and Speidel.

The change in entropy of the reaction (for standard conditions)
equals

$$\Delta S_{298} = \Delta S - \Delta C_p \ln T + \Delta C_p \ln 298.$$

To find ΔC_p it is necessary to know the molar heat capacities of the participants in the reactions. For Si(solid) and SiO$_2$
(solid) the heat capacity is found from tables; for SiO(gas),
Kubaschewski and Evans proposed the mean value between the
heat capacities of CO(gas) and S$_2$(gas). Having obtained for ΔC_p
the value 2.8 and substituted in the expression for S$_{SiO(gas)}$ all
known values, we obtain

$$S_{SiO\,(gas)} = \Delta S + 6.45 \lg T - 8.6.$$

By substituting in this equation the corresponding values for ΔS
(29.05, 37.1, 13.2, and 27.72), we obtain the following values
for S$_{SiO(gas)}$ (at 25°C): 41.3 (according to data of Tombs and
Welch), 48.75 (according to data of Schäfer and Hörnle), 25.35
(according to data of Tombs and Welch on reduction of SiO$_2$ with
hydrogen), and 39.75 cal/deg (according to data of Grube and
Speidel).

The entropy for gaseous compounds may be obtained by a
completely different method, for example by empirical formulas
and also from spectral data. Kubaschewski and Evans (1958a)
give an empirical formula for the entropy of a diatomic gas with
a molecular weight M between 20 and 30

$$S_{298} = 53.8 + 0.043M - 240M^{-1}.$$

According to this formula, the entropy of gaseous silicon monoxide equals 50.2 cal/deg. Kubaschewski and
Evans consider this value in general more reliable than that obtained from the investigation of equilibria. The
spectral data also confirm the empirically calculated value of the entropy rather than that obtained from oxidation—reduction equilibria. From spectal data for SiO(gas), Kelley (1950) obtained the value S$_{298}$ = 50.55 cal
per deg. Kubaschewski and Evans regard this as the most reliable value.

We may now find a more accurate value for the change in entropy of reaction (2) and then, knowing the
change in free energy of this reaction from the study of the equilibrium, find its heat effect. The corresponding data are given in Table 29.

The study of reduction equilibria by different authors led to results which differed strongly between each
other. At the same time, when one value of ΔS was taken, values of ΔH$_{298}$ were obtained which differed little
between each other. The average of the four values is 83,450 cal/mole, but Kubaschewski and Evans prefer to
take the value obtained from the data of Schäfer and Hörnle, which is rounded off to 82,500 cal/mole.

As a result of thermodynamic examination of all the material available up to the end of 1957,
Kubaschewski and Evans give the following data for the thermodynamic properties of silicon monoxide:

Gaseous SiO.

$$\Delta H_{298} = -22,200 \pm 2500 \quad cal/mole,$$

$$S_{298} = 50.55 \pm 0.1 \ . \ cal/deg.$$

Table 29. Thermodynamic Characteristics of the Reaction $\frac{1}{2}$ SiO$_2$(cristobalite) + Si(solid) = SiO(gas)

T,°K	ΔS	ΔF	ΔH_T	ΔH_{298}	Investigators
1750	38.2	18,750	80,600	84,650	Tombs and Welch (study of the vapor pressure of SiO over the mixture Si + SiO$_2$)
1420	38.8	24,300	79,400	82,450	Schäfer and Hörnle (study of the vapor pressure of SiO over the mixture Si + SiO$_2$)
1710	38.3	10,000	75,500	79,450	Tombs and Welch (study of the reduction of SiO$_2$ with hydrogen)
1620	38.4	21,400	83,600	87,300	Grube and Speidel (study of the reduction of SiO$_2$ with hydrogen)

Amorphous SiO.

$$\Delta H_{298} = -99,500 \pm 5000 \ \text{cal/mole},$$

$$S_{298} = 11.1 \pm 1.5 \ \text{cal/deg}.$$

Boiling point 2070°C.

$$L_{\text{evap}}(298° \text{K}) = 77,300 \pm 6000 \ \text{cal/mole},$$

$$L_{\text{evap}}(\text{b.p.}) = 66,000 \pm 6000 \ \text{cal/mole}.$$

The calculations given above were aimed at finding the thermodynamic characteristics of silicon monoxide under standard conditions. The thermodynamic properties of silicon monoxide at high temperatures were examined by Brewer and Edwards (1954).

Regardless of whether SiO exists as an individual compound in a solid state or not, to understand the processes in which silicon monoxide participates it is possible to consider the thermodynamics of the reaction

$$\text{SiO (solid)} = \text{SiO (gas)},$$

i.e., to assume that this compound exists under certain conditions. By using the data of Tombs and Welch, Brewer and Edwards obtained for this reaction a thermal effect (in the temperature region of 1573-1920°K), ΔH = 58,550 cal/mole and a change in entropy ΔS^0 = 25.45 en. units. For the change in free energy, consequently, we have ΔF^0 = 58,550 − 25.45 T. From this equation it follows that the pressure of SiO(gas) reaches 1 atm (ΔF = 0) at T = 2300°K.

A somewhat different expression for ΔF of the process examined is obtained if we use the heat capacity of crystalline silicon monoxide. Brewer and Edwards give the following values for the heat capacity of crystalline SiO and the entropies of crystalline and amorphous monoxide:

$$C_p(\text{SiO cr}) = 10.57 + 2.98 \cdot 10^{-3}T - 2.72 \cdot 10^5 T^{-2}.$$

For SiO(cr), S$_{298}$ = 6.4 en. units; for SiO(amorph), S$_{298}$ = 7.3 en. units. Hence, we have for SiO(amorph), S$_{2000}$ = 31.1 en. units. By taking for gaseous SiO, S$_{2000}$ = 66.04 en. units, we obtain for the change in entropy of the process analyzed

$$\Delta S° = 66.04 - 31.1 = 34.94 \ \text{en. units}.$$

The equation derived above,

$$\Delta F° = 58,550 - 25.45T$$

may be close to 2000°K. For higher temperatures, Brewer and Edwards recommend taking ΔH = 77,530 cal per mole, and then the change in free energy in the high-temperature region will be

$$\Delta F° = 77,530 - 34.94T.$$

Finally, the heat of the reaction

$$SiO \text{ (solid)} = SiO \text{ (gas)}$$

may be found if we know the heats of formation of these substances at 298°K: for SiO(gas), $\Delta H^0 = -21,800$ cal per mole, and for SiO(solid), $\Delta H^0 = -104,600$ cal/mole (Brewer, 1954). The heat of the reaction required $\Delta H_{298}^0 = -21,800 - (-104,600) = +82,800$ cal/mole.

For a temperature other than 298°K,

$$\Delta H_T^\circ = \Delta H_{298}^\circ + \int\limits_{298}^{T} \Delta C_p dT,$$

$$\Delta H_T^\circ = \Delta H_{298}^\circ + (H_T^\circ - H_{298}^\circ)_{SiO(gas)} - (H_T^\circ - H_{298}^\circ)_{SiO(solid)},$$

$$\Delta H_T^\circ = \Delta H_{298}^\circ + (H_T^\circ - H_0^\circ)_{SiO(gas)} - (H_{298}^\circ - H_0^\circ)_{SiO(gas)} - (H_T^\circ - H_{298}^\circ)_{SiO(solid)}$$

For 2000°K we have

$$\Delta H_{2000} = 82,800 + 8.18 \cdot 2000 - 6.955 \cdot 298 -$$

$$\int\limits_{298}^{2000} (10.57 + 2.98 \cdot 10^{-3}T - 2.72 \cdot 10^5 \cdot T^{-2}) \, dT =$$

$$= 82,800 + 16,360 - 2070 - 23,090 = 74,000 \text{ cal/mole.}$$

This value is close to the value given above for $\Delta H_{2000} = 77,530$ cal/mole.

If we deny the existence of solid silicon monoxide, then, instead of the reaction

$$SiO \text{ (solid)} = SiO \text{ (gas)}$$

it is necessary to examine the reaction

$$^1/_2 Si \text{ (liq)} + ^1/_2 SiO_2 \text{ (liq)} = SiO \text{ (gas).}$$

The free energy of this reaction may be calculated from appropriate data for Si(liq) and SiO_2(liq). At temperatures close to 2000°K the calculations give:

$$\Delta F = \Delta H_{2000} - T\Delta S_{2000} = 71,465 - 33.01T.$$

This expression is similar to that obtained on the assumption that solid silicon monoxide exists.

Solid silicon monoxide should most likely be regarded as amorphous. The heat of evaporation of the amorphous monoxide, i.e., the thermal effect of the reaction

$$SiO \text{ (amorph)} = SiO \text{ (gas),}$$

may be found by using the data of Wartenberg (1949) on the heats of solution of silicon and the amorphous monoxide. From these data, for the thermal effect of the reaction Si(solid) + $\frac{1}{2}O_2$(gas) = SiO(amorph), we obtain $\Delta H_{298} = -104,000$ cal/mole. Hence it is possible to determine the required heat of evaporation of amorphous SiO, $L_{298} = 104,000 - 22,200 = 81,800$ cal/mole.

To calculate the entropy of amorphous SiO it is necessary to resort to appropriate empirical rules, such as the tables of Latimer (1951), which make it possible to calculate the entropies of crystalline solids. According to these tables, hypothetical crystalline silicon monoxide should have an entropy of 8.6 cal/deg · mole (at 298°K). Inorganic compounds of the general formula AB have an entropy of fusion of about 5 cal/deg · mole. The conversion of a solid into an amorphous state involves a smaller entropy than fusion and it may be assumed that the entropy of this conversion for silicon monoxide is 2.5 cal/deg · mole. Then the entropy of evaporation of silicon monoxide for 298°K will be $50.55 - 8.6 - 2.5 = 39.45$ cal/deg · mole, and the free energy of the evaporation of silicon monoxide: $\Delta F = 81,800 - 39.45\,T$.

Thermodynamics of Disproportion of $SiO(amorph) = \frac{1}{2} SiO_2(solid) + \frac{1}{2} Si(solid)$.
The free energy of the disproportionation of silicon monoxide to solid crystalline silica (most likely cristobalite) and silicon may be obtained as the difference in ΔF of evaporation of SiO and ΔF of formation of gaseous SiO from $SiO_2(solid)$ and $Si(solid)$. Thus, for the disproportionation of SiO, we obtain $\Delta F = -700 + 3.75\,T$ cal/mole.

To find the free energy of disproportionation of amorphous silicon monoxide to amorphous silica and silicon, we use the free energy of conversion of cristobalite to silica glass. ΔF (cristobalite-glass) $= 700 - 0.55\,T$ (for room temperature). The free energy of disproportionation required will then be $\Delta F = 3.2\,T$. The free energy of disproportionation found is a small value.

Conclusions drawn on the stability of this compound on the basis of the thermodynamic characteristics of silicon monoxide obtained naturally cannot be conclusive. The change in free energy obtained for the reaction

$$SiO\,(amorph) = {}^{1}/_{2}\,SiO_2\,(glass) + {}^{1}/_{2}\,Si\,(solid).$$

$\Delta F = 3.2\,T$ indicates that amorphous silicon monoxide should be stable relative to silicon and silica at any temperature. However, this is contradictory to all observations of the decomposition of the monoxide in attempts to obtain it in a solid state.

Largely on the basis of the statement of Gel'd and Kochnev (1948), who considered that at 1250-1350°C the equation

$$SiO\,(amorph) = {}^{1}/_{2}SiO_2\,(glass.) + {}^{1}/_{2}Si\,(solid)$$

is shifted to the left, Kubaschewski and Evans (1958b) put forward the hypothesis that ΔF of this reaction will equal zero at 1150-1200°C and for the free energy, instead of the expression $\Delta F = 0 + 3.2\,T$ they propose $\Delta F = -4600 + 3.2\,T$. This equation may be obtained by assuming that the heat of formation of $SiO(amorph)$ equals $-99,500$ and not $-104,000$ cal/mole. The difference between these values lies within the limits of error of Wartenberg's (1949) measurements. Then the free energy of evaporation of $SiO(amorph)$ also changes. For standard conditions it is given by the equation

$$\Delta F_{298} = 77,200 - 39.45\,T,$$

and the complete equation for the free energy of evaporation of silicon monoxide assumes the form

$$\Delta F = 78,900 + 12.65\,\lg T - 76.25\,T,$$

which corresponds to the following equation for the vapor pressure:

$$\lg P_{SiO}\,(mm\ Hg) = -17,250\,T^{-1} - 2.77\,\lg T + 19.55.$$

ΔF of evaporation becomes equal to zero at about 2350°K, which is the boiling point. The heat of evaporation at this temperature, $L_{evap} = 78,900 - 5.5\,T = 66,000$ cal/mole. The entropy of evaporation, according to the data presented, equals 28 cal/deg, which is a little higher than that required by Trouton's rule.

For the reaction

$$SiO\,(amorph) = SiO\,(gas)$$

Gel'd and Kochnev give (1300°C), $\Delta F = 32,500$ cal/mole. However, Kubaschewski and Evans, using the data of the Soviet authors, give the somewhat different value of 31,100 cal. Then ΔF of the reaction $\frac{1}{2}SiO_2(cristobalite) + \frac{1}{2}Si(solid) = SiO(gas)$ will be 28,000 cal/mole.

Brewer and Edwards attempted to calculate the thermodynamic stability of solid SiO, having determined beforehand the entropy of this substance, using approximate formulas. They used Latimer's (1951) rule and

Table 30. Heat of Formation of Gaseous SiO at 298°K

Equilibrium studied	ΔH_{298}, cal /mole	Investigators
$\frac{1}{2}$ Si(solid) + $\frac{1}{2}$ SiO$_2$(solid) → SiO(gas)	$-18,212 \pm 960$	Gel'd and Kochnev, 1948
SiO$_2$(solid) + H$_2$(gas) → SiO(gas) + H$_2$O(gas)	$-15,291 \pm 2000$	Grube and Speidel, 1949
$\frac{1}{2}$ SiO$_2$(solid) + $\frac{1}{2}$ Si(solid) → SiO(gas)	$-21,411 \pm 574$	Schäfer and Hörnle, 1950
SiO$_2$(solid) → SiO(gas) + $\frac{1}{2}$ O$_2$(gas)	$-21,159 \pm 2000$	Brewer and Mastick, 1951
SiO$_2$(solid) + H$_2$(gas) → SiO(gas) + H$_2$O(gas)	$-26,004 \pm 2400$ (low-temperature data), $-21,269 \pm 1500$ (high-temperature data)	Tombs and Welch, 1952

obtained a value for the entropy of solid SiO of 6.4 en. units at 298°K. Using the value of the heat capacity of solid SiO found by Potter (1907) of 9.19 cal/mole · deg, and assuming that the temperature dependence of the heat capacity of the solid monoxide corresponds to that for solid isotropic crystalline substances, Brewer and Edwards arrived at the following expression for the heat capacity of crystalline SiO:

$$C_p = 10.57 + 2.98 \cdot 10^{-3}T - 2.72 \cdot 10^5 T^{-2}.$$

In accordance with this expression, the entropy of crystalline monoxide at 1200°C equals 22.4 en. units. Assuming that amorphous SiO has an entropy 0.9 en. units greater than that of crystalline monoxide [the same difference in entropy is found for the change SiO$_2$(cristobalite) → SiO$_2$ (amorph)], for amorphous SiO at 1200°C we have S = 23.3 en. units.

For the reaction

$$^1/_2 \text{Si (solid)} + {}^1/_2 \text{SiO}_2 \text{ (cristobalite)} = \text{SiO (amorph)}$$

Brewer and Edwards give ΔS^0 = 1.5 ± 3.0 en. units, ΔH = 1.0 ± 3.0 kcal/mole, and ΔF = −0.9 ± 6.6 kcal/mole. The thermodynamic functions obtained for this reaction are again too indefinite to draw any conclusions on the stability of silicon monoxide at 1200°C.

To determine the region of stability of silicon monoxide, Brewer and Greene (1957) studied the system Si − SiO$_2$ by high-temperature thermal analysis with induction heating, which made it possible to reach a temperature of 3000°C. The corresponding method was described by Eastman, Brewer, Bromley, Gilles, and Lofgren (1950) and Brewer and Zavitsanos (1957). The latter authors noted that heating could be achieved by electron bombardment.

Brewer and Greene took an intimate mixture of metallic silicon and quartz with excess of the latter relative to the stoichiometry of Si + SiO$_2$. The mixture was heated up to 1500°C. The authors always observed a thermal effect corresponding to the fusion of silicon (close to 1415°C) and exactly the same effect was obtained with pure silicon. These experiments show that SiO$_2$ does not react with Si at the melting point of silicon and silicon monoxide is not formed. Hence, the conclusion is drawn that solid SiO must be thermodynamically unstable at all temperatures. Thus, in their work Brewer and Greene disprove the statement of Brewer himself and Edwards that SiO(solid) may be stable and not disproportionate at temperatures above 1400°C.

In considering a series of solid monoxides of group IV elements: CO, SiO, GeO, SnO, PbO, TiO, ZrO, and ThO, Brewer and Greene report that the first four compounds are not stable in the solid state at all temperatures. Thus, solid CO, obtained by condensation of gaseous CO, is not stable at all temperatures relative to graphite and solid CO$_2$.

According to Brewer and Zavitsanos (1957), the same may be said of GeO, and according to Platteeuw and Meyer (1956), of SnO. Thus, in group IV, all elements from carbon to lead form compounds MeO, which are only stable in the gaseous state. Lead forms a stable solid compound corresponding to the composition MeO. In the fourth group, the stability of solid TiO has been established definitely, but solid ZrO and ThO are thermodynamically unstable at least as pure phases.

Table 31. Heats of Formation of Gaseous SiO for High Temperatures

T, °K	ΔH_T, cal/mole
0	−21695
298	−21411
2000	−36135
2500	−37527
3000	−38954

Thermal Properties of SiO at High Temperatures.
The heat of formation of gaseous SiO at temperatures when silicon is in the form of vapor, i.e., the thermal effect of the reaction

$$Si \text{ (gas)} + {}^1\!/_2 O_2 \text{ (gas)} = SiO \text{ (gas)},$$

may be calculated from the equation

$$\Delta H_T = \Delta H_0^\circ + (H_T - H_0^\circ)_{SiO \text{ (gas)}} - (H_T^\circ - H_0^\circ)_{Si \text{ (gas)}} - {}^1\!/_2 (H_T^\circ - H_0^\circ)_{O_2 \text{ (gas)}}.$$

The heat of formation of SiO(gas) under standard conditions was obtained from data on equilibria described above (see p. 126). Brewer and Edwards, critically examining the results of the investigation of the equilibria

$$Si \text{ (solid)} + SiO_2 \text{ (solid)} = 2SiO \text{ (gas)},$$

$$SiO_2 \text{ (solid)} + H_2 \text{ (gas)} = SiO \text{ (gas)} + H_2O \text{ (gas)},$$

give for the heat of formation of silicon monoxide ΔH_{298} the values given in Table 30. Similar values were obtained from spectral investigations.

Taking the energy of atomization of SiO(gas) obtained from the spectral data of Bonhoeffer (1928), (−170 kcal), and also the energy of atomization of O_2 and the heat of sublimation of Si (at 25°C), we obtain for the heat of the reaction

$$Si \text{ (solid)} + {}^1\!/_2 O_2 \text{ (gas)} = SiO \text{ (gas)}$$

$$\Delta H_{298} = -170 + 59.1 + 87.5 = -23.4 \text{ kcal/mole.}$$

As a result of examining the data obtained by various authors for the heat of formation of SiO, Brewer and Edwards consider that the best data are those found by Schäfer and Hörnle, i.e., $\Delta H_{298} = -21,411$ cal/mole.

Table 31 gives the thermodynamic properties of silicon monoxide for high temperatures. The data for silicon and oxygen required for the calculations were taken from the tables of Stull and Sinke (1956).

The first detailed spectral investigation of silicon monoxide was made by Bonhoeffer (1928). He used a Tammann furnace, the carbon tube of which contained silica. The furnace was heated to 1500°C and the absorption spectrum plotted of the gases inside the tube (a hydrogen discharge tube was the radiation source). To demonstrate that the absorption bands observed corresponded to the monoxide and not silicon dioxide, a long quartz tube was placed inside the furnace and the absorption spectrum of the gases inside this tube investigated. The bands observed in the first case were then not observed. Bonhoeffer estimated the energy of the bond between silicon and oxygen atoms in a molecule of gaseous silicon monoxide as 175,000 cal/mole.

Later, Barrow (1955) made a spectroscopic investigation of thermal equilibria involving gaseous suboxides and subsulfides of silicon and also confirmed the existence of SiO as a gas.

The structure of a molecule of silicon monoxide existing as a vapor has not yet been established conclusively. There is the possibility that this compound has a ring structure such as $(SiO)_3$, but evidence against this is provided by the magnitude of the entropy of gaseous silicon monoxide. The following empirical equation for calculating entropy applies to polyatomic gases:

$$S_{298} = 39.0 + 0.34M - 6.2 \cdot 10^{-4} M^2.$$

For silicon monoxide this formula gives S_{298} = 72 en. units or per SiO group, ${}^{72}\!/_3$ = 24 cal/deg · mole, which differs considerably from the entropy found experimentally.

Table 32. Thermodynamic Functions for SiO(gas) from Spectral Data (cal/mole · deg)

Function	298 °K	2000 °K	2500° K	3000° K
$\dfrac{F_T^\circ - H_0^\circ}{T}$				
Translation contribution	−32.30	−41.76	−42.86	−43.76
Electronic contribution	0.00	0.00	0.00	0.00
Vibration contribution	0.00	− 1.05	− 1.34	− 1.61
Rotation contribution	−11.27	−15.05	−15.49	−15.85
Sum	−43.57	−57.86	−59.69	−61.22
$\dfrac{H_T^\circ - H_0^\circ}{T}$				
Translation contribution	4.965	4.965	4.965	4.965
Electronic contribution	0.00	0.00	0.00	0.00
Vibration contribution	0.00	1.23	1.36	1.46
Rotation contribution	1.987	1.987	1.987	1.987
Sum	6.955	8.18	8.317	8.412
$S_T^\circ = \left(\dfrac{H_T^\circ - H_0^\circ}{T}\right)_{\text{Sum}} - \left(\dfrac{F_T^\circ - H_0^\circ}{T}\right)_{\text{Sum}}$	50.52	66.04	68.01	69.63
Correction for anharmonicity . . .	0.00	0.00	0.03	0.04
S_T°	50.52	66.04	68.04	69.67

The great importance of silicon monoxide in the high-temperature chemistry of silicates and many metallurgical processes accounts for the appearance of many papers on the investigation of the properties of gaseous silicon monoxide. The determination of the thermodynamic properties of SiO at high temperatures is facilitated by the fact that this substance is a gas and therefore spectroscopic methods may be used.

The summary of Herzberg (1949) contains data characterizing the molecular properties of gaseous SiO, which make it possible to calculate the reduced thermodynamic potential and the entropy of this substance in the high-temperature region. Only the singlet ground state is considered in these calculations. The next higher electronic state has a wave number of 42,835 cm^{-1}, which corresponds to $(hc\omega/kT) = 20.6$ at 3000°K, and the contribution to the distribution function $e^{-(hc\omega/kT)}$ is too small to be considered in the calculations.

For a single vibration of a diatomic molecule SiO, the characteristic frequency ($\omega = \omega_e - 2\omega_e x_e$) corresponds to 1230 cm^{-1}. In the calculation of the rotational contribution, for the moment of inertia I we adopt the value $38.6 \cdot 10^{-40}$ g · cm^2. Table 32 gives the correction for anharmonicity for the entropy, but this correction is very small.

The quantum-statistical calculations of Dashevskii and Khitrik (1948) also led to results which agree well with data from spectral investigations. In the calculations of the entropy of gaseous SiO, these authors used the formula for the entropy of diatomic molecules

$$S = 4.575 \left({}^7/_2 \lg T + {}^3/_2 \lg M - \lg P + \lg \frac{g_0 I}{n} \right) + S_E\left(\frac{\theta}{T}\right) + 175.37,$$

where M is the molecular weight, P is the pressure in atm, g_0 is the statistical sum of the lowest energy level, I is the moment of inertia in g · cm^2, and n is the symmetry number. Assuming that M = 44.06, P = 1 atm, $\theta = \omega(h/k) = 1780$,

$$I = \frac{m_{1,\,Si} \cdot m_{2,\,O}}{m_{1,\,Si} + m_{2,\,O}} d^2 = 14.0 \cdot 10^{-39} \text{ g} \cdot \text{cm}^2$$

[where $m_{1,Si}$ = 28.06/(6.06 · 10^{23}), $m_{2,O}$ = 16/(6.06 · 10^{23}), and d = 1.51 A], we obtain for 1873°K, S = 62.93 en. units. If we assume that $C_{P_{SiO}}$ = $C_{P_{CO}}$ = 6.6 + 1.2 · 10^{-3}T, then

$$S_{298} = S_{1873} - \int_{298}^{1873} \frac{(6.6 + 1.2 \cdot 10^{-3}T)}{T} dT = 48.93 \text{ cal/deg} \cdot \text{mole.}$$

This value is close to the most reliable value adopted by Kubaschewski and Evans, S_{298} = 50.55 cal/deg· · mole.

Table 32 gives the thermodynamic functions of gaseous SiO obtained from spectroscopic data. The translation, vibration, and rotation contributions are given for the function $F_T - H_0/T$. For the reasons given above, the electronic contribution is always taken as zero.

Determination of the Thermodynamic Properties of Silicon Oxides by Explosion in a Bomb. The explosion method, which began with the work of Nernst, has long been used for determining the thermal properties of gaseous systems at high temperatures (3000°K and above). In describing ways of determining heat capacity and heats of dissociation by the explosion method, Gurvich and Shailov (1955) limited themselves to the examples of hydrogen and water vapor only. The explosion method is only beginning to be used for studying gaseous molecules of oxides of metallic elements.

The great possibilities offered by the use of the explosion method for studying high-temperature processes and properties of substances at high temperatures are demonstrated in the work of Medvedev (1958). However, considerable development of the procedure is still required before explosions in a bomb can be used widely in high-temperature chemistry of oxide systems. Medvedev (1958) showed that under the conditions of explosion in a spherical bomb, the combustion products at the moment at which the maximum pressure is reached are in complete thermodynamic equilibrium. This opens great possibilities for the investigation of the thermodynamic properties of many gaseous compounds of metals with oxygen. In particular, by the explosion method it is possible to carry out investigations at high temperatures (3000°K and above), where substances which are quite unusual for low-temperature conditions (BO, SiO, etc.) are formed and exist stably.

Bergman and Medvedev (1959) applied the method of explosion in a spherical bomb to the study of vapors of involatile substances for the first time. The authors carried out explosions of mixtures of disilylethane $SiH_3CH_2CH_2SiH_3$ with oxygen (sometimes with other fuel gases added).

Under the conditions of explosion of substances which form involatile compounds, the vapor pressure of the latter may considerably exceed the equilibrium pressure of the saturated vapor, and there is the danger that in the time between ignition of the combustible gas mixture and the attainment of the maximum pressure (approximately 0.01 sec), the supersaturated silicon oxides formed cannot condense completely. Special experiments showed that under the conditions of explosion in a spherical bomb there is practically equilibrium condensation of supersaturated vapors. This makes it possible to use the explosion method for investigating "condensed phase—gas" equilibria and to determine the heat of sublimation of involatile substances.

In the work of Bergman and Medvedev, the maximum explosion pressure was measured with a mechanical gage described by Medvedev, Korobov, and Baibuz (1958). The rise in pressure during the explosion and the time were recorded with an optical system and a nine-loop oscillograph.

Bergman and Medvedev carried out two groups of experiments: 1) with mixtures containing excess oxygen, and 2) with mixtures containing considerable amounts of reducing gases, namely hydrogen or carbon monoxide. In the latter case, silicon monoxide was formed as a result of the explosion and, in the first case (with excess oxygen), gaseous silicon dioxide was formed. Depending on the composition of the starting mixture, various temperatures, sometimes exceeding 3000°K, were obtained in the explosions. The starting mixture was made up by the successive introduction of the components into the preliminarily evacuated bomb. The partial pressures of the separate components were measured with a mercury manometer and a cathetometer.

Bergman and Medvedev first determined the magnitude of the heat losses in the bomb in the time interval between ignition of the mixture and the moment that the maximum pressure is attained. These losses appear in the thermal balance of the explosion and the accuracy of the results depends to a considerable extent on the accuracy of their determination. It is particularly important to take into account the fact that in explosions of mixtures whose combustion products include substances in a condensed state, the thermal losses must be great. These authors derived an empirical formula for calculating the thermal losses.

Mixtures of disilylethane with a large excess of oxygen were investigated in the first series of experiments. The relatively low temperatures (2100-2500°K) must have promoted the formation of silicon dioxide in a state of very highly supersaturated vapor. The calculations were carried out with two assumptions: 1) the condensation of SiO_2 did not occur and the silicon dioxide formed was in a supersaturated vapor state up to the moment that the maximum pressure was reached; 2) complete equilibrium was reached between the supersaturated vapor and the condensed phase. The second assumption was found to be correct, i.e., condensation of the supersaturated silica vapor occurred under the experimental conditions used.

The second series of explosions were aimed at reaching higher temperatures (>2300°K), and for this purpose, carbon monoxide (up to 10%) was added to the mixture instead of excess oxygen. These experiments also confirmed the validity of the assumption that equilibrium was reached between the supersaturated silicon dioxide vapor and the condensed phase. The results of the experiments showed that the heat of sublimation of silica must be greater than 130 kcal/mole. It would apparently be more accurate to adopt the heat of sublimation of silica obtained by Porter and his co-workers (136 kcal).

In the next two series of experiments, excess carbon monoxide or hydrogen was taken. High temperatures (2800-3200°K) were reached here, but the medium was reducing and the combustion product under these conditions was found to be silicon monoxide. The authors showed that there was condensation of the SiO vapor, which was in a supersaturated state in the first moment. The thermal balance was satisfactory if $\Delta H_0 = 86$ kcal per mole was taken as the heat of sublimation of silicon monoxide; this value of ΔH_0 was also adopted by the authors of the work examined.

As was pointed out above, at the present time there is a dispute as to whether solid SiO exists and, therefore, it may be better as yet to talk of the heat of the following reaction rather than the heat of sublimation of SiO:

$$Si\,(solid) + SiO_2\,(solid) = 2SiO\,(gas).$$

From their experimental data, Bergman and Medvedev were able to calculate the heat of formation of silicon monoxide in the condensed state. The value they obtained, $\Delta H_0^0 = -108.5$ kcal/mole, is in good agreement with the results of direct calorimetric determination carried out by Wartenberg (1951), who gave the value -104 kcal/mole, converted to 0°K.

The most important result of the work of Bergman and Medvedev is the establishment of the fact that under the conditions of an explosion in a spherical bomb, in the time between the ignition of the combustible gas mixture and the attainment of the maximum pressure, which is measured in hundredths of a second, the supersaturated vapors of silicon oxides formed are able to condense and this makes it possible to use the explosion method to investigate high-temperature heterogeneous equilibria. Moreover, this method makes it possible to find the heats of sublimation of refractory oxides, which are difficult to determine.

BIBLIOGRAPHY

Barrow, R. F. Proc. Roy. Soc. (London), Ser. A 224: 374 (1954).

Barrow, R. F. Trans. Faraday Soc. 51:1480 (1955).

Bergman, G. A., and V. A. Medvedev. Trans. State Inst. Appl. Chem. 42:158-172 (1959), Goskhimizdat, Leningrad.

Bonhoeffer, K. F. Z. Phys. Chem. 131: 369 (1928).

Brady, G. W. J. Phys. Chem. 63(7):1119-1120 (1959).

Brewer, L. Chem. Rev. 52:1 (1953).

Brewer, L., and R. K. Edwards. J. Phys. Chem. 58(4): 351-358 (1954).

Brewer, L., and F. T. Greene. Phys. Chem. Solids 2: 286-288 (1957).

Brewer, L., and D. F. Mastick. J. Chem. Phys. 19: 834 (1951).

Brewer, L., and G. M. Rosenblatt. Chem. Rev. 61(3): 257-263 (1961).

Brewer, L., and P. Zavitsanos. Phys. Chem. Solids (1957), pp. 2284-2285.

Cremer, E., A. Faessler, and H. Krämer. Naturwissensch. 46: 377 (1959).

Dadape, V. V., and J. L. Margrave. Abstracts of Scientific Papers Presented at the Eighteenth Internat. Congress of Pure and Appl. Chem. (1961), pp. 103-104.

Dashevskii, Ya. V., and S. I. Khitrik. Stal', No. 10: 892 (1948).

Eastman, E. D., L. Brewer, L. A. Bromley, P. W. Gilles, and N. L. Lofgren. J. Am. Chem. Soc. 72: 2248 (1950).

Ewles, J., and R. F. Youell. Trans. Faraday Soc. 47(10): 1060-1064 (1951).

Fuller, C. S. Seventeenth Internat. Kongress Reine Angewandte Chemie, Plenarvorträge, Vol. I, Anorganische Chemie (1960), pp. 271-299.

Gel'd, P. V. High-Temperature Reduction Processes (1951), pp. 10-55.

Gel'd, P. V., and O. A. Esin. High-Temperature Reduction Processes (1957), Metallurgizdat, Moscow.

Gel'd, P. V., and M. K. Kochnev. Zh. Prikl. Khim. 21: 1249 (1948).

Geller, S., and C. D. Thurmond. J. Am. Chem. Soc. 77: 5285 (1955).

Grube, G., and H. Speidel. Z. Elektrochem. 53: 339 (1949).

Günther, K. G. Glastech. Ber. 31(1): 15 (1958).

Gurvich, A. M., and Yu. Kh. Shailov. Thermodynamic Investigations by the Explosion Method and Calculations of Combustion Processes, Izd. MGU (1955).

Hass, G. J. Am. Ceram. Soc. 33: 353 (1950).

Hass, G., and C. D. Salzberg. J. Opt. Soc. Am. 44(3): 181-187 (1954).

Herlet, A., and G. Reich. Z. Angew. Phys. 9(1): 14, 23 (1957).

Herzberg, G. Vibration and Rotation Spectra of Polyatomic Molecules [Russian translation], IL, Leningrad (1949).

Hoch, M., and H. L. Johnston. J. Am. Chem. Soc. 75(21): 5224-5225 (1953).

Honig, R. E. RCA Rev. 18: 2 (1957).

Howarth, L. E., and W. G. Spitzer. J. Am. Ceram. Soc. 44(1): 26-28 (1961).

Humphrey, G. L., S. S. Todd, J. P. Coughlin, and E. G. King. U. S. Bureau of Mines, Bulletin No. 4888 (1952).

Inuzuka, H., and M. Ageha. J. Japan Ceram. Assoc. 50: 105 (1942) [see: Chem. Abstr. 44: 8080d (1950)].

Jacobs, G. Comptes Rend. 236: 1369-1371 (1953).

Kaiser, W., H. L. Frisch, and H. Reiss. Phys. Rev. 112: 1546 (1958).

Kelley, K. K. U. S. Bureau of Mines, Bulletin No. 477, (1950), p. 113.

Kubaschewski, O., and E. L. Evans (eds.). Metallurgical Thermochemistry, 3rd ed., Pergamon Press, Inc., New York (1958), p. 195.

Kubaschewski, O., and E. L. Evans (eds.). Metallurgical Thermochemistry, Assessment of Standard Values Silicon Monoxide, Pergamon Press, Inc., New York (1958), pp. 390-396.

Latimer, W. M. J. Am. Chem. Soc. 73: 1480 (1951).

Medvedev, V. A. Zh. Fiz. Khim. 32(8): 1851-1858 (1958).

Medvedev, V. A., V. V. Korobov, and V. F. Baibuz. Zh. Fiz. Khim. 32(12) (1958).

Nesmeyanov, A. N., and L. P. Belykh. Zh. Fiz. Khim. 34: 841 (1960).

Nesmeyanov, A. N., and L. P. Belykh. Zh. Fiz. Khim. 34: 1032 (1960).

Nesmeyanov, A. N., and L. P. Firsova. Zh. Fiz. Khim. 34(9): 1907-1910 (1960).

Olett in: J. Elliott (ed.). Phys. Chem. Steelmaking, Proc. Dedham, Mass. (1956), Technology Press, M.I.T., Cambridge, Mass. (1958), pp. 18-26.

Platteeuw, J. C., and G. Meyer. Trans. Faraday Soc. 52: 1066 (1956).

Porter, R. F., W. A. Chupka, and M. G. Inghram. J. Chem. Phys. 23(1): 216-217 (1955).

Potter, H. N. Trans. Am. Electrochem. Soc. 12: 191, 215, 223 (1907).

Primak, W., L. H. Fuchs, and P. Day. J. Am. Ceram. Soc. 38: 135 (1955).

Ramstad, H. F., and F. D. Richardson (with an appendix by P. J. Bowles). Trans. Metallurg. Soc. of AIME 221(5): 1021-1028 (1961).

Ruff, O. Trans. Am. Electrochem. Soc. 68: 87 (1935).

Ruff, O., and M. Konschak. Z. Elektrochem. 32: 515 (1926).

Ruff, O., and P. Schmidt. Z. Anorg. Chem. 117: 172 (1921).

Schäfer, H., and R. Hörnle. Z. Anorg. Allgem. Chem. 263:261 (1950).

Schick, H. L. Chem. Rev. 60(4):331-362 (1960).

Sosman, R. B. Trans. Brit. Ceram. Soc. 54:655 (1955).

Stull, D. R., and G. C. Sinke. Thermodynamic Properties of the Elements (1956).

Tombs, N. C., and A. J. Welch. J. Iron Steel Inst. 172:69 (1952).

Warren, B. E., H. Krutter, and O. Morningstar. J. Am. Chem. Soc. 19:202 (1936).

Wartenberg, H. Z. Elektrochem. 53:343 (1949).

Wartenberg, H. Z. Anorg. Allgem. Chem. 265:186 (1951).

CHAPTER VIII

EVAPORATION OF OXIDES OF ALKALINE EARTH ELEMENTS AND ENERGY CHARACTERISTICS OF GASEOUS RO MOLECULES

In this and the next two chapters, we give a review of results accumulated up to the present time from the investigation of the evaporation of oxides. The review requires some preliminary comments.

Contemporary technology uses such high temperatures that even the most refractory substances evaporate and are converted into a gaseous state. Thereupon there is dissociation and association and the evaporation of refractory oxides, carbides, nitrides, etc., is not as simple as the evaporation of typical organic compounds, for example.

Gaseous oxygen compounds (as will be shown below) are often unsaturated, very reactive particles, which can react energetically with other solid and liquid substances. The evaporation of oxides may consist of not only the conversion to the gaseous state of the oxides themselves or their partial or complete decomposition products, but also other substances formed as a result of the interaction of the given oxide with the material of the containers, refractories, etc. The behavior of the material at very high temperatures may be described completely only if we take into account its reaction with the surrounding gaseous substances.

The requirements of high-temperature technology have prompted investigations of evaporation processes and the search for methods of studying gaseous molecules which exist at high temperatures. The great successes in the investigation of the evaporation of oxides have been achieved largely as a result of the use of mass spectrometry with instruments especially designed for these purposes. The Knudsen or Langmuir method of studying evaporation in combination with a mass spectrometer, which makes it possible to determine the composition of gaseous molecules, has led to the most important results available to modern science.

Some other methods of investigating molecules in high-temperature gases such as spectroscopy, electron diffraction, etc., have also helped to accumulate evidence on the state and structure of gaseous oxides.

The study of the evaporation of oxides on a large scale has begun in the last 5-7 years. Investigations are being carried out in several laboratories in the USA; in the Soviet Union the evaporation of oxides is studied in Leningrad University (S. A. Shchukarev and G. A. Semenov). Quite a lot of data have been accumulated during this time, and their generalization is an urgent problem.

As will be seen from the contents of Chapters VIII, IX, and X, at the present time the evaporation of almost all oxides, including the most refractory, has been studied to some extent. For many elements (for example aluminum and silicon), the formation of gaseous lower oxides (AlO, Al_2O, and SiO) is characteristic. The investigation of the gaseous molecules helps to elucidate the nature of the solid compounds of the same composition.

The reverse process, i.e., the condensation of oxides from their vapors, is of great interest. Fibrous crystals, "whiskers," are obtained in this way and they will probably soon be of important practical use. A special chapter (XI) is devoted to methods of preparing "whiskers" and a brief description of their properties.

Methods have recently been proposed for synthesizing complex oxygen compounds (for example, spinels) by transfer of the reagents through the gas phase. Substances with very special properties are obtained by this method.

In our review we devoted some space to the energy characteristics of oxide molecules existing in the form of vapor.

Oxides of alkaline earth elements find great practical application in high-temperature technology, but their evaporation processes have not been studied at all adequately. The composition of the gas phase above these oxides is still not definitely clear. The results of determinations by different methods of the dissociation energies and heats of sublimation of oxides of alkaline earth metals give values which do not agree or are contradictory in a number of cases.

The vapors above oxides of alkaline earth metals include gaseous MO molecules, metal vapor, and oxygen (molecular and atomic) and for beryllium and barium, polymeric gaseous molecules such as $(BaO)_2$, $(BeO)_3$, etc., are also found. The quantitative proportions of the molecules listed in the vapors of various oxides differ and have not yet been established conclusively.

1. Results of Investigating Evaporation of Beryllium, Magnesium, Calcium, Strontium, and Barium Oxides

Beryllium Oxide. The investigation of the evaporation of beryllium oxide is of great interest in view of its wide use in high-temperature technology.

The first investigation of the evaporation of beryllium oxide was made by Erway and Seifert (1951), who used Knudsen's effusion method with a tungsten cell with radioactive isotopes for determining the amount of substance effusing. The vapor pressure values obtained agreed well with the values proposed if we assume that decomposition into the constituent elements occurs during evaporation. The authors did not determine the molecules existing in the vapor and did not consider the reaction of beryllium oxide with tungsten. Using the data of Erway and Seifert, Brewer (1953) calculated the dissociation energy of the gaseous beryllium oxide molecule. The maximum value for this energy must be 125 kcal/mole.

Nesmeyanov and Firsova (1959) measured the vapor pressure of beryllium oxide in the temperature range of 1830-2300°C by the effusion method, using beakers of tungsten or sintered oxides, and also by evaporation from an open surface in vacuum. The amount of material evaporating from the effusion cell or from the surface of the samples in vacuum was determined as the weight of condensate collected and from the loss in weight of the container, from which the oxides evaporated. It was assumed that the condensation coefficient for beryllium oxide is close to unity and that evaporation occurs without a change in the molecular composition.

As a result of processing the experimental results by the method of least squares, the following equation was obtained for the temperature dependence of the vapor pressure of beryllium oxide:

$$\log P_{\text{BeO}} \text{ (atm)} = 8.16 - \frac{33,200}{T} \text{ (2103—2573° K).}$$

A detailed investigation of the evaporation of beryllium oxide was recently carried out by Chupka, Berkowitz, and Giese (1959). The authors used the effusion method with a tungsten Knudsen cell and the effusion vapors were analyzed in a mass spectrometer. The authors observed polymeric molecules in the vapor of beryllium oxide, and this is of particular interest in connection with the fact that molecules in the vapor of magnesium, calcium, and strontium oxides are single. This may be connected with the tendency of beryllium toward covalence.

In the temperature region of 1900-2400°K, the vapor of beryllium oxide consists predominantly of Be and O atoms and a certain amount of $(BeO)_3$ and $(BeO)_4$. Small amounts of the following molecules were also detected: O_2, BeO, $(BeO)_2$, $(BeO)_5$, $(BeO)_6$, WO_2, WO_3, and the complex molecules $WO_x(BeO)_y$, where x = 1,2 and y = 1, 2, 3.

The tendency to form polymeric complex groups is a general characteristic of oxygen compounds of beryllium. Hypotheses have been put forward that $Be(OH)_2^{2+}$, $Be(OH)_3^{3+}$, and $Be(OH)_4^{4+}$ exist in aqueous solutions. Careful investigation of the hydrolysis of Be^{2+} ions by electrometric titration led to the conclusion that the main hydrolysis product is the ion $Be(OH)_3^{3+}$, which has a ring structure in which the beryllium atoms have tetrahedral coordination

$$
\begin{array}{c}
\text{H} \\
\text{O} \\
(H_2O)_2Be \quad\quad Be(H_2O)_2 \\
| \quad\quad\quad | \\
\text{HO} \quad \text{OH} \\
\text{Be} \\
(H_2O)_2
\end{array}
$$

It may be surmised that the polymeric molecules in the vapor also have a cyclic structure. Certain considerations on the nature of the beryllium—oxygen bond confirm this hypothesis.

Beryllium oxide shows appreciable volatilization in the presence of water vapor, probably because of the formation of the stable gaseous molecules $Be(OH)_2$. In studying this phenomenon, Grossweiner and Seifert(1952) give for the reaction

$$BeO \text{ (solid)} + H_2O \text{ (gas)} \rightleftarrows Be(OH)_2 \text{ (gas)}$$

the free energy value $\Delta F^0_{1673} = 29.0$ kcal/mole.

Hutchison and Malm (1951) observed that beryl is less reactive toward water than BeO and that the volatility of beryllium in the form of the hydrate is less than when the pure oxide is used. This indicates that the combination of BeO with another oxide reduces its chemical potential, and this leads to the given decrease in volatility.

A detailed investigation of the reaction of water vapor with beryllium oxide and beryllium aluminates, namely crysoberyl $BeO \cdot Al_2O_3$ and the compound $BeO \cdot 3Al_2O_3$, was made by Young (1960). Young was unable to determine the molecular weight of the gaseous beryllium hydroxide and adopted the single molecule $Be(OH)_2$ only for simplicity. There is the possibility that polymeric molecules may have been present in the vapor of the hydroxides.

As for other oxides of alkaline earth elements, for beryllium oxide there is a considerable discrepancy between the values obtained for the dissociation energy of gaseous BeO by spectroscopy and by study of the evaporation process. Thus, in his summary of dissociation energies obtained from spectral data, Herzberg (1949) gives for $D_0(BeO)$ the values 69 and 85 kcal/mole, and Gaydon (1949) gives 101 kcal.

In work with a tungsten cell, it is impossible to ignore the reaction of BeO with W. Ackermann and Thorn (1958) showed that in the vapor emerging from a tungsten cell the ratio of the amount of beryllium oxide to tungsten was 0.92.

Magnesium Oxide. The evaporation of magnesium oxide has been studied less than the evaporation of other oxides of alkaline earth elements. The first important work on the evaporation of magnesium oxide was carried out by Brewer and Porter (1954). These authors used two methods, namely, measurement of the saturated vapor pressure by Knudsen's effusion method and through the change in the intensity of MgO bands on heating of the solid oxide in a King furnace by the method developed by Brewer, Gilles, and Jenkins (1948).

On the basis of special calculations, Brewer and Porter assumed that under their experimental conditions magnesium oxide evaporates practically completely in the form of MgO molecules, while the equilibrium pressure of atomic magnesium must be negligible. The dissociation energy of gaseous magnesium oxide calculated by the authors was very high (about 113 kcal/mole).

However, soon after this work of Brewer and Porter, the latter author, together with Chupka and Inghram (Porter, Chupka, and Inghram, 1955), made a mass spectrometric study of the vapor over magnesium oxide and came to a directly opposite conclusion on the sublimation products of magnesium oxide. From measurements of the ionization currents of Mg^+ and MgO^+ they assumed that solid magnesium oxide evaporates in vacuum at 1950°K practically completely in the form of atomic magnesium and oxygen.

In the work of 1955, Porter and his co-workers give a completely different value for the dissociation energy of the MgO molecule, stating that this value must be less than 95 kcal/mole. The authors do not discuss the reasons for such a sharp discrepancy, and do not indicate possible sources of error in the calculations carried out in 1954. In the mass spectrometric investigations, the amount of atomic magnesium in the vapor over magnesium oxide was considerably higher than the amount of oxygen. The authors themselves point out that this indicates considerable reduction of the magnesium oxide in Knudsen cells. Ackermann and Thorn (1958) consider that the reducing agent may be tantalum, from which the heater surrounding the oxide cell was made.

If we assume that magnesium oxide decomposes completely on evaporation to magnesium and oxygen, it is possible to calculate the vapor pressure of atomic magnesium in an effusion cell containing magnesium oxide. This calculation was made by Ackermann and Thorn (1961), knowing the standard free energy of formation of solid magnesium oxide and the vapor pressure of metallic magnesium. In giving the equation for the pressure of magnesium over magnesium oxide

$$\log P_{Mg}(\text{atm.}) = 7.36 - \frac{26{,}100}{T} \, ,$$

Ackermann and Thorn point out that the calculations by which this equation was obtained should be regarded as "formal exercises" rather than quantitative determinations. Further investigations of the evaporation of magnesium oxide are obviously required.

Altman and Searcy (1961) also consider that in the evaporation of magnesium oxide there is dissociation to the elements and the major components of the vapors are Mg, O, and O_2. This conclusion was based on the results of experiments both with a Knudsen effusion cell (in the range of 1800-2100°K) and by the flow method. It was observed that the loss in weight in a stream of argon corresponded to evaporation with decomposition to the elements, while in a stream of oxygen the loss in weight was lower. These data indicate that the amount of gaseous MgO molecules must be only about 1% at 2030°K.

Having pointed out that the literature gives different values for the heat of dissociation of gaseous magnesium oxide (from 80 to 120 kcal), Scheffee and Henderson (1960) consider that the most probable value is about 85 kcal/mole obtained by Chupka et al.

Calcium Oxide. No detailed and accurate investigations of the evaporation of calcium oxide have been described in the literature. It has been pointed out that under neutral conditions calcium oxide sublimes, giving gaseous calcium and a small amount of CaO.

The investigation of Claassen and Veenemans (1933) of the evaporation of calcium oxide deposited on platinum was not accurate, but if we correct their data, taking into account the fact that the evaporation proceeds mainly with the formation of the gaseous elements, we obtain for the vapor pressure the value $7.0 \cdot 10^{-10}$ atm at 1700°K, which is somewhat less than the calculated value of $8.7 \cdot 10^{-10}$ atm. The latter value was calculated from data on the vapor pressure of calcium obtained by Priselkov and Nesmeyanov (1954) and thermodynamic data for calcium oxide and metallic calcium. It is difficult to draw any definite conclusions on the presence of gaseous CaO molecules from a comparison of these vapor pressure values.

According to the mass spectrometric observations of Pel'khovich (1960), the ratio of Ca^+ to CaO^+ is approximately the same as the corresponding ratio of Sr^+ to SrO^+ if in both cases the oxides are evaporated in contact with platinum at the same temperature. The investigations of the flame spectra by Huldt and Lagerqvist (1954) and also Veits and Gurvich (1957a) show that the ratios $Ca^+ : CaO^+$ and $Sr^+ : SrO^+$ are approximately the same in equilibrium flames and that barium oxide differs from the oxides examined. As calcium oxide is less volatile than strontium oxide, the ratio of metal to oxide in vapor in equilibrium with calcium oxide will be less than the corresponding ratio in the case of strontium oxide, while the dissociation energy of calcium oxide must be less than that for strontium oxide.

Strontium Oxide. None of the other oxides of alkaline earth metals has produced such disputes on the character of the evaporation as strontium oxide. By spectral methods, Huldt and Lagerqvist (1950) and James (1954) determined the concentration of free strontium atoms in flames and came to the conclusion that the dissociation energy D(SrO) is about 5 eV. This comparatively high value indicates the stability of SrO molecules and therefore it may be surmised that strontium oxide must evaporate predominantly in the form of SrO molecules. From the low value they obtained for the evaporation coefficient, Morgulis, Gavrilyuk, and Kulik (1955) state that SrO vapor is almost undissociated.

Calculations of the vapor pressure based on the effusion data of Moore, Allison, and Struthers (1950) on the assumption that only SrO was present in the vapor, led to values which indicated that this assumption was incorrect and that both SrO and Sr atoms must be present in the vapor. Appropriate calculations led to the conclusion that approximately 30% of the vapor consisted of gaseous strontium. More accurate calculations showed that the quantitative ratio Sr : SrO is more likely to be about 0.45.

Mass spectrometric investigations of the vapor above strontium oxide by Plumlee and Smith (1950), Aldrich (1951), Pel'khovich (1960), Bickel and Holroyd (1954), and Porter, Chupka, and Inghram (1955), also showed that there are SrO molecules and Sr in the vapor. The work of Porter, Chupka, and Inghram (1955) led these authors to the conclusion that strontium oxide vapor consists largely of strontium atoms and a mixture of oxygen atoms and molecules. For the ratio Sr : SrO under neutral conditions at 2100°K, they give the value $6 \cdot 10^3$, and under reducing conditions, $3 \cdot 10^2$.

Others of the investigators above, working with a mass spectrometer (Pel'khovich, Aldrich, Bickel, and Holroyd) with the same ionization potential as Porter and his co-workers, give a value close to 50 for the Sr : SrO ratio. Ackermann and Thorn (1961), reviewing the results of investigations of the evaporation of strontium oxide, came to the conclusion that the thermochemical data must be more reliable, while the results of Porter and his co-workers are regarded as unreliable, and particularly emphasize the reducing action of the material of the effusion cell (tantalum), which distorts the results.

Veits and Gurvich (1957b) showed that Porter, Chupka, and Inghram incorrectly calculated the equilibrium constant of the reaction

$$SrO \rightleftarrows Sr + \frac{1}{2}O_2.$$

The dissociation constant $K_p = 1 \cdot 10^6$ given by these authors could not be related to the observed ionization currents at all. Veits and Gurvich consider that the ratio of the ionization currents observed by Porter and his co-workers leads to a value for the equilibrium constant of the given reaction of $1.8 \cdot 10^3$. This value gives the more probable value for the dissociation energy of the SrO molecule of 103 kcal/mole instead of 88 kcal per mole, if $K_p \cong 10^6$ is taken.

The great contradictions in the data from the investigation of strontium oxide vapor may be connected with the presence in the vapor (especially under oxidizing conditions) of complex molecules such as Sr_2O_2 and even of higher molecular weight. In particular, this suggestion has been put forward by Charton and Gaydon (1956).

Barium Oxide. The first investigations of the evaporation of barium oxide were made in the thirties by Claassen and Veenemans (1933) and Blewett, Liebhafsky, and Hennelly (1939). The latter authors considered that the partial pressure of barium in equilibrium with the solid oxide was less than 1% of the total vapor pressure of the oxide, i.e., the gas phase contained mainly oxide molecules. In the first mass spectrometric investigations of Aldrich (1951), Bickel and Holroyd (1954), and Pel'khovich (1960), the ratio of the ion intensities of barium and barium oxide was 0.5, 1, and 0.2, respectively. These figures, which indicate that there is not a predominance of BaO molecules in the vapor, cannot be regarded as accurate. Thermodynamic calculations showed that these ratios must be reduced by a factor of 10^2-10^3.

In 1955, Inghram, Chupka, and Porter (1955) made a thorough mass spectrometric investigation of the evaporation of barium oxide and their data, which agree well with the results of flame studies, apparently give quite an accurate picture of the evaporation of barium oxide. These authors took some precautions which made it possible to avoid the errors they made in the study of the evaporation of strontium oxide. In particular, they studied the temperature dependence of the Ba^+ ion current, and also they determined experimentally the ionization cross sections of Ba and BaO.

In contrast to all the other oxides of alkaline earth elements, barium oxide evaporates predominantly in the form of BaO molecules. Moreover, surprisingly, the following complex gaseous compounds of barium and oxygen are formed: Ba_2O, Ba_2O_2, BaO, and Ba_2O_3. Under reducing conditions, Ba_2O predominates; under neutral conditions, BaO; and under oxidizing conditions, Ba_2O_3. Comparison of the heat of formation of Ba_2O with the

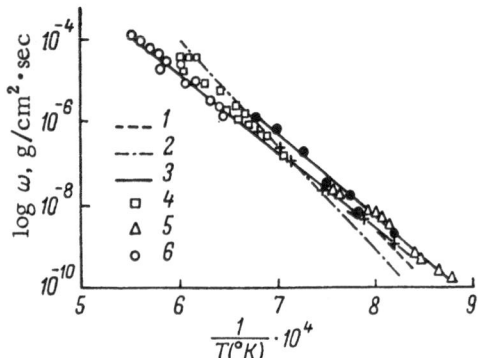

Fig. 94. Temperature dependence of the rate of evaporation of BaO according to the data of different authors. 1) Leverton and Shepherd; 2) German; 3) Claassen and Veenemans; 4) Blewett, Liebhafsky, and Hennelly; 5) Nikonov; 6) Nikonov and Otmakhova.

value calculated (see Brewer and Mastick, 1951) on the basis of the linear ion model shows that the formation of the bond must involve the unpaired electron of the Ba^+ ion. The most suitable electronic model of Ba_2O_3 must include peroxide oxygen rather than the Ba^{3+} ion. The amount of polymeric molecules is relatively low, namely, less than 0.7% under neutral conditions, but their presence is indisputable.

The evaporation of oxides of alkaline earth metals has been studied in connection with the fact that these compounds are of interest as thermoelectronic emitters, operating at high temperatures.

Nikonov and Otmakhova (1961) studied the evaporation of barium oxides and a series of chalcogenides of alkaline earth metals (Nikonov, 1956). These authors used the effusion method with a platinum and nickel cell and an aluminum target, covering the temperature region of 1100-1600°K. The starting materials were barium and strontium carbonates containing radioactive Ba^{140} and Sr^{89}.

For the rate of evaporation of barium oxide, Nikonov and Otmakhova obtained data which are represented well by the formula

$$\log \omega \left(g/cm^2 \cdot sec \right) = 7.7 - 20{,}000/T.$$

Figure 94 gives the data of various authors for the temperature dependence of the rate of evaporation of barium oxide. The agreement of the results on the evaporation of BaO obtained in different scientific institutes confirms the accuracy of the values given by Soviet scientists for the rate of evaporation of BaO.

Some data on the evaporation of barium oxide were obtained in connection with investigations of the cathodes of electronic tubes. In the so-called "impregnated cathode" (Levi, 1953), the pores of a tungsten emitter are completely filled with a fused mixture of barium oxide and aluminum oxide with the composition $5BaO \cdot 2Al_2O_3$. In accordance with data from the phase diagram of the system $BaO-Al_2O_3$ (see Toropov and Galakhov, 1952), this composition is a eutectic mixture of $Ba_3Al_2O_6$ and $BaAl_2O_4$. Rittner, Rutledge, and Ahlert (1957, 1958) consider that at 1190°C, the compound $Ba_3Al_2O_6$ reacts with tungsten by the reaction

$$^2/_3 Ba_3Al_2O_6 \, (\text{solid}) + ^1/_3 W \, (\text{solid}) = ^1/_3 BaWO_4 \, (\text{solid}) + ^2/_3 BaAl_2O_4 (\text{solid}) + Ba \, (\text{gas}). \qquad (1)$$

Rittner and his co-workers consider that in addition to the evaporation of barium, there is also the evaporation of barium oxide in accordance with the reaction

$$^1/_2 Ba_3Al_2O_6 \, (\text{solid}) = ^1/_2 BaAl_2O_4 \, (\text{solid}) + BaO \, (\text{gas}).$$

For the quantitative determination of the rate of evaporation of barium oxide and its partial vapor pressure, Rutledge, Milch, and Rittner (1958) proposed a simple method. Vapor containing a mixture of gaseous Ba and BaO was directed onto a tungsten wire and the time required to reach the maximum thermionic emission of the tungsten wire was determined. Calibration was first carried out with vapor of known composition. Using this method, it was found that the BaO content of the vapor over barium aluminate with the composition $5BaO \cdot 2Al_2O_3$ is 44% (the rest is Ba).

As was to be expected, calculation of the composition of the vapor from thermodynamic data gave a value which differed markedly from this figure.

By using data from appropriate handbooks, it is possible to calculate the partial vapor pressures of barium and barium oxide:

$$\log P_{Ba} \, (\text{mm Hg}) = -20{,}360/T + 8.56,$$

$$\log P_{BaO} \, (\text{mm Hg}) = -21{,}960/T + 8.87.$$

These equations make it possible to determine the fraction of BaO in the vapor. In actual fact, by taking the difference in these equations, we obtain

$$\log (P_{Ba}/P_{BaO}) = 1600/T - 0.31,$$

whence, at 1190°C, we have $P_{Ba}/P_{BaO} = 6.1$.

According to thermodynamic calculations, the barium oxide content of the vapor must be only 16.5%, while the barium oxide content found experimentally (by the method of Rutledge et al.) was 44%.

If we limit ourselves to the results of mass spectrometric investigations, which make it possible to determine in general the composition of the vapor and give indications of which molecules predominate and which are present in small amounts, we obtain the picture given by Table 33. Polymeric molecules are ignored here and the ratio between the metal atoms and oxide molecules in the vapor is given.

Apparently including other data (in addition to mass spectrometric), Ackermann and Thorn (1961) state that the fraction of oxide molecules in the gas phase for beryllium oxide at 2200°K is 0.2, for strontium oxide at 1600°K, approximately 0.4, and for barium oxide at 1400°K, approximately 1. The fraction of oxide molecules for calcium oxide at 1700°K is small and is quite insignificant for magnesium oxide (at 2000°K).

Medvedev (1961) generalized all possible data from the study of oxide vapors, covering work on pressure and also spectroscopic work, and calculated the composition of the evaporation products of oxides of alkaline earth metals. He considers that the evaporation of beryllium oxide occurs mainly in the form of beryllium atoms. At low temperatures, magnesium oxide evaporates mainly in the form of magnesium atoms, but at high temperatures it is in the form of MgO. At low temperatures the vapor of calcium oxide contains Ca and CaO in approximately identical amounts, but at 2500-3000°K the pressure of CaO is approximately twice that of Ca. In the opinion of Medvedev, strontium oxide evaporates at all temperatures mainly in the form of SrO. The evaporation of barium oxide at all temperatures up to 3000°K is almost exclusively in the form of BaO.

The composition of the evaporation products, expressed as the corresponding partial pressures, taken from the work of Medvedev, are given in Table 34 to give an idea of the quantitative ratios of the components of the vapors of alkaline earth oxides.

2. Energy Characteristics of Gaseous Oxides of Alkaline Earth Elements

The dissociation energy, i.e., the energy required to break down the molecule to its component atoms, is the most important characteristic of gaseous molecules. A knowledge of this value makes possible various thermodynamic calculations of processes in which the corresponding oxides participate.

In the years before mass spectrometry was used widely, the vapors of alkaline earth element oxides were studied by measuring the equilibrium dissociation constants in flames and on the basis of measurements of the saturated vapor pressure by the effusion method.

The first group includes the work of the school of Lagerqvist (Huldt, Lagerqvist, 1950) and also Veits and Gurvich (1956b) and James (1954). Measurements of the saturated vapor pressure by the effusion method were made, for example, by Barrow and Drummond (1951), Brewer and Porter (1954), and others.

Veits and Gurvich (1956b) developed a simple experimental method of determining the dissociation energy by determining the equilibrium constant of the dissociation of metal oxides in flames on condition that a zone of thermal equilibrium exists in them.

For determining the equilibrium dissociation constant of a metal oxide in a flame, fed with a fuel mixture of definite composition, a solution of a salt of the metal of known composition is introduced. The partial pressure of the metal vapor P_{Me} in the flame is determined by spectroscopic measurement of the intensity of

Table 33. Results of Mass Spectrometric Measurements of the Composition of Alkaline
Earth Metal Oxides

Temperature range, °K	Composition of vapor	Authors
2000—2400	Be : BeO=230—60	Chupka, Berkowitz, Giese, 1959
1950	Mg : MgO > 1000	Porter, Chupka, Inghram, 1955
1550—1800	Ca : CaO > 50	Pel'khovich, 1960
1400—1750	Sr : SrO > 50	Pel'khovich, 1960
2100	Sr : SrO > 300	Porter Chupka, Inghram, 1955
1150—1400	BaO	Pel'khovich, 1960
1530—1758	BaO	Inghram, Chupka, Porter, 1955

Table 34. Composition of Products of Evaporation of Alkaline Earth Metal Oxides
(according to Medvedev), in atm

T, °K	P_{MeO}	P_{Me}	P_{O_2}
	Beryllium oxide		
1500	$3.46 \cdot 10^{-15}$	$5.95 \cdot 10^{-13}$	$1.93 \cdot 10^{-14}$
2000	$1.57 \cdot 10^{-10}$	$1.26 \cdot 10^{-8}$	$3.21 \cdot 10^{-10}$
2500	$6.95 \cdot 10^{-7}$	$1.39 \cdot 10^{-5}$	$7.32 \cdot 10^{-7}$
3000	$3.15 \cdot 10^{-4}$	$1.97 \cdot 10^{-3}$	$2.00 \cdot 10^{-4}$
	Magnesium oxide		
1500	$2.57 \cdot 10^{-11}$	$2.03 \cdot 10^{-10}$	$8.29 \cdot 10^{-11}$
2000	$1.99 \cdot 10^{-6}$	$3.75 \cdot 10^{-6}$	$1.46 \cdot 10^{-6}$
2500	$1.57 \cdot 10^{-3}$	$1.27 \cdot 10^{-3}$	$4.80 \cdot 10^{-4}$
3000	$1.27 \cdot 10^{-1}$	$5.93 \cdot 10^{-2}$	$2.16 \cdot 10^{-2}$
	Calcium oxide		
1500	$5.97 \cdot 10^{-12}$	$4.72 \cdot 10^{-12}$	$6.85 \cdot 10^{-13}$
2000	$3.94 \cdot 10^{-7}$	$1.75 \cdot 10^{-7}$	$3.00 \cdot 10^{-8}$
2500	$2.74 \cdot 10^{-4}$	$9.15 \cdot 10^{-5}$	$1.65 \cdot 10^{-5}$
3000	$1.71 \cdot 10^{-2}$	$5.21 \cdot 10^{-3}$	$9.07 \cdot 10^{-4}$
	Strontium oxide		
1500	$1.07 \cdot 10^{-10}$	$4.50 \cdot 10^{-11}$	$1.48 \cdot 10^{-11}$
2000	$2.70 \cdot 10^{-6}$	$8.08 \cdot 10^{-7}$	$2.40 \cdot 10^{-7}$
2500	$1.05 \cdot 10^{-3}$	$2.71 \cdot 10^{-4}$	$7.42 \cdot 10^{-5}$
3000	$3.91 \cdot 10^{-2}$	$1.06 \cdot 10^{-2}$	$2.52 \cdot 10^{-3}$
	Barium oxide		
1500	$1.69 \cdot 10^{-7}$	$4.56 \cdot 10^{-11}$	$1.50 \cdot 10^{-11}$
2000	$4.57 \cdot 10^{-4}$	$6.58 \cdot 10^{-7}$	$1.85 \cdot 10^{-7}$
2500	$3.13 \cdot 10^{-2}$	$1.58 \cdot 10^{-4}$	$3.59 \cdot 10^{-5}$
3000	$3.80 \cdot 10^{-1}$	$5.38 \cdot 10^{-3}$	$9.51 \cdot 10^{-4}$

Note. Part of the oxygen is in the atomic state. The amount of this oxygen is readily
found from the relation $P_O = P_{Me} - 2P_{O_2}$.

Table 35. Dissociation Energies of Oxides of Alkaline Earth Elements in kcal/mole

Ground molecular state $^1\Sigma$	Measurement of equilibrium constants of dissociation in flames				Measurement of vapor pressure				Theoretical calculation from data of Gaspar and Csaviszky (1957)	Linear extrapolation by the method of Birge and Sponer (1926).
	Veits, Gurvich (1956a)	Huldt, Lagerqvist (1950)	Huldt, Lagerqvist (1950)	James (1954)	Barrow, Drummond (1951)	Brewer, Porter (1954)	Porter, Inghram, Chupka (1955)	Inghram, Chupka, Porter (1955)		
MgO	100.0	120.0	—	—	—	112.8	94.5	—	103.7	76.0
CaO	114.8	120.0	112.7	—	118.6	—	—	—	—	94.0
SrO	111.0	111.0	111.7	—	112.9	—	88.4	—	—	84.0
BaO	138.1	126.8	128.0	134.5	131.2	—	—	130.5	—	157.0

the resonance lines of the metal atoms. The partial pressure of the oxide vapors P_{MeO} may be calculated on the assumption that the reaction $Me + O \rightleftharpoons MeO$ is the only one in which the metal participates in the flame.

The dissociation energy of gaseous MeO molecules is determined from the formula

$$D_0 \, (MeO) = T \, (\Delta\Phi^* - R \ln K_p).$$

The equilibrium constant of the dissociation of the oxide is determined from data on the vapor pressures

$$K_p \, (T) = \frac{P_M \cdot P_O}{P_{MO}} \, .$$

The change in the reduced thermodynamic potential of the reaction is found from the relation

$$\Delta\Phi^* = \Phi^*_{Me} + \Phi^*_O - \Phi^*_{MeO}.$$

The values of the reduced thermodynamic potential of the metal in the vapor Φ^*_M, atomic oxygen Φ^*_O, and gaseous molecules Φ^*_{MO} are calculated by statistical thermodynamics method (see, for example, Gryaznov and Frost, 1950 and Godnev, 1956). The molecular constants of MgO, CaO, SrO, and BaO required in these calculations have been found by spectral methods, mainly by Lagerqvist and Gaydon. In the literature it is possible to find different opinions on the ground state of gaseous oxides of alkaline earth elements. Veits and Gurvich (1957a) demonstrates that the ground states of MeO molecules are singlets and not triplets, as was stated in their review by Brewer and Searcy (1958).

In the investigation of the dissociation of oxides of alkaline earth metals in flames obtained by the combustion of hydrogen compounds there is the possibility of the formation in the flame of MeOH molecules [or (MeOH)$^+$ ions]. Sugden and Wheeler (1955) consider that (MeOH)$^+$ ions, which have a filled electron shell of the inert gas type (analogous to molecules of alkali metal hydroxides and molecules of alkaline earth metal oxides), may be relatively stable. Without denying the possibility of the presence of such molecules in flames, Veits and Gurvich (1957b) consider that the numbers of them are small and therefore should not affect the calculation of the dissociation energy of MeO molecules.

Table 35 gives the dissociation energies of oxides of alkaline earth elements, obtained by various authors from measurement of the equilibrium constants of dissociation in flames, measurement of the saturated vapor pressure (by effusion and mass spectrometric methods), and some theoretical calculations.

The last column of Table 35 gives the dissociation energy of the oxides of alkaline earth elements obtained by linear extrapolation by the method of Birge and Sponer from the values of the normal vibration frequencies ω_e and the anharmonicity constants $x_e \cdot \omega_e$ of the corresponding molecules. As can be seen, this method leads to inaccurate data because of the fact that ionic bond forces, which are ignored in the linear extrapolation method, must play an important part in molecules of metal oxides.

Table 36. Molecular Constants of MgO, CaO, SrO, and BaO in cm^{-1}

Molecule	State	ν_{00}	ω_e	$\omega_e \cdot x_e$	B_e	$\alpha \cdot 10^2$	$D_0 \cdot 10^6$
MgO	$X^1\Sigma$	0	785.06	5.18	0.5743	0.50	−1.21
	$A^1\Pi$	3563.3	664.44	3.91	0.5056	0.46	−1.21
	$B^1\Sigma$	19984.0	824.08	4.76	0.5822	0.45	−1.13
CaO	$X^1\Sigma$	0	732.11	4.81	0.44447	0.335	−0.656
	$A^1\Sigma$	11548.8	716	1.6	0.4063	0.141	−0.54
SrO	$X^1\Sigma$	0	653.47	3.95	0.3379	0.21	−0.42
	$A^1\Sigma$	10885.0	619.6	0.9	0.30471	0.112	−3.2
BaO	$X^1\Sigma$	0	669.81	2.054	0.31249	0.13	−0.26
	$A^1\Sigma$	16807.1	500	1.6	0.2584	0.11	−0.28

Veits and Gurvich consider that the following should be regarded as the most accurate values of the dissociation energies of gaseous molecules of alkaline earth element oxides (in kcal/mole): $D_0(MgO) = 100.0 \pm 3$, $D_0(CaO) = 115.0 \pm 3$, $D_0(SrO) = 111.0 \pm 3$, and $D_0(BaO) = 135.0 \pm 3$. These values were obtained on the assumption that the ground electronic state of the oxide molecules is the $^1\Sigma$-state.

Until recently data on the thermodynamic functions of alkaline earth element oxides in the gas state were unknown or were known only for a limited temperature range. In connection with the success of spectral investigations, it became possible to obtain reliable data on oxide molecules at high temperatures.

Spectroscopic investigations of diatomic molecules make it possible to determine a series of molecular constants, with a knowledge of which it is possible, for example, to calculate the reduced thermodynamic potential and the entropy of the corresponding gaseous molecules.

Statistical thermodynamics gives formulas for calculating these thermodynamic functions. Thus, for the reduced thermodynamic potential at a pressure of 760 mm Hg, we have

$$\Phi_T^* = R \cdot \ln Q_B + {}^3/_2 R \ln M + {}^5/_2 R \ln T - 7.2825.$$

The entropy is calculated from the equation

$$S_T^0 = R \ln Q_B + RT \frac{\partial \ln Q_B}{\partial T} + {}^3/_2 R \ln M + {}^5/_2 R \ln T - 2.3145.$$

Here, T is the temperature in °K, R is the gas constant (1.98719 cal/mole · deg), M is the molecular weight, and Q_B and $\partial Q_B / \partial T$ are the sum through the intramolecular states and the derivative of this sum with respect to temperature. These values are determined from the molecular constants of the corresponding molecules (spectral data). The change in enthalpy $H_T^0 - H_0^0$ is obtained simply from the expression

$$H_T^\circ - H_0^\circ = T \left(S_T^0 - \Phi_T^* \right).$$

The spectra of the oxides of alkaline earth metals have been investigated by a series of authors. In recent years the most careful work has been carried out by Lagerqvist and his co-workers (Lagerqvist, Uhler, Hultin, Almkvist, 1949, 1950, 1954), and also Kovacs (Kovacs, Budo, 1953) and Dcezsi (Dcezsi, Koczkas, Mátrai, 1953). Older work has been generalized in the books of Herzberg (1949) and Gaydon (1949).

In the investigation of the emission spectra of alkaline earth element oxides, a system of bands was observed which lay over a wide range of the spectrum, from the near-ultraviolet to the infrared region.

All the systems of bands of MgO, CaO, SrO, and BaO examined belong to transitions between singlet states of these molecules, the lowest of which for all four molecules is the $^1\Sigma$-state. Some authors (for example, Brewer, 1953) considered that the ground states of these molecules must be the triplet $^3\Sigma$- or $^3\pi$-states. However, in later work of Gaydon (1955) it was shown that the complex systems of bands which could have been

Table 37. Thermodynamic Properties of Alkaline Earth Element Oxides in the Gaseous State

T, °K	Φ_T^*, cal/mole·deg	S_T°, cal/mole·deg	$H_T^\circ - H_0^\circ$, cal/mole	$\log K_p$
		Magnesium oxide		
293.16	43.685	50.816	2091	−70.2329
298.16	43.806	50.945	2129	−68.9709
400	45.930	53.257	2931	−50.1094
500	47.584	55.093	3755	−39.0419
1000	53.019	61.260	8241	−16.8311
1500	56.492	65.415	13385	−9.4104
2000	59.134	68.561	18854	−5.7132
2500	61.273	71.006	24333	−3.5060
3000	63.064	72.969	29715	−2.0413
3500	64.600	74.606	35021	−0.9987
		Calcium oxide		
293.16	45.189	52.357	2101	−81.2695
298.16	45.311	52.488	2140	−79.8227
400	47.448	54.826	2951	−58.2021
500	49.114	56.673	3780	−45.5200
1000	54.555	62.683	8128	−20.0759
1500	57.909	66.314	12608	−11.5510
2000	60.352	68.931	17158	−7.2711
2500	62.281	70.995	21785	−4.6940
3000	63.881	72.720	26517	−2.9697
3500	65.252	74.218	31381	−1.7312
		Strontium oxide		
293.16	47.590	54.827	2122	−78.4518
298.16	47.712	54.960	2161	−77.0525
400	49.873	57.340	2987	−56.1463
500	51.660	59.213	3827	−43.8840
1000	57.062	65.268	8206	−19.2838
1500	60.444	68.908	12696	−11.0418
2000	62.903	71.533	17260	−6.9038
2500	64.843	73.606	21908	−4.4118
3000	66.452	75.341	26667	−2.7436
3500	67.832	76.850	31563	−1.5440
		Barium oxide		
293.16	48.903	56.118	2115	−96.3379
298.16	49.024	56.251	2155	−94.6384
400	51.178	58.617	2976	−69.2517
500	52.858	60.480	3811	−54.3653
1000	58.336	66.502	8166	−24.5146
1500	61.700	70.117	12626	−14.5194
2000	64.142	72.706	17128	−9.4957
2500	66.065	74.729	21660	−6.4511
3000	67.652	76.396	26232	−4.3834
3500	69.005	77.824	30867	−2.8689

ascribed to transitions between triplet states of oxide molecules actually belong to molecules of the type MeOH and Me_2O_2. Veits and Gurvich (1957a) put forward weighty arguments showing that the ground states of MgO, CaO, SrO, and BaO molecules are their low $^1\Sigma$-states.

Table 36 gives the values of the molecular constants from which it is possible to calculate the thermodynamic functions. This table gives: ν_{00}, the wave number of the zero band of the system of bands; ω_e, the vibration constant; $\omega_e \cdot x_e$, the vibration constant; B_e, the rotation constant for the state corresponding to the complete absence of vibrations; α_e, the rotation constant; D_0, the dissociation energy.

Veits, Gurvich, and Rtishcheva (1958), who calculated the thermodynamic properties of gaseous magnesium, calcium, strontium, and barium oxides and also gaseous atoms of these elements and their monohydrides over a wide range of temperatures, used the "tabular" method of Gordon and Barnes (1933) in their calculations. In some work the thermodynamic functions of gaseous molecules at high temperatures were calculated assuming a harmonic oscillator, i.e., a rigid rotator, without allowance for excited electronic states. However, diatomic molecules of alkaline earth oxides have electronic states with low excitation energies, which should be taken into account in the calculation of thermodynamic functions.

The intramolecular contribution to the value of the entropy S_T^0 for MgO caused by the components of the excited electronic states equals 2.0839 en. units at 3400°K. Neglecting this component introduces quite a large error into the thermodynamic values. The method of calculating the thermodynamic functions of diatomic gases taking into account the excited states of their molecules was described in the article of Gurvich and Korobov (1956).

In the vapors of alkaline earth element oxides there is a chemical equilibrium between MeO molecules and the dissociation products of these molecules, i.e., vapors of atomic metal and atomic oxygen. The equilibrium constant of the reaction

$$MeO \text{ (gas)} \rightleftarrows Me \text{ (gas)} + O \text{ (gas)}$$

may be determined from known values of the reduced chemical potential of all participants in the reaction

$$R \ln K_p = \Phi_{Me}^\bullet + \Phi_O^\bullet - \Phi_{MeO}^\bullet - \frac{D_0}{T} ,$$

where D_0 is the dissociation energy of the diatomic gas at 0°K.

The results of the calculations of Veits, Gurvich, and Rtishcheva (1958) given in Table 37 give the thermodynamic constants for gaseous MgO, CaO, SrO, and BaO, which, in their turn, may be used in calculations on processes involving these molecules. In their calculations the authors used the values of D_0 which they themselves found (Veits and Gurvich, 1957b) from flame spectroscopic investigations (see p. 147). The data of Veits, Gurvich, and Rtishcheva are apparently regarded as the best at the present time. In particular, this is evident from the fact that in a recent review, Ackermann and Thorn (1961) gave the constants obtained by the Soviet authors.

The most important result of the study of the vapor of the oxides is the possibility of determining the heat of sublimation, the dissociation energy, and other thermodynamic values. For oxides of alkaline earth elements, all these energy characteristics are very different for the same oxide, depending on which of the investigation methods is used.

Medvedev (1961) recently compared the dissociation energies and heats of sublimation obtained for BeO, MgO, CaO, SrO, and BaO from measurements of vapor pressure (effusion method without a mass spectrometer), investigation of equilibria in a flame, and the mass spectrometric method. Medvedev used the results of the work of a large number of authors and with the exception of some mass spectrometric studies, their results could be processed only by the second law of thermodynamics, while the results of all the other work were recalculated using the third law of thermodynamics through the equation

$$\Delta H_0 = T \left(\Delta \varphi^\bullet - R \ln K_p \right).$$

Table 38. Comparison of Results of Determining Heats of Sublimation and Dissociation Energies of Alkaline Earth Metal Oxides by Different Methods

Method	Temp. range, °K	$\Delta H_{subl.}$, kcal/mole	D_0, kcal/mole	Authors
Beryllium oxide				
Effusion method				
Knudsen	2250-2413	Authors did not detect		Erway, Seifert, 1951
Langmuir	2103-2573	BeO molecules in the vapor		Firsova, 1959
Mass spectrometry				
(Knudsen)	2000-2400	171.9	106.4	Chupka, Berkowitz, Giese, 1959
Magnesium oxide				
Boiling point	2723-2903	144	93	Ruff, Schmidt, 1921
Effusion method				
Knudsen	2040-2200	129	108	Brewer, Porter, 1954
	2410	120	117	Huldt, Lagerqvist, 1954
Flame spectroscopy	2366-2536	100 ± 4	136.9	
	1750-2400	102 ± 2	134.9	Bulewicz, Sugden, 1959
Mass spectrometry				
(Knudsen)	1950	145	92	Porter, Chupka, Inghram, 1955
Calcium oxide				
Langmuir...................	1617-1728	140	112	Claassen, Veenemans, 1933
Flame spectroscopy	2240-2430	139	133.2	Huldt, Lagerqvist, 1954
	2452-2562	137.3	114.9 ± 4	Veits and Gurvich, 1956b
Mass spectrometry				
(Langmuir)................	1550-1800	174	78	Pel'khovich, 1960
Strontium oxide				
Langmuir.................	1495-1635	128	111	Claassen, Veenemans, 1933
Langmuir.................	1290-1650	128 ± 4	111	Moore, Allison, Struthers, 1950
Flame spectroscopy	2240-2430	125.5	113.2	Huldt, Lagerqvist, 1954
	3080-3210	125.5	113.2 ± 4	Veits, Gurvich, 1956b
Mass spectrometry				
(Langmuir)................	1400-1750	154	84.7	Pel'khovich, 1960
Mass spectrometry				
(Knudsen)	2100	153	85.7	Porter, Chupka, Inghram, 1955
Barium oxide				
Langmuir.................	1223-1475	102.3 ± 0.5	132.7	Claassen, Veenemans, 1933
Langmuir.................	1200-1550	102.5 ± 2	132.5	Hermann, 1937
Knudsen	1200-1800	100.7 ± 0.6	134.3	Blewett, Liebhafsky, Hennelly, 1939
Flame spectroscopy	2240-2430	106	129	Huldt, Lagerqvist, 1954
	–	100.5	134.5	James, 1954
	3080-3210	98	137 ± 4	Veits, Gurvich, 1956b
Mass spectrometry	1150-1250	114.3	120.7	Pel'khovich, 1960
(Langmuir)................	1260-1400	94.3	140.7	Pel'khovich, 1960
	1173-1473	102	133.0	Shchukarev, Semenov, 1957
Mass spectrometry				
(Knudsen)	1530-1758	102	133.0	Pel'khovich, 1960
				Inghram, Chupka, Porter, 1955

Table 39. Most Reliable Values of Heats of Sublimation, Dissociation Energies,
and Heats of Formation of Gaseous Alkaline Earth Metal Oxides, kcal/mole

Oxides	$\Delta H_{0\ subl}$	D_0	$D_{298.15}$	$\Delta H^0_{f_0}$	$\Delta H^0_{f_{298.15}}$
BeO	171.9	106.4±3	107.4	30.2	30.2
MgO	135.9	101±2	102.0	—7.0	—7.1
CaO	137.1	115±4	115.9	—13.9	—14.2
SrO	126.1	112±2	112.9	—13.8	—14.2
BaO	101.0	134±3	134.9	—32.1	—32.6

The Σ states were adopted as the ground electronic states of alkaline earth metal oxides.

From the heat of sublimation ΔH_{subl} it is easy to obtain the dissociation energy D_0 for oxides with the formula MeO by using the simple reaction

$$D_0(MeO) = \Delta H_{0\ subl}(Me) - \Delta H_{0\ subl}(MeO) - \Delta H^0_{f_0}(MeO_{cr}) + \Delta H^0_{f_0}(O_{gas}),$$

where $\Delta H^0_{f_0}$ is the heat of formation; all the values refer to 0°K. Of the values of D_0 given in Table 38, the results of Veits and Gurvich are reliable, as these authors carried out investigations in several flames over a wide range of temperatures. For magnesium oxide, the work of the Soviet authors is in good agreement with the results of Bulewicz and Sugden. It was shown by Brewer and also by Veits and Gurvich that the value of D_0 (MgO) obtained by Huldt and Lagerqvist is incorrect.

The dissociation energies of calcium and strontium oxides obtained by flame spectroscopy by Huldt and Lagerqvist, and also Veits and Gurvich, agree well with each other. For barium oxide, the Swedish investigators apparently give a low value and the better agreement between the data of Veits and Gurvich and James indicates that the value of D_0 for BaO is close to 135 kcal/mole.

In the literature on the evaporation of oxides (see, for example the review article of Ackermann and Thorn, 1961), values are often given for the dissociation energies of alkaline earth metal oxides which have been estimated by linear extrapolation of vibration energy levels. The following values were obtained by this method: $D_0(BeO) = 90$, $D_0(MgO) = 76$, $D_0(CaO) = 94$, $D_0(SrO) = 84$, $D_0(BaO) = 157$ kcal/mole. However, in giving these data, Gaydon (1949) himself points out that this method does not give reliable results in the case of alkaline earth metals. The completely different value $D_0(BeO) = 111$ kcal/mole was obtained by Lagerqvist by basing the calculation on the dissociation limit of the $A^1\Pi$-state. However, the error here may be ±10 kcal per mole.

The comparison by Medvedev of the values obtained by different methods for the dissociation energies of alkaline earth metal oxides shows that in general there is quite good agreement between the results of measurements of vapor pressures and investigations of equilibria in flames. At the same time, mass spectrometric investigations lead to markedly different values for all oxides except BaO. It is possible that the high values of the heats of sublimation in mass spectrometric investigations are connected with the dissociation of oxide molecules on ionization. Even for the most stable oxide BaO, Inghram and his co-workers and Pel'khovich found that approximately half of the BaO$^+$ ions dissociate on ionization. For the less stable molecules of MgO, CaO, and SrO, dissociation may proceed to an even greater degree. Table 38 gives the results of determinations of the heats of sublimation and dissociation energies by all three of the methods examined, namely, effusion, mass spectrometry, and flame spectroscopy.

As a result of examining these data, Medvedev considers that for beryllium oxide it is possible to adopt the values obtained by mass spectrometry. The best values for the dissociation energies of magnesium and calcium oxides were obtained from investigations of equilibria in flames. For strontium oxide it is possible to accept both data from measurements of the vapor pressure of strontium oxide and also the results of investigating equilibria in flames. For barium oxide we can accept the values of $D_0(BaO)$ obtained by measuring the vapor pressure by the normal method and with a mass spectrometer.

Table 39 gives the most reliable values for the heats of sublimation, dissociation energies, and heats of formation of gaseous oxides of alkaline earth metals.

BIBLIOGRAPHY

Ackermann, R. J., and R. J. Thorn Congrès Internat. de Chim. Pure et Appl., Paris, 1957; Mémoires, Presentés, á la Section de Chimie Minérale, Paris (1958), pp. 667-684 .

Ackermann, R. J., and R. J. Thorn. Progr. Ceram. Sci. 1:50-51 (1961).

Aldrich, L. T. J. Appl. Phys. 22:1168 (1951).

Almkvist, G., and A. Lagerqvist. Arkiv Fysik. 2:233 (1950).

Altman, R. L., and A. W. Searcy. Abstracts of Scientific Papers at the XVIII International Congress of Pure and Applied Chemistry (1961), p. 100.

Barrow, R., and L. Drummond. Trans. Faraday Soc. 47:1275 (1951).

Bickel, P. W., and L. V. Holroyd. J. Chem. Phys. 22:1793 (1954).

Birge, R., and H. Sponer. Phys. Rev. 28:259 (1926).

Blewett, J. P., H. A. Liebhafsky, and E. F. Hennelly. J. Chem. Phys. 7:478 (1939).

Brewer, L. Chem. Rev. 52:48 (1953).

Brewer, L., Gilles, and Jenkins. J. Chem. Phys. 16:797 (1948).

Brewer, L., and D. F. Mastick. J. Chem. Phys. 19:834 (1951).

Brewer, L., and R. Porter. J. Chem. Phys. 22:1867 (1954).

Brewer, L., and A. Searcy. Usp. Khim. 27(8):969 (1958).

Bulewicz, E. M., and T. M. Sugden. Trans. Faraday Soc. 55:720 (1959).

Charton, M., and A. G. Gaydon. Proc. Phys. Soc. (London) 69A:520 (1956).

Chupka, W. A., J. Berkowitz, and C. F. Giese. J. Chem. Phys. 30(3):827-834 (1959).

Claassen, A., and C. F. Veenemans. Z. Phys. 80:342 (1933).

Dcezsi, J., E. Koczkas, and F. Mátrai. Acta Phys. Acad. Sci. Hung. 3:95 (1953).

Erway, N. D., and R. L. Seifert. J. Electrochem. Soc. 98:83 (1951).

Gaspar, R., and P. Csaviszky. Acta Phys. Acad. Sci. Hung. 5:65 (1955).

Gaydon, A. Dissociation Energies and Spectra of Diatomic Molecules [Russian translation], IL, Moscow (1949).

Gaydon, A. Proc. Roy. Soc. A231:437 (1955).

Godnev, I. N. Calculation of Thermodynamic Functions from Molecular Data (1956).

Gordon, A., and C. Barnes. J. Chem. Phys. 1:297 (1933).

Grossweiner and Seifert. J. Am. Chem. Soc. 74:2701 (1952).

Gryaznov, V. M., and A. V. Frost. Statistical Methods of Calculating Thermodynamic Values (1950).

Gurvich, L. V., and V. V. Korobov. Zh. Fiz. Khim. 30:2794 (1956).

Hermann, G. Z. Physik. Chem. B35:298 (1937).

Herzberg, G. Spectra and Structure of Diatomic Molecules [Russian translation], IL, Moscow (1949).

Huldt, L., and A. Lagerqvist. Arkiv Fysik. 2:333 (1950).

Huldt, L., and A. Lagerqvist. Z. Naturforsch. 9a:991 (1954).

Hultin, M., and A. Lagerqvist. Arkiv Fysik. 2:471 (1950).

Hutchison, C. A., and J. G. Malm. J. Am. Chem. Soc. 71:1338 (1951).

Inghram, M. G., W. A. Chupka, and R. F. Porter. J. Chem. Phys. 23:2159 (1955).

James, L. Thesis, Cambridge University, Cambridge, England (1954).

Kovacs, G., and A. Budo. Ann. Phys. 12:17 (1953).

Lagerqvist, A., E. Lind, and R. Barrow. Proc. Phys. Soc. A63:1132 (1950).

Lagerqvist, A., U. Uhler, M. Hultin, and G. Almkvist. Arkiv Fysik 1:159, 477 (1949); 2:233, 471 (1950); 8:83 (1954).

Leverton, W. F., and W. G. Shepherd. J. Appl. Phys. 23:787 (1952).

Levi, R. J. Appl. Phys. 24:233 (1953).

Medvedev, V. A. Zh. Fiz. Khim. 35(7):1480-1488 (1961).

Moore, C. E. Atomic Energy Levels, Nat. Bur. Std. (U. S. A.) Circ. 467 (1958).

Moore, C. E., H. W. Allison, and J. D. Struthers. J. Chem. Phys. 18:1572 (1950).

Morgulis, N. D., V. M. Gavrilyuk, and A. E. Kulik. Dokl. Akad. Nauk SSSR 101: 479 (1955).

Nesmeyanov, A. N., and L. P. Firsova. Izv. Akad. Nauk SSSR, Otd. Tekhn. Nauk, Met. i Toplivo 3: 150-151 (1959).

Nikonov, B. P. Tr. Nauchn.-Issled. Inst., Min. Radio-Tekhn. Prom. SSSR 1: 29, 3 (1956).

Pel'khovich, I. Problems in Cathode Electronics (1960).

Plumlee, R. H., and L. P. Smith. J. Appl. Phys. 21: 811 (1950).

Porter, R. F., W. A. Chupka, and M. G. Inghram. J. Chem. Phys. 23: 1347 (1955).

Priselkov, I. A., and A. N. Nesmeyanov. Dokl. Akad. Nauk SSSR 95: 1207 (1954).

Rittner, E. S., W. C. Rutledge, and R. H. Ahlert. J. Appl. Phys. 28(12): 1468 (1957); 29: 744 (1958).

Ruff, O., and P. Schmidt. Z. Anorg. Chem. 117: 172 (1921).

Rutledge, W. C., A. Milch, and E. S. Rittner. J. Appl. Phys. 29(5): 834-839 (1958).

Scheffee, R. S., and C. B. Henderson. Kinetics, Equilibria, and Performance of High-Temperature Systems, pp. 1-4, in: G. S. Bahn and E. E. Zukoski (eds.).

Shchukarev, S. A., and G. A. Semenov. Zh. Neorgan. Khim. 2: 1217 (1957).

Sugden, T. M., and R. C. Wheeler. Discussions Faraday Soc. 19: 76 (1955).

Toropov, N. A., and F. Ya. Galakhov. Dokl. Akad. Nauk SSSR 82(1): 69 (1952).

Veits, I. V., and L. V. Gurvich. Dokl. Akad. Nauk SSSR 108: 659 (1956).

Veits, I. V., and L. V. Gurvich. Opt. i Spektroskopiya 1: 22 (1956).

Veits, I. V., and L. V. Gurvich. Opt. i Spektroskopiya 2: 145 (1957).

Veits, I. V., and L. V. Gurvich. Zh. Fiz. Khim. 31: 2306 (1957).

Veits, I. V., L. V. Gurvich, and N. P. Rtishcheva. Zh. Fiz. Khim. 32(11): 2532-2542 (1958).

Young, W. A. J. Phys. Chem. 64(8): 1003-1006 (1960).

CHAPTER IX

EVAPORATION OF OXIDES OF GROUP III ELEMENTS
(INCLUDING RARE EARTHS)

Aluminum oxide is one of the most important materials of high-temperature technology. Its behavior at high temperatures must be well known, and this explains the great interest in elucidating the state of aluminum oxide in vapors. However, despite the series of studies of the composition of the vapor over aluminum oxide and the determination of the thermodynamic characteristics of gaseous compounds of aluminum with oxygen, much remains unclear.

Unfortunately, none of the oxides studied has presented such difficulties to investigators as aluminum oxide. The results obtained by some authors are disputed by others. Many contradictions found in the literature remain unresolved. Undoubtedly in the near future there will appear new investigations which will give more definite data on the high-temperature chemistry of the Al—O system. We consider it necessary to give here a more or less detailed review of the present state of the problem.

It should be pointed out that interest in gaseous oxygen compounds of aluminum grew because of the requirements of single crystal technology. Methods were recently proposed for growing crystals of alumina in the form of fibers ("whiskers") and plates (see Chapter XI). Here the formation of crystals proceeds through gaseous compounds of aluminum and naturally a knowledge of the composition of the vapor and the thermodynamic properties of the molecules in the vapors is very important. At the present time it is considered that the following gaseous oxides of aluminum exist: Al_2O, AlO, Al_2O_2, Al_2O_3.

1. Question of Existence of Solid Lower Oxides of Aluminum

Attempts were made long ago to determine whether aluminum suboxides exist in a solid state. Baur and Brunner (1934a), who investigated the system $Al—Al_2O_3$, stated that there exists the compound Al_8O_9, melting at 2323°C with a eutectic between Al_2O_3 and Al_8O_9.

Beletskii and Rapoport (1951) obtained light blue acicular crystals by heating a mixture of aluminum oxide, carbon, and silica in vacuum above 1800°C. Similar crystals were obtained when a mixture of aluminum oxide and metallic aluminum was heated to 1900-2000°C at normal pressure. These crystals were found on top of the charge of briquettes, i.e., they were deposited from the gas phase.

Beletskii and Rapoport consider that they obtained the solid suboxide Al_2O. Because of the small amount of the substance, it was impossible to carry out a chemical analysis of the crystals, but x-ray diffraction and crystal—optical investigations indicated characteristics which do not apply to compounds that could have been formed here (for example, AlN, Al_4C_3, or Al_5C_3N). The interesting experiments of Beletskii and Rapoport still do not definitely prove the existence of aluminum suboxides under normal conditions.

The formation of the solid suboxides of aluminum Al_2O and AlO is indicated by the work of Hoch and Johnston (1954). These compounds were obtained from a mixture of metallic aluminum and alumina (α-form) directly in a high-temperature x-ray camera and from the appearance of x-ray lines different from those characteristic of Al_2O_3 it was possible to determine whether a new compound was formed. In accordance with the stoichiometry of the equations

$$4Al + Al_2O_3 \rightleftarrows 3Al_2O,$$

$$Al + Al_2O_3 \rightleftarrows 3AlO$$

in the mixtures studied the ratio of aluminum to oxide was 4:1 and 1:1.

Up to 1000°C there was no reaction between the alumina and the metal and new lines appeared only at 1100°C. Hoch and Johnston assigned the new x-ray lines observed in the region of 1100-1500°C to the compound Al_2O. At 1500°C there appeared lines of what, in their opinion, was the compound AlO, which exists together with Al_2O up to 1600°C. Above 1700°C only AlO is observed. At 1900°C, a sample (i.e., a mixture of alumina and aluminum) weighing 15 mg volatilized completely, indicating the high vapor pressure of AlO at this temperature. On cooling, even as rapidly as possible, the compounds formed in the high-temperature region dissociate to Al and Al_2O_3, i.e., are unstable at normal temperatures.

The diffraction lines of both suboxides correspond to a cubic structure with lattice constants of 4.98 A for Al_2O at 1110°C and 5.67 A for AlO at 1700°C.

The existence of aluminum suboxides in a condensed state is confirmed by the experiments of Cochran (1955), who determined the melting point of alumina—metallic aluminum mixtures. Thus, for a mixture containing 50 wt.% alumina, the melting point on a tungsten backing in an argon atmosphere was 1945°C. This depression of melting point (as compared to pure aluminum oxide) was ascribed by Cochran to the formation of the suboxide, but he was unable to construct a phase diagram of the system $Al-Al_2O_3$ as the melting point of the mixtures did not change regularly with the composition.

2. Evaporation of Aluminum Oxide

Gaseous AlO has been known for a long time from spectroscopic investigations (see, for example, Herzberg, 1949a). In 1943, Zintl et al. (Zintl, Krings, Branning, 1943) described a method of determining aluminum in aluminum alloys by evaporation of the suboxide, to which was assigned the formula AlO. It was also observed that a mixture of aluminum oxide and metallic aluminum sublimes in vacuum from which it was surmised that there occurs the reaction

$$Al + Al_2O_3 \rightleftarrows 3AlO.$$

Even earlier, Zintl et al. (Zintl, Morawietz, Gastinger, 1940) stated that they obtained this compound by the reaction

$$Al_2O_3 + B \rightleftarrows 2AlO + BO.$$

By heating a mixture of alumina and silicon at 1800°C in a high vacuum, Grube, Schneider, Esch, and Flad (1949) obtained a volatile sublimate, whose analysis corresponded to the composition Al_2O. The authors suggested that the formation of the volatile oxide Al_2O proceeds by the reaction

$$Al_2O_3 + 2Si \rightleftarrows Al_2O + 2SiO.$$

Cochran (1955) studied the action of gaseous aluminum on alumina by placing a plate of sintered pure alumina over a boat of titanium carbide containing aluminum. In the temperature region studied (1500-1600 °C), the total loss in weight corresponded approximately to the reaction

$$4Al\,(gas) + Al_2O_3\,(solid) \rightarrow 3Al_2O\,(gas).$$

Cochran did not detect the compound AlO in a spectroscopic investigation, either. The absorption spectrum was studied for vapor obtained from a mixture of alumina and aluminum heated to 2200°C in a tantalum carbide boat. The emission spectrum was observed in an aluminum arc in an oxygen atmosphere. In neither case were bands characteristic of AlO observed.

White (1955) also observed bands of Al_2O in the spectrum of burning aluminum. In 1951 there was published the extensive investigation of Brewer and Searcy (1951) on the gas phase over the system $Al-Al_2O_3$. Brewer and Searcy observed that under reducing conditions, Al_2O_3 evaporates to give gas of a different composition from under neutral or oxidizing conditions. They observed that the volatility of Al_2O_3 in the presence of aluminum is 100 times as great as in its absence. From thermodynamic considerations they ascribed this increase in volatility to the formation of gaseous Al_2O.

In a comparative study of the vapor pressure (using a tungsten Knudsen cell) under reducing (Al_2O_3 + Al) and almost neutral (Al_2O_3) conditions, Brewer and Searcy concluded that under reducing conditions the main components of the vapor are Al and Al_2O, and under neutral conditions, AlO and atomic oxygen.

The next work of Brewer and Searcy is open to criticism, as these authors did not take into account the fact that the material of the effusion cell (tungsten) may reduce aluminum oxide to form volatile tungsten oxides WO_3, WO_2, etc. Ackermann and Thorn (1956) showed that this reduction proceeds so rapidly that at 2582°K the ratio of alumina to tungsten distilling is 1.16. In investigating the equilibrium

$$Al_2O_3 \ \text{(solid)} \rightarrow 2AlO \ \text{(gas)} + O \ \text{(gas)}$$

Brewer and Searcy obtained for the dissociation energy of AlO the value 6.0 ± 0.4 eV. Although this value is much more probable than that obtained by spectral methods (see p. 163), in the experiments of Brewer and Searcy the composition of the vapor over the solid aluminum oxide remained unknown and, in particular, it was not established experimentally that it did not contain gaseous aluminum oxide Al_2O_3. Therefore, the value found for the dissociation energy of AlO may include a substantial error.

Brewer and Searcy consider that the discrepancy between the values for the dissociation energy of AlO calculated by linear extrapolation (linear extrapolation for the $x^2\Sigma$-state of the AlO molecule leads to the value 4.16 eV) and found by them experimentally, is explained by the fact that the lowest known state of AlO is not the ground state of this molecule. However, Veits and Gurvich (1956) demonstrated that this hypothesis is incorrect, and that, as in the case of AlF and some other molecules, this discrepancy is caused by the ionic character of the chemical bond.

In 1955, Porter, Schissel, and Inghram (1955) studied the evaporation of the system $Al-Al_2O_3$ in a zirconium dioxide crucible which, in its turn, was placed in a tantalum crucible (Knudsen cell). The vapor formed at high temperature was identified with a mass spectrometer. The temperature range investigated was limited to 1500-1800°K. The main components of the vapor were Al_2O molecules and Al atoms. The amount of all the other gaseous oxides, which could be AlO, Al_2O_2, and Al_2O_3, was at least a factor of 1000 less than Al_2O. In a similar mass spectrometric investigation of the vapor effusing from a tungsten cell containing Al_2O_3 (i.e., close to neutral conditions), the same authors again identified Al and Al_2O. The ratio $Al_2O^+ : Al^+$ varied from about 1 for the system $Al-Al_2O_3$ to 0.1 in the system $W-Al_2O_3$. The ratio of the intensities of AlO^+ to Al^+ was found to be less than 0.1, i.e., AlO is not the main component of the vapor under the experimental conditions of these authors.

In 1959, a new investigation of the evaporation of aluminum oxide was carried out in the Chicago laboratory of Inghram (see De Maria, Drowart, Inghram, 1959; Drowart, De Maria, Burns, Inghram, 1960). The authors used tungsten and molybdenum Knudsen cells which contained pure aluminum oxide. Measurements were carried out up to 2500°K, i.e., the melting point of alumina was exceeded.

The following ions were found in the ionized vapor: O^+, O_2^+, Al^+, AlO^+, Al_2O^+, and $Al_2O_2^+$. Moreover, in the tungsten cell the ions WO^+, WO_2^+, and WO_3^+ were found, and in the molybdenum cell, MoO^+, MoO_2^+, and MoO_3^+. The mass spectrometer was capable of determining high molecular weights (up to 600), but no more complex ions $Al_xO_y^+$ were detected. Likewise, no complex molecules containing both Al and W (or Mo) were detected. The authors used ionizing electrons of low energy (less than 15 eV), and hence they could state that the ions O^+, Al^+, AlO^+, Al_2O^+, and $Al_2O_2^+$ were obtained by simple ionization of the molecules AlO, Al_2O, and Al_2O_2 and O and Al atoms. The Al_2^+ ion observed arose as a fragment of the Al_2O molecule.

To obtain absolute values of the vapor pressures, the authors calibrated the instrument by several methods: 1) with silver; 2) from the O : O_2 ratio; 3) with molybdenum; and, 4) by self-calibration. In the latter method a weighed amount of aluminum oxide (about 10 mg) was completely evaporated from a molybdenum or tungsten cell and the time dependence of the size of the Al^+, AlO^+, and Al_2O^+ peaks was recorded. From the number of Al atoms in the starting sample it was possible to calculate the pressure corresponding to unit intensity.

Table 40. Partial Pressures of Al, O, AlO, Al_2O, and Al_2O_2 above Al_2O_3, in atm

T, °K	Al	O	AlO	Al_2O	Al_2O_2
			Neutral conditions		
2000	$6.7 \cdot 10^{-9}$	$7.7 \cdot 10^{-9}$	$2.5 \cdot 10^{-10}$	$6.4 \cdot 10^{-11}$	$1.3 \cdot 10^{-13}$
2300	$7.9 \cdot 10^{-7}$	$9.1 \cdot 10^{-7}$	$7.1 \cdot 10^{-8}$	$2.9 \cdot 10^{-8}$	$1.3 \cdot 10^{-10}$
2500	$9.0 \cdot 10^{-6}$	$1.0 \cdot 10^{-5}$	$1.1 \cdot 10^{-6}$	$5.0 \cdot 10^{-7}$	$3.1 \cdot 10^{-9}$
			In a tungsten cell		
2000	$7.1 \cdot 10^{-9}$	$6.3 \cdot 10^{-9}$	$2.2 \cdot 10^{-10}$	$9.4 \cdot 10^{-11}$	$1.0 \cdot 10^{-13}$
2300	$1.0 \cdot 10^{-6}$	$7.8 \cdot 10^{-7}$	$6.8 \cdot 10^{-8}$	$5.4 \cdot 10^{-8}$	$1.5 \cdot 10^{-10}$
2500	$1.2 \cdot 10^{-5}$	$8.7 \cdot 10^{-6}$	$1.4 \cdot 10^{-6}$	$8.7 \cdot 10^{-7}$	$4.1 \cdot 10^{-9}$

The evaporation from a tungsten or molybdenum cell occurred under reducing conditions. To obtain the partial pressures under neutral conditions, the authors, using certain thermodynamic data, first calculated the partial pressures of aluminum and oxygen which should be produced under neutral conditions. It was assumed that reducing conditions (due to the presence of tungsten or molybdenum) have little effect on the reaction

$$Al_2O_3 \text{ (solid or liq)} = 2Al \text{ (gas)} + 3O \text{ (gas)},$$

and, therefore, the pressures of Al and O may be calculated from thermodynamic data. Table 40 gives the partial pressures of aluminum, atomic oxygen, and AlO, Al_2O, and Al_2O_2 molecules under neutral conditions (obtained by the method given above) and in a tungsten cell. The quantitative ratios of the gaseous molecules remain the same in the two cases. Under neutral conditions the partial pressure of aluminum somewhat exceeds that of atomic oxygen. In a tungsten cell there is naturally the reverse relation.

AlO and Al_2O molecules are present in almost the same amount, but the concentration of AlO is still slightly greater even in a tungsten cell, i.e., under reducing conditions. The amount of dimeric molecules of Al_2O_2 is considerably lower, but at a high temperature (2500°K) their partial pressure almost reaches the partial pressure of oxygen or aluminum observed at 2000°K.

An important result of the work of Drowart, De Maria, Burns, and Inghram is the establishment of the fact that under neutral or weakly reducing conditions the vapor over alumina contains both the monoxide and the compound Al_2O. It is also important that the authors did not detect molecules with the composition Al_2O_3.

Molecules with the structure Al_2O_2 are of great interest. Here there are two possibilities: a linear molecule O=Al—Al=O, or a cyclic molecule, $Al\diagup\!\!\!\!\diagdown Al$. The authors were inclined to favor the cyclic structure, emphasizing the difference from the dimer B_2O_2 which has an analogous composition, and for which the linear structure O=B—B=O is adopted. The values of the reduced thermodynamic potential for the Al_2O_2 molecule, $\Phi_{Al_2O_2} = -(F^0 - H_0^0)/T$, calculated from spectral data and molecular constants for linear and cyclic structures, are quite close, namely, 81.47 and 80.86 kcal/mole · deg, respectively, at 2000°K.

Thus, the reduced thermodynamic potential is not sensitive to the structure of the Al_2O_2 molecule. The ring structure is also supported by comparison of the dimerization energies of Al_2O_2 and B_2O_2 molecules. While the heat of the reaction

$$Al_2O_2 \text{ (gas)} \rightarrow 2AlO \text{ (gas)}$$

is 135 kcal/mole, the corresponding value for B_2O_2 is only 115 kcal/mole.

Drowart, De Maria, Burns, and Inghram give the following values for the atomization energies of the gaseous molecules: $D_0^0(AlO) = 115 \pm 5$ kcal/mole, $D_0^0(Al_2O) = 245 \pm 7$ kcal/mole, and $D_0^0(Al_2O_2) = 365 \pm 7$ kcal per mole.

Table 41. Loss in Weight of a Sapphire Single Crystal in Vacuum

T, °K	Loss in weight, $g/cm^2 \cdot sec$	Vapor pressure, atm
1493	$3.43 \cdot 10^{-10}$	$4.50 \cdot 10^{-11}$
1564	$1.523 \cdot 10^{-9}$	$2.06 \cdot 10^{-10}$
1614	$2.445 \cdot 10^{-9}$	$3.28 \cdot 10^{-10}$

The volatility of aluminum oxide is strongly affected by the metals in contact with the Al_2O_3, such as the tungsten, molybdenum, and tantalum forming the vessel containing the oxide. In connection with this, Sears and Navias (1959) made an interesting attempt to study the evaporation of aluminum oxide without a metal in contact with the latter.

An alumina rod was held with a tungsten wire, the point of contact with which was always kept cold. A special screen protected the rod from the bombarding electrons with which the furnace was heated. When a tungsten heating tube was used, after heating at 1900°C for an hour, there were no deposits or sublimate in the cold parts. With a tantalum heater appreciable sublimate was observed even at 1600°C.

The vapor pressure over aluminum oxide calculated from weight loss was $5 \cdot 10^{-7}$ mm Hg (at 1900°C in a tungsten heater). This value is less by a factor of 4000 than that obtained by Brewer and Searcy (1951), who had the alumina in contact with a metal.

Sears and Navias showed that the rate of evaporation of aluminum oxide which is not in contact with tungsten (the tungsten was at some distance), was less by a factor of three than that of aluminum oxide in a tungsten crucible. The vapor pressure over aluminum oxide determined experimentally was less by a factor of 20 than that calculated from thermodynamic data on the assumption that Al_2O_3 decomposes to the elements on evaporation.

In the work of Walker, Efimenko, and Lofgren (1961), the rate of evaporation of aluminum oxide was determined in vacuum and in water vapor. The measurements were made through the loss in weight (Langmuir's method).

Up to 1600°C, Walker and his co-workers used an induction furnace. Experiments with molten alumina (2055 + 10°C) were carried out with a solar reflector.

The evaporation of solid alumina was studied with a single crystal of sapphire (fire-polished), suspended from a platinum wire. It was assumed that the vapor consists of AlO (Table 41).

The vapor pressure obtained was five times that of Brewer and Searcy, who worked with a Knudsen cell. Walker and his co-workers consider that up to 1600°C there is no reaction between solid aluminum oxide and water vapor, though they were unable to carry out experiments because of the strong effect of the water vapor on the platinum support.

The evaporation of alumina in an atmosphere of water vapor (2-25 mm H_2O) in the presence of oxygen was investigated with a solar reflector at a temperature close to the melting point of alumina (2055 ± 10°C). An appreciable reaction was observed between liquid alumina and water vapor. The authors were unable to determine the molecules present in the vapor, but there was the possibility of the presence of AlOH.

Ackermann and Thorn (1958) showed that in explaining the evaporation of alumina, it is impossible to ignore the gaseous tungsten oxides, which are inevitably present in work with a tungsten effusion cell. The complete description of the process requires a knowledge of the free energy of formation of gaseous Al_2O, Al, WO_3, AlO, and WO_2. Ackermann and Thorn considered the following reactions which may occur in a tungsten cell containing aluminum oxide:

$$\tfrac{1}{2}Al_2O_3 \, (liq) = Al \, (gas) + \tfrac{3}{2}O,$$

$$Al_2O_3 \, (liq) = Al_2O \, (gas) + 2O,$$

$$W \, (solid) + 3O = WO_3 \, (gas),$$

$$W \, (solid) + 2O = WO_2 \, (gas).$$

Moreover, they considered that gaseous AlO and WO may be formed.

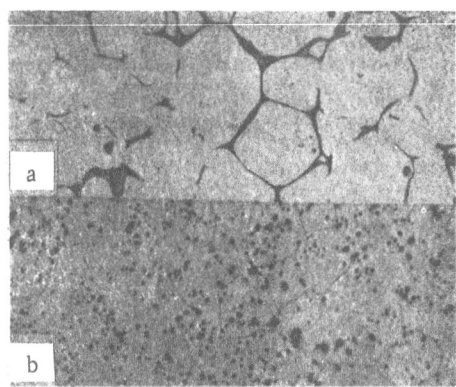

Fig. 95. Photomicrograph of refractory with the composition 0.19 CaO + 0.81 ZrO$_2$. a) Fired at 1900°C. in a furnace with an Al$_2$O$_3$ tube; b) fired at 2100°C in a furnace with a molybdenum element (etched section).

Brewer and Searcy quote a free energy for the reaction

$$^4/_3\text{Al (liq)} + ^1/_3\text{Al}_2\text{O}_3 \text{ (solid)} = \text{Al}_2\text{O (gas)},$$

but their value is unreliable, as they used an open effusion cell. The heat of formation of gaseous Al$_2$O may be obtained from the mass spectrometric measurements of Porter and his co-workers, who studied this reaction. Ackermann and Thorn, using this heat of formation and the absolute value of the entropy of Al$_2$O, obtained the following expression for the free energy of formation of gaseous Al$_2$O:

$$\Delta F^{\circ}_{\text{form}} = -47,200 - 9.27T \text{ cal/mole.}$$

In the work of De Maria, Drowart, and Inghram (1959), it was shown that it is necessary to take into account the presence of AlO and possibly WO in the gas phase. From the work of these authors, it follows that AlO must be quite stable and the dissociation energy of this compound given in the spectroscopic literature (0.95 eV) is incorrect. This is supported by Lagerqvist (1957), who has criticized the old determinations of the dissociation energy of AlO. In calculating the free energy of formation of AlO, Ackermann and Thorn used the dissociation energy of this molecule given by Herzberg (1949a) (3.7 eV), and considered it probable. Using the molecular constants given by Herzberg, and taking the vapor pressure from their own work in 1957, Ackermann and Thorn obtained the following expression for the free energy of formation of AlO:

$$\Delta F^{\circ}_{\text{form}} = 38,000 - 13.17T \text{ cal/mole.}$$

On the basis of the values obtained for the free energies of formation of Al$_2$O and AlO, Ackermann and Thorn showed that the ratio of the amount of Al$_2$O to AlO should be 7, while the mass-spectrometric investigations of De Maria, Drowart, and Inghram give 0.6 for this ratio.

Thus, at the present time, there are contradictions between the spectroscopic dissociation energy and the data of De Maria, Drowart, and Inghram. In the absence of other investigations, preference must go to the mass-spectrometric observations of the latter authors. However, since these observations were made under conditions which were far from equilibrium, a low concentration of AlO might be expected for equilibrium conditions, and then Ackermann and Thorn would be right in excluding this substance from their thermodynamic analysis.

Aluminum suboxide in a vapor state will transfer oxide from an alumina refractory to a material heated in the furnace. This may adversely affect the quality of the object heated if the introduction of alumina leads to a change in the structure, density, etc.

Rhodes (1961) showed that sintering of a zirconium refractory with the composition 0.19 CaO + 0.81 ZrO$_2$ proceeds differently, depending on the furnace in which it is heated. If a sample of the given composition on a tungsten backing is heated at 1900°C in a hydrogen atmosphere in a corundum tube with an external winding, heating for 1000 min gives a material whose structure contains a second phase, which is localized in the spaces between grains of cubic CaO–ZrO$_2$ crystals. Rhodes considered that vapor containing aluminum oxides penetrated the still porous, unfired material, and then the aluminum oxide reacted with the solid solution of the CaO–ZrO$_2$ system to form a liquid phase. The latter was responsible for liquid-phase sintering with the formation of the two-phase structure which is characteristic of this type of sintering. The amount of the second phase was estimated (from polished sections) as 12%, and it was surmised that it was a calcium aluminate glass (Hafner, Kreidl, Weidel, 1958).

If the same zirconium–calcium solid solution was sintered in a furnace with a molybdenum heater inside an alumina tube, the temperature of which was only 1200°C, and all evaporation of aluminum was prevented, the fired material had a completely different structure and a polished section showed only one phase (Fig. 95).

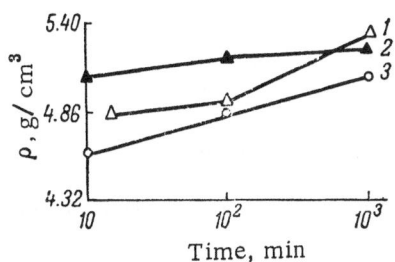

Fig. 96. Relation of the relative density of refractory with the composition $0.19\,CaO + 0.81\,ZrO_2$ to sintering time in a hydrogen atmosphere at $1900°C$. 1) Furnace with alumina tube; 2) with $0.75\,wt.\%\,Al_2O_3$ added, furnace with molybdenum element; 3) furnace with molybdenum element.

When $0.75\,wt.\%\,Al_2O_3$ was added to the charge before firing and the firing carried out in a molybdenum heating element, the same structure as in the first case, with interlayers of glass between grains of zirconium—calcium solid solution was obtained. This experiment shows that vapor evaporates from an alumina tube flushed with hydrogen and is deposited on the material fired. The increases in density of the samples on sintering were different in the three cases described. Figure 96 shows that samples with $0.75\,wt.\%\,Al_2O_3$ added had the highest density. The density of samples on which aluminum oxide vapors acted increased more slowly in the first 100 secs than subsequently. This indicates the peculiar action of the vapor, whose composition was not examined in Rhodes' article.

The observations described deserve great attention as aluminum oxide is the normal material for high-temperature furnaces and other units.

The work of Veits, Gurvich, and Medvedev on the determination of the dissociation energy of gaseous AlO, Al_2O, and Al_2O_3 was of great importance in elucidating the nature of gaseous aluminum oxides.

Very different values have been given for the dissociation energy of AlO. The study of predissociation in spectra normally leads to very low values. Thus, Herzberg (1949a) considered that $D_0^0(AlO)$ was about 3.75 eV on the basis of predissociation in the $B^2\Sigma$-state of AlO, while Rosen (1951) considered that $D_0^0(AlO)$ must be less than 0.93 eV from predissociation in the $X^2\Sigma$-state. The latter value cannot be correct, as spectral investigations showed that in the spectra of stars, AlO is present together with such stable molecules as TiO, ZrO, and BO. This indicates that the dissociation energy of AlO must be close to the dissociation energy of these molecules, i.e., 5-7 eV.

Linear extrapolation for the $X^2\Sigma$-state of the AlO molecule leads to the value $D_0^0(AlO) = 4.16$ eV, which may be regarded only as the lower limit of the value sought, as the ionic forces of the bond must play a substantial part in the AlO molecule (see Veits and Gurvich, 1956). Veits and Gurvich found the dissociation energy of AlO, having determined the equilibrium constant of the dissociation of AlO,

$$K_p(T) = \frac{P_{Al} \cdot P_O}{P_{AlO}}$$

in an acetylene—oxygen flame by the method developed previously by these authors.

The partial pressure of atomic aluminum was found from measurement of the intensity of the Al line at $\lambda = 3961.5$ A. In addition to AlO, the flame may have contained Al_2O and Al_2O_3 molecules. The number of these molecules (which is proportional to the partial pressure) may be estimated from the equations

$$P_{Al_2O} = \frac{P_{Al}^2 \cdot P_O}{K_p(Al_2O)}, \quad P_{Al_2O_3} = \frac{P_{Al}^2 \cdot P_O^3}{K_p(Al_2O_3)},$$

The thermodynamic values required for calculations of the equilibrium constant $K_p(Al_2O_3)$ were taken from the summary of Huff, Gordon, and Morrell (1950); the value of $K_p(Al_2O)$ was calculated from the data of Brewer and Searcy, and also Porter, Schissel, and Inghram. It was found that the partial pressures of Al_2O_3 and Al_2O are several orders less than ΣP_{Al} (the total pressure of all the aluminum compounds in the vapor), and, consequently, P_{AlO} may be calculated as the difference between the total vapor pressure and the vapor pressure of metallic aluminum, i.e.,

$$P_{AlO} = \Sigma P_{Al} - P_{Al}.$$

Having determined experimentally the value of $K_p(AlO)$ for a temperature T, it is possible to find the dissociation energy of AlO from the relation

Table 42. Results of Calculations of the Experiments
of Brewer and Searcy

$D_0^0(AlO)$, kcal/mole	$\Delta H_{0\,subl}^0$ (Al_2O_3), kcal/mole	$P_{Al_2O_3} \cdot 10^6$, atm	$P_{AlO} \cdot 10^6$, atm	$P_O \cdot 10^6$, atm
138	180 ± 3	1.3	8.4	2.6
137	178 ± 2.5	2.3	7.3	2.2
136	177 ± 3	3.0	6.3	1.9
134	176 ± 3	4.3	4.7	1.4

Note. The partial pressures of Al_2O_3, AlO, and O
are given for 2325°K.

$$D_0 = T(\Delta\Phi - R \ln K_p),$$

where $\Delta\Phi$ is the change in the reduced thermodynamic potential of the reaction $Al + O \rightleftharpoons AlO$

$$\Delta\Phi = \Phi_{Al}^* + \Phi_O^* - \Phi_{AlO}^*.$$

The values of the reduced thermodynamic potential for Al, O, and AlO may be calculated by the usual methods of statistical thermodynamics. For the dissociation energy of AlO, Veits and Gurvich give the value 133.5 ± 3 kcal/mole (5.8 eV). The change in the reduced thermodynamic potential $\Delta\Phi = 27.7$ kcal/mole· · deg.

Somewhat later, Gurvich and Veits (1958), having corrected their data of 1956, gave for the dissociation energy of AlO the value 137 ± 4 kcal/mole. They consider that the sharp discrepancy between the values of $D_0^0(AlO)$ they obtained and the data of Herzberg (1949a), Rosen (1946), and Howell (1945) is explained by the incorrect interpretation of the spectra of AlO molecules. This conclusion is confirmed by the work of Lagerqvist and his co-workers (Lagerqvist, Nillson, Barrow, 1956).

Gurvich and Veits determined the dissociation energy of a series of diatomic molecules MeO for group III elements. They stated that the calculation of the equilibrium of the reaction of these metals with gases of flames made it possible for them to show that the partial pressures of other possible compounds such as Al_2O_3 and Al_2O do not exceed 5% of the value of $P_{\Sigma M}$, and they consider that the presence of Al_2O and Al_2O_3 cannot affect the determination of the dissociation energy of AlO.

In an article entitled "Method of Calculating Effusion Experiments in the Case of Dissociation of the Evaporation Products," Medvedev (1958) demonstrated that effusion measurements may be used for calculating heats of dissociation also in the case where dissociation products are present in the vapor together with the main evaporation products.

Medvedev recalculated the results of Brewer and Searcy's experiment on the vapor pressure of aluminum oxide. The latter assumed that the evaporation products of aluminum oxide are AlO and O, and calculated $D_{298}^0(AlO) = 138$ kcal/mole, which corresponds to $D_0^0(AlO) = 136.8$ kcal/mole. However, Medvedev showed, as a result of his calculations, that under the conditions of Brewer and Searcy's experiments (temperature 2300-2600°K and pressure 10^{-4}-10^{-5} atm) the evaporation products must also contain a considerable amount of Al_2O_3. In the calculations, the thermodynamic properties of gaseous aluminum compounds were taken to be the same as in the work of Gurvich and Veits (1958).

The thermodynamic functions of aluminum oxide in the condensed state were calculated with the values of the heat capacity obtained by Ginnings and Furukawa (1953), and also from the summary of Kelley (1949). Medvedev carried out calculations for various values of $D_0^0(AlO)$. The results he obtained are given in Table 42.

The partial pressures of molecular oxygen and Al_2O in all the experiments were relatively low and are not given in the table. Thus, the main conclusion of Medvedev is the demonstration of the existence of Al_2O_3 molecules in the vapor.

In their review, Ackermann and Thorn (1961) considered the work of Veits and Gurvich (1956), Gurvich and Veits (1958), and Medvedev (1958), which we examined above (p. 163). Ackermann and Thorn were mainly interested in the question of the presence of gaseous Al_2O_3 molecules, whose presence in the vapor above liquid aluminum oxide was demonstrated by the Soviet scientists. Ackermann and Thorn pointed out that Veits, Gurvich, and Medvedev, in their calculations, used thermodynamic data taken from the tables of Huff, Gordon, and Morrell (1950). However, the numerical values in these tables were based on the heat of evaporation of Al_2O_3 according to the measurements of Ruff and Konschak (1926), who placed aluminum oxide in a tantalum crucible and carried out the experiments at 2800°K. However, at this temperature there is considerable reduction of alumina by tantalum, which leads to a large amount of gaseous tantalum oxides. Because of this, the resulting pressure obtained was too high and the heat of evaporation was too low, as the system studied did not correspond to simple evaporation of liquid Al_2O_3 as such.

Ackermann and Thorn (1958) consider that Al_2O has a cyclic structure with Al–O and Al–Al distances of 1.6 and 2.8 A, respectively. The vibration frequencies of the Al_2O molecule are taken to be the same as for ethylene oxide, and only the difference in the weights of CH_2 and Al is considered. By assuming a nondegenerate ground state, we find for the change in entropy of the reaction

$$^5/_3Al \ (liq) + {}^1/_3Al_2O_3 \ (solid) = Al_2O \ (gas)$$

$\Delta S = 37.82$ en. units.

The entropy of this reaction may also be obtained by using the data from the mass spectrometric measurements of Porter, Schissel, and Inghram. The ratio of the ionization currents of Al_2O and Al was 2.5. By using the data of Baur and Brunner (1934a) for the vapor pressure of aluminum, we obtain for the entropy of the reaction given the value 35.21 en. units. The entropy values obtained by different methods are quite close. Ackermann and Thorn preferred the second value.

By taking for the heat of formation of Al_2O from aluminum and alumina the value obtained by Porter, Schissel, and Inghram (86.9 kcal/mole), we obtain for the free energy of the reaction

$$^5/_3Al \ (liq) + {}^1/_3Al_2O_3 \ (solid) = Al_2O \ (gas)$$
$$\Delta F^0_{form} \cong 86,900 - 35.21T \ cal/mole.$$

Then the free energy of formation of Al_2O(gas) will be

$$\Delta F^0_{form} \ (Al_2O_{gas}) \cong -47,200 - 9.27T \ cal/mole.$$

In accordance with the proposed structure of Al_2O, for the temperature-dependent part of the free energy $F^0 - H^0_0$ we obtain

$$f^\circ (Al_2O) = -R'T \ (7 \log T + 649T^{-1} - 7.51),$$
$$R' = 2.303R.$$

The equation for f^0 of atomic oxygen and aluminum at temperatures above 2000°C may be obtained by using the data from the summary of Moore (1958)

$$f^\circ (O) = -R'T \ (^5/_2 \log T - 48.69T^{-1} + 31,193T^{-2} + 1.169),$$
$$f^\circ (Al) = -R'T \ (^5/_2 \log T - 46.67T^{-1} + 1254 \ T^{-2} + 1.333).$$

By taking as the free energy of dissociation of molecular oxygen

$$\Delta F^\circ_D (O) = 121,945 - 32.22T \ cal/mole,$$

we find for the standard free energy of dissociation of Al_2O

$$\Delta F^\circ_D (Al_2O) \cong 252,100 - 2.288T \log T - 51.92T \ cal/mole.$$

In the final expression, the terms with T^{-1} and T^{-2} may be neglected, as they are insignificant.

In the conclusion of the examination of work on the evaporation of aluminum oxide, Ackermann and Thorn (1961) wrote: "It is quite obvious that the evaporation of alumina from a tungsten cell is a complex process, which includes four or even more chemical equilibria. Therefore it is very important to be sure of the reliability of the values of the free energies of formation. However, before further calculations can be undertaken, some quantitative experiments must be carried out. It is necessary to remeasure the absolute effusion rate of evaporation of gaseous Al, Al_2O, AlO, etc., from a tungsten cell. It is necessary to study the extent of

Table 43. Thermodynamic Properties of Some Aluminum Compounds (According to
Baer, Geene, Smith, and Wortman) at 3000°K

Compound	Aggregate state	H_T, kcal/mole	S_T, cal/mole· deg	Coefficients of heat capacity equation $C_p = \alpha + \beta T + \frac{\gamma}{T^2}$		
				α	β	$\gamma \cdot 10^{-7}$
Al	Gas	196.880	50.842	5.011	$-0.6000 \cdot 10^{-3}$	—
AlH	Gas	195.337	62.842	8.908	0.1628	0.3086
AlO	Gas	211.640	71.397	8.938	$0.8664 \cdot 10^{-2}$	0.2914
Al_2O	Gas	279.296	92.571	13.895	$0.2907 \cdot 10^{-3}$	0.7246
Al_2O_3	Liquid	118.276	90.145	31.15	2.0	—
Al_2O_3	Solid	91.571	78.634	27.43	3.06	0.847

the reaction of alumina with tungsten to determine whether the partial pressures depend on the surface of contact between liquid alumina and tungsten. The free energy of formation of gaseous tungsten oxides must be measured more accurately."

Thermodynamic calculations with computers are being used increasingly in practice. Such an example is the work of Baer, Geene, Smith, and Wortman (1960), who used a high-speed digital computer to calculate the enthalpy and entropy of some aluminum compounds at 3000°K together with other substances. In Table 43 we give the thermodynamic characteristics of some aluminum compounds taken from the article of Baer and his co-workers.

Without giving an exhaustive review of investigations of the evaporation of boric anhydride, we should note only certain facts, mainly for comparison with aluminum oxide.

Boric anhydride is a readily fusible substance and solid boric acid compounds are of no interest for high-temperature technology. However, gaseous compounds of boron have a very high stability and are not decomposed at temperatures of 2500-3000°K. It is particularly interesting that boron compounds containing hydrogen are found in vapors above 2000°K.

The vapor pressure of boric anhydride was studied by Speiser, Naiditch, and Johnston (1950), Inghram, Porter, and Chupka (1956), Scheer (1957), Soulen and Margrave (1956), and Searcy and Meyers (1957). From the data of these authors, for the heat of sublimation of B_2O_3 at 0°K values are obtained in the range of 97.2-102.8 kcal/mole. Evans, Prosen, and Wagman (1959) obtained somewhat different values: $\Delta H_{1500} = 84$ kcal per mole, and $\Delta H_0^0 = 96.6$ kcal/mole.

Nesmeyanov and Firsova (1959) studied the evaporation of liquid borid anhydride using Knudsen cells of molybdenum or tantalum. The following temperature dependence was obtained for the vapor pressure of boric anhydride (it is assumed that evaporation occurs without a change in the molecular composition):

$$\log P \,(\text{atm}) = 6.56 - \frac{16800}{T} \quad (1299—1515°\,\text{K}).$$

The vapor pressure of boric anhydride was also studied by Akishin and Gorokhov (1961). Berl and Renish (1959), referring to work carried out in the National Bureau of Standards (USA), give the following expression for the vapor pressure of boric anhydride:

$$\log P \,(\text{atm}) = \frac{18850}{T} - 2.9 \log T + 17.23\,478.$$

According to this equation, the boiling point of boric anhydride must be 2560°K.

In the evaporation of B_2O_3 in the presence of water vapor, the main component of the vapor is the compound HOBO. Margrave and his co-workers (Margrave, Soulen, Leroi, Greene, Randell, 1958) consider that this compound has the composition $H_2B_2O_4$. For the monomer HOBO, Evans and his co-workers give the heat of formation $\Delta H_{form}^0 = -140$ kcal/mole.

The compound B_2O_3 as a vapor was obtained by Inghram et al. by heating a mixture of B_2O_3 and boron. Searcy and Meyers obtained B_2O_2 in studying the effusion of gases formed by heating a mixture of MgO + B. B_2O_2 was also obtained by Scheer. As the mean of the determinations of these authors we can take as the heat of formation of gaseous B_2O_2 105 kcal/mole. Preliminary mass spectrometric determinations by Chupka gave for the dissociation energy of B_2O_2(gas) to form 2BO(gas) the value 115 kcal/mole.

As a further development of the work described in Chapter VII on the use of the explosion method for studying gaseous oxides, Rusin and Tatevskii (1961) proposed a method with simultaneous determination of the explosion pressure and absorption spectrum. With this method it was possible to establish conclusively the presence of BO_2 molecules in the combustion products of boron. Even in 1958, on the basis of a study of flames, Mal'tsev and Tatevskii put forward the hypothesis that the fluctuation bands they observed belonged to BO_2 particles. Further spectral investigations (Mal'tsev, Matveev, and Tatevskii, 1961a, 1961b) convinced the authors that the hypothesis was correct. For conclusively establishing the nature of the fluctuation bands and the composition of the combustion products of boron, a method was developed based on the study of the relation of the absorption at the band at 5470 A to the concentration of oxygen, water, and hydroxide for explosion in a bomb with central ignition.

Rusin and Tatevskii consider that B_2O_3, B_2O_2, BO, BO_2, and HBO_2 may exist at the moment of maximum pressure on ignition of a mixture of CO_2, O_2, and H_2 with diborane added in a bomb. All these molecules apart from BO_2 were detected in mass spectrometric investigations. It is possible that BO_2 is formed only at the high temperatures developed in the bomb (above 3000°K). Rusin and Tatevskii calculated the concentrations of the molecules listed above and obtained compatible results if it was assumed that there were present BO_2 molecules (and not B_2O_3), which were formed in accordance with the equilibrium

$$HBO_2 + OH \rightleftarrows BO_2 + H_2O.$$

Under the conditions of experiments with a bomb, the main boron-containing combustion products at 3080°K were HBO_2 and BO_2.

The work of Rusin and Tatevskii shows the great promise of the use of explosions in a bomb for studying high-temperature processes.

3. Evaporation of Gallium, Indium, and Thallium Oxides

Gallium oxide dissociates in vacuum and, therefore, the vapor pressure of this substance cannot be determined by methods which involve heating the substance in high vacuum. Shchukarev, Semenov, and Rat'kovskii (1961) determined the saturated vapor pressure of Ga_2O_3 in the temperature range of 1523-1682°C by the flow method in an oxygen atmosphere. The temperature dependence of the saturated vapor pressure of gallium oxide is described well by an equation derived on the assumption that Ga_2O_3 is monomeric in the vapor:

$$\log P \, (\text{mm Hg}) = -27.098 T^{-1} + 13.339.$$

The enthalpy and entropy of sublimation of gallium oxide, calculated from the slope of the line giving the relation $\log P = f(1/T)$ were found to equal 126 kcal/mole and 49 kcal/mole·deg, respectively. Assuming that the heat capacities of Al_2O_3 and Ga_2O_3 have the same temperature dependence, Shchukarev and his co-workers calculated the thermodynamic properties of Ga_2O_3(solid) under standard conditions.

Enthalpy of sublimation . . . . . 137 kcal/mole
Entropy of sublimation . . . . . . 57 kcal/mole·deg
Enthalpy of formation . . . . . . −121 kcal/mole
Entropy of formation . . . . −15 kcal/mole·deg

In studying the evaporation of indium oxide, Shchukarev, Semenov, Rat'kovskii, and Perevozchikov (1961) also used the flow method in an oxygen atmosphere. For determining the amount of oxide evaporating, they used a radiochemical method employing the isotope In[114].

Table 44. Dissociation Energy of Some Compounds of Group III Elements, kcal per mole

Compound	Data of Gurvich and Veits	Data of Howell
GaO	115.6 ± 3	—
InO	103 ± 5	26
TlO	< 90	0
GaOH	< 108	—
TlOH	< 70	—

Table 45. Partial Pressure of Components and Enthalpy of the Reaction Y_2O_3(solid) $\rightarrow$ 2YO(gas) + O(gas)

T, °K	$P_{YO} \cdot 10^6$, atm	$P_O \cdot 10^6$, atm	ΔH_O^0, kcal/mole
2500	2.70	0.53	489.0
2600	8.63	1.69	—
2700	24.0	4.73	—

The vapor pressure of indium oxide in the temperature range of 1290-1490°C is expressed by the equation

$$\log P\,(\text{mm Hg}) = -27.791\,T^{-1} + 14.353.$$

The authors estimated the accuracy of their data as 6%. The enthalpy of sublimation of In_2O_3 was found to equal 118 kcal per mole and the entropy of sublimation 42 kcal/mole · deg.

Using the method they developed for studying the equilibrium of the dissociation of molecules in flames, Gurvich and Veits (1958) determined the dissociation energies of GaO, InO, and TlO, which are the predominant components of the vapors under their experimental conditions. Working with flames containing hydroxyl, Gurvich and Veits observed the molecules GaOH and TlOH, for which the dissociation energies were also found (Table 44).

The sharp discrepancy in the data of Howell (1945) is explained by the inaccurate interpretation of the spectra of the molecules. From a comparison of the stabilities of the molecules BO, AlO, GaO, and InO, it can be stated that Howell's conclusion on the low dissociation energy of InO is inaccurate.

4. Evaporation of Rare Earth Oxides

Interest in investigating the composition and stability of compounds present in vapors over rare earth oxides has increased particularly in recent years, when these substances have become more available for wide investigations and their practical use has become possible. It should be noted particularly that rare earth elements are formed in nuclear reactions and are of interest as reactor materials. Increasing interest has arisen in rare earth oxides from the point of view of high-temperature technology. This necessitates their thermodynamic investigation, but only the first steps in this direction have been taken.

For thermodynamic knowledge of oxides, it is necessary to study their evaporation and also characterize the chemical interactions in the system metal—oxygen, as this makes possible a deeper understanding of the nature of homogeneous phases of variable composition. Such phases are known to be formed by a series of rare earth oxides.

The investigation of the evaporation of rare earth oxides began very recently. In his review, Brewer (1953) reported only that Wartenberg (1930) observed appreciable evaporation of Ce_2O_3 at 2600°K and Mott (1918) as early as 1918 determined the boiling points of Y_2O_3 and La_2O_3 and found that they equaled 4570° and 4470°K, respectively. Conclusions on the energy characteristics of molecules of rare earth oxides in the vapor state, based on spectral investigations, were contradictory, though there was no doubt of the existence of gaseous oxides. From the values of the dissociation energies of the molecules MeO, Herzberg came to the conclusion that the following reaction is of great importance for yttrium, lanthanum, and cerium:

$$\text{Me (solid)} + Me_2O_3 \text{ (solid)} = 3MeO \text{ (gas)}.$$

However, Gaydon (1949) gives for the dissociation energy of LnO molecules values about 50 kcal lower, and then the reaction given could not play a substantial role. Only study of the evaporation of the oxides made it possible to find the dissociation energy of gaseous compounds of rare earth oxides, which was apparently beyond the scope of spectral methods.

The study of the thermal properties of gaseous molecules of rare earth oxides should shed light on some peculiarities in the properties of solid sesquioxides of these elements. Huber, Head, and Holley (1960) found an anomalously high heat of formation of solid oxides Ln_2O_3. In connection with this, the enthalpy of formation per oxygen atom for the compounds Ln_2O_3 is greater than normal and only for Li_2O and some oxides of group II elements are the values observed close to those for rare earth oxides.

Gaseous oxides of rare earths (especially monoxides), which have a similar electronic structure, still show a considerable difference in bond energies. All these circumstances prompt new investigations of these compounds at high temperatures.

The study of the evaporation of rare earth oxides was begun later than for oxides of other elements. Some of the first of these investigations were begun in Leningrad University in the inorganic chemistry department (Prof. S. A. Shchukarev and G. A. Semenov). In the USA, apart from the small amount of work on the evaporation of lanthanum oxide in Chicago University (Inghram, Chupka, and Porter), the investigations of the evaporation of rare earth oxides have been concentrated mainly in the Cryogenics Laboratory in Columbus (Goldstein, Walsh, White, and Dever). The results given below from the investigation of the evaporation of rare earth oxides show that this is only the beginning of great work.

Yttrium Oxide. In the present section we include the results of investigations of the evaporation of yttrium and scandium oxides, as these oxides are similar to rare earth oxides with respect to evaporation processes.

The evaporation of yttrium oxide was studied by Walsh and White, and also Ackermann and Thorn. A brief report on the results of this work may be found in a review article of Ackermann and Thorn (1961). Walsh and White measured the rate of effusion of the vapor over solid Y_2O_3 from a tungsten cell in the range of 2500-2700°K. By means of vacuum balances, Ackermann and Thorn also measured the effusion rate of evaporation of yttrium oxide (probably in the form of a mixture of YO and O) in the range of 2509-2720°K. The latter authors showed that there is quite an appreciable reaction between yttrium oxide and tungsten at 2800°K and the loss of tungsten from the effusion cell due to this reaction is 10% of the weight of the yttrium oxide evaporating.

The quantitative proportions of atomic yttrium, oxygen, and molecular YO in the vapor are known only approximately, as there are no accurate data on the vapor pressure of metallic yttrium. Table 45 gives the data of White and Walsh for the reaction

$$Y_2O_3 \text{ (solid)} \rightarrow 2YO \text{ (gas)} + O \text{ (gas)}.$$

Ackermann and Thorn consider that the partial pressure of gaseous yttrium is one-tenth of the partial pressure of yttrium monoxide.

For the dissociation energy of the YO molecule, Walsh and White give the value 7.25 ± 0.30 eV, and Ackermann and Thorn, 7.1 ± 0.3 eV. Spectroscopic determinations of the dissociation energy of YO molecules lead to inaccurate and unreliable results. Herzberg (1949a) gives the value 9.0 eV; Gaydon (1949) gives the possible limits of variations in the dissociation energy, and these reach almost 30% (7 ± 2 eV).

Taking the value for the heat of formation of solid Y_2O_3 obtained recently by Huber et al. (Huber, Head, Holley, 1960), and using the tabular values of Coughlin (1954), it is possible to give the temperature dependence of the free energy of formation of Y_2O_3 (solid) in the temperature range of 1800÷2000°K,

$$\Delta F^0 = -460,000 + 74.76 \, T \text{ cal/mole.}$$

The free energy of formation of the gaseous YO molecule may be calculated from a knowledge of its dissociation energy and the reduced thermodynamic potential and free energy of formation of solid Y_2O_3, using data from the effusion measurements on yttrium oxide and the vapor pressure of metallic yttrium. For the range of 2500-2700°K, Ackermann and Thorn obtained the following equation for the free energy of formation of the gaseous molecule YO:

$$\Delta F^0 = -11,000 - 12.9 \, T \text{ cal/mole.}$$

This equation in its turn makes it possible to determine the partial pressure of YO in effusion experiments with Y_2O_3. At 2500°K this partial pressure will equal $1.6 \cdot 10^{-6}$ atm. The partial pressure of the monoxide arising as a result of the reaction

$$Y_2O_3 \text{ (solid)} + {}^1\!/_2 W \text{ (solid)} = 2YO \text{ (gas)} + {}^1\!/_2 WO_2 \text{ (gas)},$$

may be estimated as $5 \cdot 10^{-7}$ atm. It is evident that in accurate investigations it is necessary to take account of the reaction between Y_2O_3 and tungsten.

Scandium Oxide. Ackermann and Thorn, pointing out the absence of quantitative data on the evaporation of scandium oxide, reported only that the volatility of this oxide is several times as great (at 2500 °K) as the volatility of Y_2O_3. A particularly low volatility is characteristic of yttrium oxide.

Cerium Oxide. The evaporation of cerium oxides was studied by Shchukarev and Semenov (1961) with a mass spectrometer by the method described below (see p. 185).

The evaporation of oxides of this element is a complex process. Cerium dioxide CeO_2 dissociates on heating in vacuum with the evolution of a large amount of oxygen and gradual conversion to the oxide Ce_2O_3. At 1200°C and a pressure of about 10^{-7} mm Hg, the dissociation product is close to the composition Ce_4O_7. The molecules CeO_2 and CeO were detected in the vapor above this oxide. As the preseparation evaporated, the ratio $CeO : CeO_2$ in the gas phase increased in connection with the approach of the solid phase to Ce_2O_3. The molecules CeO_2 and CeO were also found in the vapor over cerium oxide Ce_2O_3, but their ratio is quite different from that found at the beginning of the process when the composition of the solid is close to CeO_2. It may be surmised that the evaporation of Ce_2O_3 proceeds in accordance with the two schemes

$$Ce_2O_3 \text{ (solid)} \rightarrow 2CeO \text{ (gas)} + O \text{ (gas)},$$
$$Ce_2O_3 \text{ (solid)} \rightarrow CeO \text{ (gas)} + CeO_2 \text{ (gas)}.$$

Walsh, Goldstein, and White (1958) briefly reported their experiments with cerium dioxide. Like Semenov, they observed that serium dioxide is not stoichiometric, losing oxygen preferentially to form a compound corresponding to the composition $CeO_{1.8}$. The pressure of oxygen over CeO_2 at 1700°C equals approximately 10^{-5} atm. On the other hand, the same authors report that Ce_2O_3 evaporates stoichiometrically and that this oxide is apparently more volatile than Nd_2O_3 and La_2O_3. We give some data below on gaseous CeO in examining the recent (since 1961) work of these authors.

Lanthanum Oxide. On the basis of spectroscopic data, it has long been known that the monoxide LaO exists as a vapor. With a mass spectrometer, Chupka, Inghram, and Porter (1956) showed that at 1800°K, the monoxide LaO and atomic oxygen and lanthanum are in equilibrium with lanthanum oxide; it was also shown that the saturated vapor pressure at this temperature is $2.2 \cdot 10^{-7}$ atm. Using simultaneously mass spectrometric and effusion data, the same authors studied the reaction

$${}^1\!/_3 La \text{ (gas)} + {}^1\!/_3 La_2O_3 \text{ (solid)} = LaO \text{ (gas)}$$

in the region of 1650-1900°K and found for the dissociation energy of the molecule LaO the value 8.15 ± 0.35 eV (this was later corrected and the value 7.98 eV given). From this high dissociation energy it follows that LaO molecules must be in a predominant amount over solid La_2O_3.

Walsh, Goldstein, and White (1960), like Chupka and his co-workers, used a tungsten cell to measure the rate of effusion of the vapor formed over La_2O_3 in the temperature region of 2230-2440°K. The first authors calculated the dissociation energy of the molecule LaO. The value obtained, 8.02 ± 0.17 eV, agrees within the limits of intrinsic errors with the value given by Chupka and his co-workers. Taking for the dissociation energy of the molecule LaO the value 7.98 eV and the reduced thermodynamic potential of LaO obtained from spectroscopic data, Ackermann and Thorn arrived at the following equation for the free energy of formation of gaseous LaO:

$$\Delta F = 36,000 - 9.6T \text{ cal/mole.}$$

Inghram and Drowart (1962), referring to the work of Chupka, Inghram, and Porter, give the heats of the reactions:

$$^1/_3 La \text{ (gas)} + {}^1/_3 La_2O_3 \text{ (solid)} = LaO \text{ (gas)},$$
$$LaO \text{ (gas)} = La \text{ (gas)} + O \text{ (gas)}.$$

Depending on the calculation method (using the second or the third law of thermodynamics), the heats of formation have slightly different values. For the first reaction: $\Delta H_{298}^0 = 89 \pm 9$ kcal/mole (second law) and $\Delta H_{298}^0 = 82 \pm 7$ kcal/mole (third law); for the second reaction: $\Delta H_{298}^0 = 182 \pm 10$ (second law) and $\Delta H_{298}^0 = 189 \pm 8$ kcal/mole (third law).

A large amount of work to investigate the evaporation of rare earth element oxides was carried out by Walsh and his co-workers (Walsh, Dever, Goldstein, White, 1961). They studied: 1) the evaporation of lanthanum and neodymium oxides; 2) the heats of sublimation of neodymium, praseodymium, gadolinium, terbium, dysprosium, holmium, erbium, and lutecium, and 3) isomolecular exchange of oxygen in metal−monoxide systems. The authors used a time-of-flight mass spectrometer, which is described in the collection "Advances in Mass Spectrometry" (1958), and also in articles by Wiley (Wiley, McLaren, 1955; Wiley, 1956). A tantalum or tungsten Knudsen cell was used.

Isotherms of the rate of evaporation of lanthanum and neodymium oxides showed that there is a change in the composition of the solid phase as the gaseous products are liberated. At the beginning of the process, when fresh oxides are present, there is an increased rate of evaporation, which falls until the composition of the products reaches $La_2O_{2.96}$ and $Nd_2O_{2.96}$. Oxides of this composition evaporate at a constant rate. The following particles were observed in a mass spectrometer at all temperatures: LnO, Ln, and O (Ln = La or Nd). LnO molecules were the predominant components. Metal ions were formed as fragments of LnO molecules.

Lanthanum and neodymium oxides reacted with the material of the Knudsen cell (tantalum or tungsten). However, this did not interfere with the determination of accurate thermodynamic characteristics of gaseous LnO molecules as the formation of tantalum and tungsten oxides occurs as a result of the diffusion of oxygen through the walls of the cell and the evolution of the molecules TaO, TaO_2, etc., from the outer surface of the cell. The loss of tantalum and tungsten is 0.1 mole per mole of Ln_2O_3, though no gaseous molecules of oxide compounds of tantalum and tungsten were detected in the effusion vapors.

Walsh, Goldstein, and White observed in a tantalum effusion cell new materials to which they assigned the formula $LnTaO_4$. These compounds are probably formed by the reaction

$$3Ln_2O_3 \text{ (solid)} + Ta \text{ (solid)} = 5LnO \text{ (gas)} + LnTaO_4 \text{ (solid)}.$$

Such compounds are obtained with lanthanum, neodymium, and cerium. In the last case, the compound may have a more complex composition. No corresponding yttrium compound was observed.

The evaporation of La_2O_3 and Nd_2O_3 is represented by the following equation:

$$^1/_2 Ln_2O_3 \text{ (solid)} = LnO \text{ (gas)} + {}^1/_2 O \text{ (gas)}.$$

From data on the rates of evaporation, and using the required thermodynamic characteristics of the participants in the reaction, Walsh and his co-workers calculated the thermal effects of the reaction given:

$$\text{for La } \Delta H_{2140} = 209.2 \pm 4.2 \text{ kcal/mole},$$
$$\text{for Nd } \Delta H_{2307} = 204.5 \pm 4.4 \text{ kcal/mole}.$$

The corresponding values for absolute zero will be

$$\text{for La } \Delta H_0^0 = 217.6 \pm 6.6 \text{ kcal/mole},$$
$$\text{for Nd } \Delta H_0^0 = 216.3 \pm 6.2 \text{ kcal/mole}.$$

The heat of formation and dissociation energy of lanthanum and neodymium monoxides are characterized by the following figures:

$$\Delta H_0^0 (LaO) = -29.8 \pm 4 \text{ kcal/mole}, \quad D_0^0 (LaO) = 8.08 \pm 0.2 \text{ eV},$$
$$\Delta H_0^0 (NdO) = -30.0 \pm 6 \text{ kcal/mole}, \quad D_0^0 (NdO) = 7.18 \pm 0.3 \text{ eV}.$$

In calculations of the thermodynamic properties of gaseous molecules of rare earth monoxides, it is necessary to use spectral data. In these data the greatest uncertainty is connected with the electronic contribution, which, in the case of rare earth compounds, is unusually high as a result of the possibility of multiplicity of the ground states and excitation at high temperatures.

Walsh, Dever, and White carried out a mass spectrometric investigation of isomolecular exchange of oxygen of the general form

$$Ln' \text{ (gas)} + Ln''O \text{ (gas)} \rightleftarrows Ln'O \text{ (gas)} + Ln'' \text{ (gas)},$$

and were able to draw conclusions on the electronic contribution to the entropy of gaseous monoxides. These authors studied the equilibrium constants of the following reactions:

$$Ce \text{ (gas)} + LaO \text{ (gas)} = La \text{ (gas)} + CeO \text{ (gas)}, \tag{1}$$
$$Pr \text{ (gas)} + LaO \text{ (gas)} = La \text{ (gas)} + PrO \text{ (gas)}, \tag{2}$$
$$Nd \text{ (gas)} + PrO \text{ (gas)} = Pr \text{ (gas)} + NdO \text{ (gas)}. \tag{3}$$

The heat of any of these reactions is the difference in the dissociation energies of the two monoxides participating in the reactions. Thus, the study of reactions of this type is a means of checking the accuracy of the values obtained for the dissociation energies using spectroscopic data.

Having obtained the equilibrium constants for the three reactions given, Walsh and his co-workers calculated the values of ΔH and ΔS of these reactions. The following values were found for ΔH and ΔS: for reaction (1), $\Delta H_{1870} = 1.05 \pm 0.20$ kcal/mole, $\Delta S_{1870} = 0.33 \pm 0.12$ kcal/mole $\cdot$ deg; for reaction (2), $\Delta H_{1913} = 15.8 \pm 0.4$ kcal/mole, $\Delta S_{1913} = 3.6 \pm 0.2$ kcal/mole $\cdot$ deg; for reaction (3), $\Delta H_{1910} = 6.9 \pm 0.6$ kcal/mole, $\Delta S_{1910} = 1.5 \pm 0.3$ kcal/mole $\cdot$ deg. By adding reactions (2) and (3) it is possible to obtain the change in enthalpy and entropy of the reaction

$$Nd \text{ (gas)} + LaO \text{ (gas)} = NdO \text{ (gas)} + La \text{ (gas)}.$$

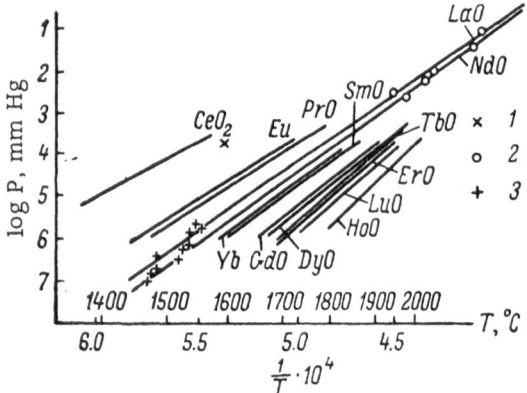

Fig. 97. Vapor pressure above rare earth oxides according to Shchukarev and Semenov. For LaO according to the data of different authors: 1) Inghram et al.; 2) White et al.; 3) Shchukarev and Semenov.

$\Delta H_{1910} = 22.7 \pm 0.7$ kcal/mole, $\Delta S_{1910} = 2.1 \pm 0.4$ kcal per mole $\cdot$ deg.

These values make it possible to determine the difference in the dissociation energies of the molecules LaO and NdO. It was found that $D_0^0(LaO) - D_0^0(NdO) = 23.5 \pm 1$ kcal/mole. This difference is in good agreement with the difference obtained from determinations of the vapor pressures over lanthanum and neodymium oxides (20.8 ± 9 kcal/mole).

The investigations of the exchanges of oxygen do not make it possible to determine the dissociation energies of the monoxides, as they give only the differences in these energies. For one of the oxides it is necessary to take a dissociation energy determined by other methods. It was assumed that $D_0^0(LaO) = 8.08 \pm 0.2$ eV (see above), and then $D_0^0(NdO) = 7.06 \pm 0.2$ eV, $D_0^0(CeO) = 8.03 \pm 0.2$ eV, $D_0^0(PrO) = 7.40 \pm 0.3$ eV.

Table 46. Relative Composition of Vapor Over Rare Earth Oxides
with Various Ionizing Potentials

Ln_2O_3	45 V			10 V		
	Ln^+	LnO^+	LnO_2^+	Ln^+	LnO^+	LnO_2^+
La_2O_3	0.16	1.0	—	—	1.0	—
CeO_2	0.28	0.6	1.0	—	1.0	0.95
Ce_2O_3	0.18	0.91	1.0	—	1.0	0.3
Pr_2O_3	0.36	1.0	0.2	—	1.0	0.01
Nd_2O_3	0.25	1.0	—	—	1.0	—
Sm_2O_3	1.0	0.99	—	1.0	0.77	—
Eu_2O_3	1.0	0.2	—	1.0	0.15	—
Gd_2O_3	0.3	1.0	—	1.0	0.83	—
Tb_2O_3	0.2	1.0	0.01	0.9	1.0	—
Dy_2O_3	0.5	1.0	—	1.0	0.59	—
Ho_2O_3	1.0	0.77	—	1.0	0.45	—
Er_2O_3	1.0	1.0	—	1.0	0.72	—
Yb_2O_3	1.0	0.01	—	1.0	—	—
Lu_2O_3	1.0	0.9	—	1.0	0.60	—

Table 47. Equilibrium Partial Vapor Pressures Over Ln_2O_3 at 2000°K (in atm)
According to Data of Panish

Ln_2O_3	P_{Ln}	P_{LnO}	P_{LnO_2}	$I_{Ln^+} : I_{LnO^+}$
Pr_2O_3	$< 1 \cdot 10^{-8}$	$1 \cdot 10^{-7}$	$< 5 \cdot 10^{-9}$	< 0.1
Nd_2O_3	$< 5 \cdot 10^{-9}$	$5 \cdot 10^{-8}$	—	< 0.1
Sm_2O_3	$\approx 3 \cdot 10^{-9}$	$\approx 3 \cdot 10^{-9}$	—	$0.5-1.0$
Eu_2O_3	$7 \cdot 10^{-8}$	$9 \cdot 10^{-9}$	—	8.0 ± 2

The results of the work show that the electronic contribution to the thermodynamic properties of the monoxides examined must be considerable. Thus, for LaO(gas), the electronic component of the entropy at 1900°K is estimated at 1.38 cal/mole · deg, and for NdO(gas), 4.4 ± 0.5 cal/mole · deg.

Europium Oxide. On evaporation, Eu_2O_3 almost completely dissociates to gaseous elements. In this case, it would seem that there must be the lowest volatility, but this is not observed. It is possible that gaseous europium monoxide has a very low dissociation energy, or that on evaporation there is the formation of EuO in a condensed state at the interphase.

In 1961 Shchukarev and Semenov reported their results from investigation of all the oxides of rare earth elements apart from thulium. The results of the work of Shchukarev and Semenov are given in Table 46. For the different oxides the measurements were carried out at temperatures such that the vapor pressure of the main component was of the order of 10^{-5} mm Hg. Figure 97 gives data on the vapor pressure of rare earth element monoxides according to the measurements of Semenov and other authors.

Shchukarev and Semenov observed an important rule, which has been confirmed by other investigators. The volatility of rare earth oxides in high vacuum, and also the composition of their vapors vary considerably for oxides of different elements of this group. The general tendency in the lanthanides is for a fall in the stability of gaseous monoxides with a change from lanthanum to lutecium compounds and in mass spectrometric measurements this appears as an increase in the ratio $Ln^+ : LnO^+$. This rule clearly shows periodicity: the ratio $Ln^+ : LnO^+$ is greatest for oxides of those elements which show a valence of +2 (for example, europium and ytterbium) and have the lowest enthalpy of sublimation of the metal; these oxides also have the highest volatility.

Oxides of elements showing a valence of +4 give molecules of gaseous dioxides. Thus, CeO_2, PrO_2, and TbO_2 were observed in the vapors. Terbium oxide Tb_2O_3 was evaporated from a tungsten ribbon, as with the use

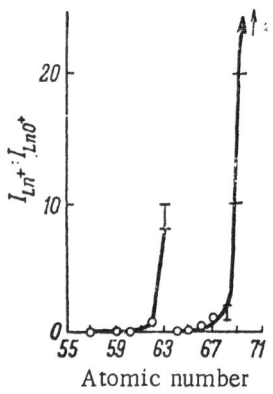

Fig. 98. Relation of the ratio of the ionization currents $I_{Ln^+} : I_{LnO^+}$ to the atomic number of the rare earth element in the evaporation of Ln_2O_3 (according to Panish).

of an iridium ribbon, with which the experiments were carried out, the ion currents of Ir^+ and TbO_2^+ could not be resolved by the mass spectrometer used. Shchukarev and Semenov observed no more complex molecules such as Ln_2O_3 or polymeric $(LnO)_n$.

Shchukarev and Semenov consider that the composition of the gas phase on evaporation of rare earth oxides is determined by the competition of the two reactions

$$Ln_2O_3 \text{ (solid)} = 2LnO \text{ (gas)} + O \text{ (gas)},$$

$$Ln_2O_3 \text{ (solid)} = 2Ln \text{ (gas)} + 3O \text{ (gas)}.$$

A decrease in the dissociation energy of LnO in combination with an increase in the heats of formation of Ln_2O_3 theoretically should lead to: 1) a decrease in the volatility of the oxides, and 2) an increase in dissociation to the elements. This rule holds in general, but there are exceptions.

The heats of formation of rare earth oxides show a tendency to increase with an increase in the molecular weight, whereupon the heats of sublimation of the elements fall. For the heats of formation the increase is generally monotonic, while the fall in the heats of sublimation shows sharp periodicity with elements which have a possible valence of +2 showing the lowest heat of sublimation.

In two brief communications, Panish (1961) described the results of investigation of the evaporation of the oxides of 12 rare earth elements: Pr_2O_3, Nd_2O_3, Sm_2O_3, Eu_2O_3, Gd_2O_3, Tb_2O_3, Dy_2O_3, Ho_2O_3, Er_2O_3, Tu_2O_3, Yb_2O_3, and Lu_2O_3 in the temperature region of 2000-2500°K. Panish used a time-of-flight mass spectrometer and an iridium Knudsen cell. LnO molecules and Ln and oxygen atoms were detected in the vapors. PrO_2 molecules were also found above praseodymium oxide. For the oxides of metals of lower atomic weight, he obtained the equilibrium partial pressures and the ratios of the ionization currents $I_{Ln^+} : I_{LnO^+}$ (Table 47).

The evaporation of the oxides listed in Table 47 occurred under neutral conditions in accordance with the stoichiometry of the reactions

$$Ln_2O_3 \text{ (solid)} = 2LnO \text{ (gas)} + O \text{ (gas)}, \qquad (1)$$

$$LnO \text{ (gas)} = Ln \text{ (gas)} + O \text{ (gas)}. \qquad (2)$$

The residues after the experiments, which were examined by x-ray diffraction, corresponded solely to the formula Ln_2O_3.

In the yttrium group, the following values were obtained for the equilibrium partial pressures of the molecules LnO at 2300°K: LnO: TbO, DyO, ErO, LuO $\approx 10^{-7}$ atm, TuO, YbO $\approx 10^{-6}$ atm.

When thorium oxide was used for the Knudsen cell, no reaction was observed between solid Ln_2O_3 and ThO_2. The vapor pressure of ThO(gas) and ThO_2(gas) was several orders below the vapor pressure of the least volatile of the rare earth oxides studied.

Table 47 shows that with an increase in the atomic number of the element, an increasingly great part is played by the second reaction, i.e., evaporation with decomposition to the component elements. This rule becomes even more noticeable if we examine the ratio $I_{Ln^+} : I_{LnO^+}$ for all the oxides studied by Panish. Figure 98 shows the relation of the ratio $I_{Ln^+} : I_{LnO^+}$ to the atomic number. As the figure shows, the oxides Ln_2O_3 are divided into two groups, in each of which the evaporation changes from reaction (1) to reaction (2) as the atomic number increases. A sharp change in the character of the evaporation occurs after Eu and Yb, elements for which the $4f$-electron shell is half or completely filled. It is interesting that if we exclude Lu_2O_3, from the character of their evaporation the rare earth elements are divided into the cerium and yttrium groups, just as in accordance with their behavior in various separation schemes.

BIBLIOGRAPHY

Ackermann, R. J., and R. J. Thorn. J. Am. Chem. Soc. 78: 4169 (1956).

Ackermann, R. J., and R. J. Thorn. XVI Congrès Internat. de Chim. Pure et Appl., 1957, Mémoires presentés à la Section de Chimie Minèrale, Paris (1958), p. 675.

Ackermann, R. J., and R. J. Thorn. "Vaporization of oxides," Progr. Ceram. Sci., Vol. 1 (1961).

Akishin, P. A., L. N. Gorokhov, and Yu. S. Khodeev. Zh. Strukt. Khim. 2: 209 (1961).

Baer, P. G., R. Geene, H. Smith, and J. Wortman. In: Kinetics, Equilibria and Performance of High Temperature Systems, G. S. Bahn and E. E. Zukoski (eds.) (1960), p. 90.

Baur, E., and R. Brunner. Z. Elektrochem. 40: 155 (1934).

Baur, E., and R. Brunner. Helv. Chem. Acta 17: 958 (1934).

Beletskii, M. C., and M. B. Rapoport. Dokl. Akad. Nauk SSSR 80: 751 (1951).

Berl, W. G., and W. Renish. In: Thermodynamic and Transport Properties of Gases, Liquids, and Solids, U. S. Touloukian (ed.), McGraw-Hill Book Co., Inc., New York (1959), p. 247.

Brewer, L. Chem. Rev. 52: 1 (1953).

Brewer, L., and A. W. Searcy. J. Am. Chem. Soc. 73: 5309-5314 (1951).

Brukl, A., and Y. Ortner. Z. Anorg. Chem. 203: 23 (1932).

Chupka, W. A., M. G. Inghram, and R. F. Porter. J. Chem. Phys. 24: 792 (1956).

Cochran, C. N. J. Am. Chem. Soc. 77: 2190 (1955).

Coughlin, J. P. U. S. Bureau of Mines, Bulletin No. 542 (1954).

De Maria, G., J. Drowart, and M. Inghram. J. Chem. Phys. 30: 318 (1959).

Drowart, J., G. De Maria, R. Burns, and M. Inghram. J. Chem. Phys. 32(5): 1366-1372 (1960).

Evans, W. H., E. J. Prosen, and D. D. Wagman. In: Thermodynamic and Transport Properties of Gases, Liquids, and Solids, U. S. Touloukian (ed.), McGraw-Hill Book Co., Inc., New York (1959), pp. 226-235.

Gaydon, A. G. Dissociation Energies and Spectra of Diatomic Molecules [Russian translation], IL, Moscow (1949).

Ginnings, D. C., and G. T. Furukawa. J. Am. Chem. Soc. 75: 522 (1953).

Goldstein, H. W., P. N. Walsh, and D. White. J. Phys. Chem. 64: 1087 (1960).

Grube, G., A. Schneider, U. Esch, and M. Flad. Z. Anorg. Chem. 260: 120-126 (1949).

Gurvich, L. V., and I. V. Veits. Izv. Akad. Nauk SSSR, Ser. Fiz. 22(6): 673-676 (1958).

Hafner, H. C., N. J. Kreidl, and R. A. Weidel. J. Am. Ceram. Soc. 41(8): 315-323 (1958).

Herzberg, G. Spectra and Structure of Diatomic Molecules [Russian translation], IL, Moscow (1949).

Herzberg, G. Vibration and Rotation Spectra of Polyatomic Molecules [Russian translation], IL, Moscow (1949).

Hoch, M., and H. L. Johnston. J. Am. Chem. Soc. 76: 2560-2561 (1954).

Howell, H. C. Proc. Phys. Soc. 57: 32 (1945).

Huber, E. J., E. L. Head, and C. E. Holley. J. Phys. Chem. 64: 1768 (1960).

Huber, E. J., C. E. Holley, and E. L. Head. J. Phys. Chem. 61: 497 (1957).

Huff, V. N., S. Gordon, and V. E. Morrell. General Method and Thermodynamic Tables for Computation of Equilibrium Composition and Temperature of Chemical Reactions (1950).

Inghram, M., R. F. Porter, and W. A. Chupka. J. Chem. Phys. 25: 498 (1956).

Inghram, M., and J. Drowart. In: Investigations at High Temperatures [Russian translation], IL, Moscow (1962). pp. 274-306.

Kelley, K. K. Bull. Am. Mines (1949), p. 476.

Lagerqvist, A. Mém. Soc. Sci. Liège 18: 550 (1957).

Lagerqvist, A., N. E. Nillson, and R. F. Barrow. Proc. Phys. Soc. A69: 356 (1956).

Mal'tsev, A. A., V. K. Matveev, and V. M. Tatevskii. Dokl. Akad. Nauk SSSR 137(1) (1961).

Mal'tsev, A. A., V. K. Matveev, and V. M. Tatevskii. Vestn. Mosk. Univ., No. 1 (1961).

Mal'tsev, A. A., and V. M. Tatevskii. Otchet Khim. Fak. Mosk. Univ., No. 210 (1958).

Margrave, J. L., J. R. Soulen, G. E. Leroi, F. T. Greene, and S. P. Randell. XVI Congrès Internat. de Chim. Pure et Appl., 1957, Mémoirs presentés à la Section de Chimie Minérale, Paris (1958).

Medvedev, V. A. Zh. Fiz. Khim. 32: 1690 (1958).

Moore, C. E. Atomic Energy Levels, Natl. Bur. Std. (U.S.), Circ. 467 (1958).

Mott, W. R. Trans. Am. Electrochem. Soc. 34: 255 (1918).

Nesmeyanov, A. N., and L. P. Firsova. Izv. Akad. Nauk SSSR, Otd. Tekhn. Nauk, Met. i Toplivo 3: 150-151 (1959).

Panish, M. B. J. Chem. Phys. 34(3): 1079-1080 (1961); 34(6): 2197-2198 (1961).

Porter, R. F., P. Schissel, and M. G. Inghram. J. Chem. Phys. 23(2): 339-343 (1955).

Rhodes, W. H. J. Am. Ceram. Soc. 44(6): 300 (1961).

Rosen, B. Physica 12: 184 (1946).

Rosen, B. Données Spectroscopiques Concernant les Molecules Diatomiques (1951).

Ruff, O., and M. Konschak. Z. Elektrochem. 32: 515 (1926).

Rusin, A. D., and V. M. Tatevskii. Dokl. Akad. Nauk SSSR 139(3): 630-633 (1961).

Scheer, M. D. J. Phys. Chem. 61: 1184 (1957).

Searcy, A. W., and C. E. Meyers. J. Phys. Chem. 61: 957 (1957).

Sears, G. W., and L. Navias. J. Chem. Phys. 30(4): 1111-1112 (1959).

Shchukarev, S. A., and G. A. Semenov. Dokl. Akad. Nauk SSSR 141: 652 (1961).

Shchukarev, S. A., G. A. Semenov, and I. A. Rat'kovskii. Zh. Neorgan. Khim. 6(8): 1973 (1961).

Shchukarev, S. A., G. A. Semenov, I. A. Rat'kovskii, and V. A. Perevozchikov. Zh. Obshch. Khim. 31: 2090-2092 (1961).

Shchukarev, S. A., G. A. Semenov, and I. A. Rat'kovskii. Zh. Prikl. Khim. 35(7): 1454 (1962).

Soulen, J. R., and J. L. Margrave. J. Am. Chem. Soc. 78: 2911 (1956).

Soulen, J. R., P. Sthapitanonda, and J. L. Margrave. J. Phys. Chem. 59: 132 (1955).

Speiser, R., S. Naiditch, and H. L. Johnston. J. Am. Chem. Soc. 72: 2578 (1950).

Veits, I. V., and L. V. Gurvich. Dokl. Akad. Nauk SSSR 108(4): 659-661 (1956).

Walker, R. F., J. Efimenko, and N. L. Lofgren. Planetary Space Sci. 3: 24-30 (February 1961).

Walsh, P. N., D. T. Dever, H. W. Goldstein, and D. White. J. Phys. Chem. 65(8): 1400-1413 (1961).

Walsh, P. N., H. W. Goldstein, and D. White. Air Force Office of Sci. Res. Rept. (1958), pp. 58-382.

Walsh, P. N., H. W. Goldstein, and D. White. J. Am. Ceram. Soc. 43(5): 229-233 (1960).

Wartenberg, H, and W. Gurr. Z. Anorg. Allgem. Chem. 196: 381 (1930).

White, W. S. Astrophys. J. 121: 271 (1955).

Wiley, W. C. Science 124: 217 (1956).

Wiley, W. C., and J. H. McLaren. Rev. Sci. Instr. 26: 1150 (1955).

Zintl, E., W. Krings, and W. Branning. German Patent, 742,330, October 14, 1943.

Zintl, E., W. Morawietz, and E. Gastinger. Z. Anorg. Chem. 245: 8 (1940).

CHAPTER X

EVAPORATION OF HIGH-TEMPERATURE OXIDES
OF GROUP IV – VI ELEMENTS

1. Evaporation of Germanium, Titanium, Zirconium, Hafnium, and Thorium Oxides

Germanium Oxides. According to the data of Shchukarev and Semenov (1958), GeO_2 evaporating from the surface of a platinum strip (at 1000°C) gives in the vapors the germanium oxides Ge_2O_2, Ge_3O_3, GeO, and Ge_2O and small amounts of Ge_3O_2 and GeO_2. Moreover, atomic germanium and oxygen (molecular and atomic) were present in the vapor. The enthalpies obtained by these authors are given below.

	Ge^+	GeO^+	$Ge_2O_2^+$	$Ge_3O_3^+$	O_2^+	O^+
ΔH_T, kcal/mole	106 ± 5	101 ± 2	108 ± 4	113 ± 5	108	110

Semenov also determined the vapor pressure of GeO at 1065°C and obtained the value $2 \cdot 10^{-5}$ mm Hg.

The evaporation of germanium dioxide shows some similarity to the evaporation of silica, but the composition of the gas phase over GeO_2 is more complex.

Titanium Oxides. The complexity of the evaporation of oxygen compounds of titanium is connected with the presence of numerous solid titanium oxides, which often remain homogeneous over wide ranges of composition. Reports on phases existing in the titanium—oxygen system have been published by Kubaschewskii and Dench (1953), Devries and Roy (1954), Ariya et al. (1957, 1958), and others. Only TiO, Ti_2O_3, Ti_3O_5, and TiO_2 are apparently the phases that have been determined thermodynamically.

The evaporation of the phase TiO was studied by Gilles (1961). On the basis of his effusion measurements, this author gives for the dissociation energy of TiO the value 6.9 ± 0.1 eV, which confirms the spectroscopic data of Herzberg (but not Gaydon). The evaporation of TiO is incongruent and during evaporation the condensed phase is enriched in oxygen.

Groves, Hoch, and Johnston (1955) measured the rate of effusion of the vapor formed over solid TiO, Ti_2O_3, Ti_3O_5, and TiO_2. Conclusions on the composition of the vapor were based on analysis of the condensate formed and the residue after evaporation. In the evaporation of Ti_2O_3 in the region of 1971-2151°K, it was observed that the ratio $TiO:TiO_2$ in the vapor was close to unity, and hence these authors concluded that Ti_2O_3 evaporates congruently.

Experiments with Ti_3O_5 led to contradictory results: chemical analysis of the remaining condensed phase indicated congruent evaporation, while x-ray diffraction patterns of the condensate collected contained only lines characteristic of TiO_2. The evaporation of TiO_2 in a molybdenum effusion cell is accompanied by reduction of the titanium dioxide by the metallic molybdenum and, therefore, the data given by Groves in this section are incorrect.

Berkowitz, Chupka, and Inghram (1957b) identified by mass spectrometry the gaseous molecules in equilibrium with TiO, Ti_2O_3, and TiO_2, using effusion cells of molybdenum and thorium dioxide. These authors confirmed the observations of Gilles on the incongruent evaporation of TiO: in the gas phase there was a considerable amount of gaseous titanium together with TiO. The dissociation energy of TiO was found to be 6.8 eV. According to the data of Berkowitz, the evaporation of Ti_2O_3 is also incongruent. The vapor over this substance consists almost exclusively of TiO, while the solid phase is consequently enriched in oxygen. According to Gilles, the composition of this solid phase corresponds to Ti_2O_5.

In the evaporation of TiO_2 from a molybdenum effusion cell, Berkowitz and his co-workers observed a large amount of gaseous molybdenum trioxide in the vapors. TiO_2 evaporates incongruently. As a result of the preferential loss of oxygen, in the cell there remains a dark blue solid solution of TiO_{2-x}, which is always obtained on heating rutile in vacuum or in a reducing atmosphere. For the dissociation energy of TiO_2, Berkowitz and his co-workers give the value 13.5 eV.

As a result of all the above investigations, it was shown that the vapors over any titanium oxides contain only TiO and TiO_2 molecules and Ti atoms. To determine the free energy of formation of any gaseous TiO, it is possible to use spectroscopic data for the reduced thermodynamic potential of TiO (Phillips, 1952) and the heat of sublimation of titanium according to Edwards, Johnston, and Ditmars (1953). Taking 6.9 eV for the dissociation energy of TiO, we obtain the temperature dependence (over the range of 1800-2300°K) of the free energy of formation of TiO(gas)

$$\Delta F^0_{\text{form}} = 4920 - 16.64\,T \quad \text{cal/mole},$$

The mass spectrometric investigations of Berkowitz, Chupka, and Inghram give for the heat of evaporation for

$$TiO \text{ (solid)} = TiO \text{ (gas)}$$

the value $\Delta H_{298} = 139.4$ kcal/mole (according to the second law of thermodynamics) or 148.6 kcal/mole (according to the third law of thermodynamics). The thermal effect of the disproportionation

$$2TiO \text{ (gas)} = Ti \text{ (gas)} + TiO_2 \text{ (gas)}$$

at 298°K is 8.8 kcal.

The reduced thermodynamic potential of gaseous TiO_2 was calculated by Berkowitz et al., and also by Brewer and Rosenblatt (1961). It was necessary to make a series of assumptions. Because of the lack of reliable methods of calculating the electronic contribution, it was assumed to be negligible and ignored. It was assumed that the TiO_2 molecule is linear and that the interatomic distances are the same as in TiO. In the calculation of the vibration constants an analogy was drawn with NO_2, CS_2, and CO_2 molecules. Brewer and Rosenblatt give for the reduced thermodynamic potential $-(F^0 - H_{298})/T$ of gaseous TiO_2 the following values:

Temperature, °K	$-(F^0 - H_{298})/T$, cal/mole·deg
298	56.6
1000	63.0
1500	67.1
2000	70.4
2500	73.1
3000	75.4

The enthalpy difference of gaseous TiO_2, $H_{298} - H^0_0 = 2670$ cal/mole.

Taking into account thermodynamic and calculated data, Ackermann and Thorn (1961) obtained the following expression for the free energy of formation of gaseous TiO_2 in the range of 1800-2300°K:

$$\Delta F^0_{\text{form}} = -86,000 + 6.1\,T \quad \text{cal/mole}.$$

The heat of evaporation of titanium dioxide, i.e., ΔH of the process

$$TiO_2 \text{ (solid)} = TiO_2 \text{ (gas)},$$

for 298°K is 142.7 kcal/mole (according to the second law of thermodynamics) and 148.6 kcal/mole (according to the third law of thermodynamics), according to the data of Inghram and Drowart (1960) from mass spectrometric measurements.

Using the mass spectrometric results of Berkowitz et al., but taking the new data of Kelley (1960) for the enthalpy of rutile, Brewer and Rosenblatt obtained somewhat different values for the heat of evaporation of titanium dioxide, namely, 139.6 kcal/mole (according to the second law of thermodynamics) and 143.0 kcal per mole (according to the third law of thermodynamics). For the standard (at 298°K) energy of atomization of TiO_2, i.e., the reaction

$$TiO_2 \text{ (gas)} = Ti \text{ (gas)} + 2O \text{ (gas)},$$

Brewer and Rosenblatt give the value 315 kcal/mole.

Zirconium Dioxide. The system Zr−O in the condensed state was studied by Domagala and McPherson (1954). There are no definite data on the existence of a particular suboxide phase between metallic zirconium and ZrO_2, but the mutual solubility of the metal and the dioxide at high temperature is considerable. Gaseous zirconium monoxide was found spectroscopically and Herzberg gave the molecular constants of ZrO.

By means of mass spectrometric measurements, Starodubtsev and Timokhina (1949) discovered the gaseous molecule ZrO_2. Chupka, Berkowitz, and Inghram (1957) confirmed the existence of both gaseous zirconium oxides ZrO and ZrO_2.

Hoch, Nakata, and Johnston (1954) measured the rate of effusion of vapor on heating ZrO_2 and a Zr−ZrO_2 mixture in a tantalum cell. It was observed that metallic zirconium has little effect on the overall effusion rate and these authors came to the conclusion that the main component of the vapor is gaseous ZrO_2. However, Chupka and his co-workers showed that tantalum reduces ZrO_2 at temperatures above 2000°K and forms the gaseous products ZrO, TaO, and TaO_2. Thus, Hoch and his co-workers interpreted their results incorrectly.

The study of the proposed equilibrium,

$$ZrO_2 \text{ (solid)} + Zr \text{ (solid)} = 2ZrO \text{ (gas)}$$

made it possible for Chupka, Berkowitz, and Inghram to obtain the dissociation energy of gaseous ZrO, which equaled 7.8 ± 0.2 eV. This agrees well with spectroscopic data. However, this agreement must be regarded as being by chance. In view of the considerable mutual solubility of Zr and ZrO_2 at 1800-1900°C, the formula of zirconium dioxide should be written as ZrO_{2-x}.

The agreement between the spectroscopic and thermochemical data compels us to assume that the standard free energy of formation of the phase ZrO_{2-x} is more positive than $ZrO_{2.00}$ by a value almost equal to the integral free energy of solution of oxygen in the zirconium phase of equilibrium composition. These two values have opposite signs and, therefore, the use of the free energy of formation of stoichiometric ZrO_2 on the assumption that metallic zirconium has an activity close to unity does not introduce serious errors.

The dissociation energy of gaseous ZrO_2 was obtained as a result of studying the equilibrium

$$ZrO_2 \text{ (solid)} = ZrO_2 \text{ (gas)}.$$

The required heat of sublimation at 0°K was calculated by Chupka. Assuming, like Chupka, the molecular constants for gaseous ZrO_2, Ackermann and Thorn (1961) obtained for the heat of sublimation of ZrO_2 the value 175 kcal/mole instead of 186 kcal/mole given by Chupka. For the heat of sublimation of zirconium ΔH_0^0 (using the more accurate value of the reduced thermodynamic potential of solid Zr), Ackermann and Thorn give the value 145 kcal/mole and not 142.1 kcal/mole, which was proposed by Skinner and Johnston (1951). As a result of their calculations based on the work of Chupka et al., Ackermann and Thorn give for the dissociation energy of gaseous ZrO_2 the value 15.2 ± 0.2 eV (at 0°K).

In their experimental investigations of the effusion of vapor of zirconium oxides in tantalum and tungsten cells, Ackermann and Thorn established that the reaction of ZrO_2 with tungsten at 2600°K is insignificant. The results obtained with this cell made it possible for them to find the heat and entropy of sublimation of solid ZrO_2.

For the heat of sublimation at 2600°K they obtained the value 163 kcal/mole, and for the entropy of sublimation, 38.0 en. units. After appropriate conversions to 0°K, Ackermann and Thorn obtained for the dissociation energy of gaseous ZrO_2 the value 15.3 ± 0.2 eV, which is very close to the above value, which was obtained from the measurements of Chupka et al.

After some of it had evaporated, the white sample of zirconium dioxide in the effusion cell became dark gray and had the composition $ZrO_{1.97}$, though it consisted of a monoclinic phase of zirconium dioxide.

Using thermodynamic and spectroscopic data for the gaseous molecule of zirconium monoxide, Ackermann and Thorn calculated the free energy of formation of ZrO(gas). For the temperature-dependent of the free energy they gave the equation

$$f^0 = 3.885\,T - 20.592\,T \log T - 480 \text{ cal/mole.}$$

Taking for the dissociation energy of the ZrO molecule the value 7.8 eV, we obtain the following temperature dependence of the standard free energy of formation of gaseous ZrO in the temperature range of 2300-2700°K:

$$\Delta F^0_{form} = 17,800 - 16.78\,T \text{ cal/mole.}$$

The thermodynamic properties of the gaseous molecule ZrO_2 were calculated right up to 3000°K. The reduced thermodynamic potential $-(F^0 - H^0_{298})/T$ for this molecule has the following values:

Temperature, °K	$-(F^0 - H^0_{298})/T$, cal/mole·deg
298	58.6
1000	65.0
1500	69.2
2000	72.5
2500	75.3
3000	77.6

The atomization energy, i.e, ΔH_{298} for the reaction

$$ZrO \text{ (gas)} = Zr \text{ (gas)} + 2O \text{ (gas),}$$

is 347 kcal/mole.

The standard free energy of sublimation of solid zirconium dioxide ΔF_{sub} is $163,000 - 38.0\,T$ cal/mole according to Ackermann and Thorn. By summing this value with the standard free energy of formation of ZrO_2 (solid), we obtain the standard free energy of formation of gaseous ZrO_2

$$\Delta F_{form} = -95,000 + 4.9\,T \text{ cal/mole.}$$

Ackermann, Thorn, and Winslow (1961a) recently determined the vapor pressure above both zirconium dioxide of normal stoichiometry and also over the compound with a deficit of oxygen. The authors observed a fall in the volatility of zirconium dioxide as the ratio O:Zr changed from 2 to 1.96. This fall may be explained by assuming that in the zirconium dioxide lattice there arise vacancies, which were previously occupied by oxygen ions. Having determined the overall volatility of the compound $ZrO_{1.96}$, and knowing the thermodynamic properties of gaseous ZrO and ZrO_2, Ackermann and his co-workers were able to determine the "interstitial disorder" in stoichiometric zirconium dioxide. The degree of this "disorder" is characterized by the figure $2 \cdot 10^{-3}$ at 2470°C.

Hafnium Dioxide. Hafnium dioxide is the least volatile of all the oxides studied. According to the measurements of Ackermann and Thorn (1961), who used a tungsten effusion cell, the vapor pressure over hafnium dioxide at 2940°K is $3 \cdot 10^{-5}$ atm. The vapor pressures of zirconium and thorium dioxides at the same temperature are five times as great. The molecules HfO and HfO_2 are apparently present in the same amounts in the vapor over hafnium dioxide.

Spectral investigations of Krishnamurty (1951) gave for the dissociation energy of HfO the value 8.0 eV. If we take this dissociation energy and use the known thermodynamic characteristics of HfO, we obtain for the free energy of formation of gaseous HfO the equation

$$\Delta F_{form} = 29{,}600 - 14.5\,T \quad cal/mole.$$

For the free energy of formation of gaseous HfO_2, Ackermann and Thorn gave the expression

$$\Delta F^0_{form} = -81{,}000 + 4.0\,T \quad cal/mole.$$

Thorium Dioxide.   The valence four is most characteristic of crystalline compounds of thorium. On the basis of x-ray investigations, Hoch and Johnston (1954) and also Katzin (1958), also considered that solid thorium monoxide exists. Weinreich and Danforth (1952), who studied the optical properties of crystalline thorium dioxide, stated that they observed slight deviations from stoichiometry. In the opinion of these authors, the nonstoichiometry of thorium dioxide is too small to affect its thermodynamic characteristics. Kubaschewskii (1961) holds a different opinion.

In the vapor above thorium dioxide, the molecules ThO_2 and ThO were identified and are present in relatively large amounts. The spectrum of gaseous ThO was studied by Krishnamurty (1951). From his data it follows that the most probable ground state of the ThO molecule is $^3\Pi$ with values of the terms of 0.2721 and 4177 cm^{-1}, though the spectral data were found to be insufficient for determining the dissociation energy of ThO. Inghram, Chupka, and Berkowitz (1956) demonstrated the presence of the gaseous oxides ThO and ThO_2 in the vapors by mass spectrometry.

The vapor pressure over thorium dioxide was determined by Shapiro (1952), Hoch and Johnston (1954), and Ackermann, Gilles, and Thorn (1956b). Dissimilar results were obtained, and this caused disagreement between these authors. Shapiro measured the rate of evaporation of thorium dioxide from a heated tungsten wire coated with a layer of ThO_2. He considered that there was some reduction of ThO_2 by tungsten. Subsequently, Ackermann and Thorn showed by quantitative experiments that at 2765°K the weight of tungsten escaping from a tungsten diffusion cell as a result of reaction with ThO_2 is less than 2% of the weight of thorium dioxide evaporating.

Hoch and Johnston obtained a value for the vapor pressure over ThO_2, 40 times as great as that observed by Shapiro. The results of Hoch and Johnston must be regarded as incorrect, as they used for the containers metallic tantalum which is much more electronegative than tungsten. The reduction of thorium dioxide by tantalum yielded volatile tantalum oxides, which distorted the results.

By special experiments, Ackermann and Thorn showed that ThO_2 hardly reacts with tungsten; the volatility was the same for pure ThO_2 and for a mixture of ThO_2 with tungsten. In later work, Ackermann and Thorn measured the effusion rate of vapor obtained by heating thorium dioxide in a tungsten effusion cell in the temperature range of 1875-2130°K. For more accurate determination of the amount of thorium in the effusate, they used thorium dioxide containing the predominantly α-active Th^{230}. However, the weights of the samples were probably too low to ensure that a saturated vapor was produced inside the effusion cell. Ackermann and Thorn give the following equation for the vapor pressure over thorium dioxide in the range of 2500-2900°K:

$$\log P \;(atm) = 7.985 - 3.489 \cdot \frac{10^4}{T}.$$

Darnell and McCollum (1961) studied the evaporation of a mixture of liquid thorium and ThO_2 in the range of 1984-2564°K in a tungsten effusion cell lined on the inside with thorium dioxide. ThO molecules formed the main component of the vapor. The partial pressure of ThO is given by the equation

$$\log P \;(atm) = -(22{,}200 \pm 700)\,T^{-1} + (4.70 \pm 0.31).$$

The evaporation of pure thorium dioxide was studied with a tungsten effusion cell (2268-2593°K). The partial vapor pressure of $ThO_2(gas)$ was represented by the equation

$$\log P \;(atm) = -(35{,}500 \pm 1100)\,T^{-1} + (8.16 \pm 0.47).$$

The partial pressure of metallic thorium was determined by extrapolation:

$$\log P \ (\text{atm}) = -27,960 \ T^{-1} + 5.575.$$

Assuming that the difference in heat content of gaseous ThO_2, $H_{2150}^0 - H_{298}^0$ is the same as for gaseous titanium, zirconium, and hafnium dioxides, i.e., 26.4 kcal/mole, Brewer and Rosenblatt calculated the enthalpy of evaporation of thorium dioxide and obtained the value ΔH_{298} = 179.7 kcal/mole. If we adopt this enthalpy of evaporation and use Shapiro's data on the vapor pressure of ThO_2, for the reduced thermodynamic potential $(F^0 - H_{298}^0)/T$ of gaseous thorium dioxide at 2150°K we obtain the value −79.1 cal/mole · deg. If we take the same value of ΔH_{298}(evap.), but use data on the absolute vapor pressure of thorium dioxide obtained by Ackermann, Thorn, and Gilles, then for the reduced thermodynamic potential of ThO_2(gas) we obtain the values −84.8 (2800°K) and −81.4 cal/mole · deg (2150°K).

Taking as the mean value of the reduced thermodynamic potential of ThO_2(gas) at 2150°K the value −79.5 cal/mole · deg, Brewer and Rosenblatt obtained the following values for $-(F^0 - H_{298}^0)/T$ of gaseous thorium dioxide:

Temperature, °K	$-(F^0 - H_{298}^0)/T$, cal/mole·deg
298	63.4
1000	70.4
1500	74.9
2000	78.5
2500	81.6
3000	84.2

To find the atomization energy, i.e., the thermal effect of the reaction

$$ThO_2 \ (\text{gas}) = Th \ (\text{gas}) + 2O \ (\text{gas}),$$

it is possible to use the data of Huber, Holley, and Meierkord (1952) for the heat of formation of ThO_2(solid) at 298.15°K (−293.2 kcal/mole) and Darnell, McCollum, and Milne (1960) for the heat of sublimation of Th(solid) (137.3 kcal/mole). Hence, we obtain the heat of atomization of ThO_2(gas), ΔH_{298} = 370 kcal/mole.

In view of the high melting point of thorium dioxide, it is important to know the reduced thermodynamic potential of ThO_2 up to higher temperatures than given in the review of Margrave (1959). Taking data for the heat content and entropy of ThO_2(solid) from Kelley's summary, and assuming that above 2000°K the heat capacity of ThO_2(solid) remains constant and equals 22 cal/mole · deg, we obtain for the reduced thermodynamic potential $-(F^0 - H_{298}^0)/T$ of solid thorium dioxide 34.9 (2150°K) and 39.3 cal/mole · deg (2800°K). It was assumed that for ThO_2(solid) S_{298} = 15.6 cal/mole · deg (according to Osborne and Westrum, 1953).

In connection with the study of the vapor pressure of thorium, Darnell, McCollum, and Milne showed that when thorium is heated with ThO_2 there occurs the reaction

$$Th \ (\text{solid}) + ThO_2(\text{solid}) = 2ThO \ (\text{gas}). \qquad (1)$$

Thorium monoxide may also be formed by the action of oxygen on thorium in accordance with the reaction

$$Th \ (\text{solid}) + {}^1/_2 O_2 \ (\text{gas}) = ThO \ (\text{gas}).$$

Assuming that the activities of Th(solid) and ThO_2(solid) equal unity, the authors calculated the free energy of reaction (1), and this made it possible (knowing the free energy of formation of solid ThO_2 from Coughlin's review) to find the free energy of formation of ThO(gas) at 1883°K: ΔF_{form}^0 = −41.5 kcal/mole.

Darnell and his co-workers calculated the entropy of ThO(gas) by statistical methods. At 1883°K, S^0 = 79.3 en. units. This value consists of the components S_{trans}^0 = 51.56, S_{rot}^0 = 18.66, S_{el}^0 = 5.91, and S_{vib}^0 = 3.15 en. units. For the heat of formation and dissociation energy of ThO, Darnell and his co-workers give the following data: ΔH_{1883}^0 = −8.0 kcal/mole, D_0^0 = 196.1 kcal/mole.

2. Evaporation of Vanadium, Niobium, and Tantalum Oxides

Vanadium Oxides. In the study of gaseous molecules in the vapor above vanadium oxides, a series of complex compounds right up to V_6O_{14}(gas) was found, and this is evidently connected with the complexity and abundance of phases in the system vanadium—oxygen.

It is recognized that the following vanadium oxides exist: V_2O_5, V_6O_{13}, VO_2, V_2O_3, and VO (Allen, Kubaschewskii, Goldbeek, 1951); Rostoker (1958) gives still more oxides. In regions adjacent to the compounds V_2O_5, VO_2, V_2O_3, and VO, there exist homogeneous phases of variable composition, which undoubtedly affect the evaporation process.

The first investigation of the evaporation of vanadium oxides was carried out by Polyakov (1946). He used the flow method and observed some anomalies in the evaporation of vanadium pentoxide. According to his measurements, the entropy of evaporation of V_2O_5 is unusually low, namely 13 en. units.

Using a mass spectrometer, Berkowitz, Chupka, and Inghram (1957b) established that in the vapor over vanadium oxides there are the monoxide, dioxide, dimers, and more complex polymers — derivatives of the pentoxide. These polymers are formed not simply by condensation of V_2O_5, but with the participation of oxygen, for example,

$$2V_2O_5 = V_4O_8 + O_2.$$

In addition to V_4O_8 were observed V_6O_{14}, V_6O_{12}, and V_2O_4. It is interesting that no monomeric pentoxide was observed. The molecule V_4O_{10} probably has a structure similar to that of the molecule P_4O_{10}, which was studied by Hampson and Stosick (1938). The molecule P_4O_{10} has a skeleton formed from four phosphorus atoms and five oxygen atoms, while the other oxygen atoms are around the skeleton. It may be assumed that the removal of oxygen atoms outside the skeleton with the formation of the dimer of vanadium pentoxide V_4O_{10} is more favored energetically than the formation of the gaseous monomer of the pentoxide. However, this explanation is contrary to the presence of V_2O_4 in the vapor.

Under oxidizing conditions, vanadium pentoxide evaporates to form polymers, predominantly dimers, in the vapor. At the beginning of the evaporation, V_4O_{10} and V_4O_8 appear in the vapor, but as the solid phase changes its composition from V_2O_5 to approximately $VO_{2.47}$, there is a tendency for the evaporation of such molecules as V_6O_{14}, V_6O_{12}, V_4O_8, and V_2O_4 and the relative amount of these increases during evaporation.

Solid vanadium monoxide evaporates to form VO molecules and a small amount of gaseous dioxide and metal atoms. The value 6.4 eV was obtained for the dissociation energy of VO, and this is in good agreement with spectroscopic measurements if we assume that the ground state of VO is $^2\Delta$. The heat of evaporation of VO(solid) was obtained from mass spectrometric observations. Depending on the calculation method, ΔH of the reaction

$$VO \text{ (solid)} = VO \text{ (gas)}$$

had the value at 0°K of 135.3 kcal/mole (according to the second law of thermodynamics) or 129.5 kcal/mole (according to the third law of thermodynamics).

Berkowitz, Chupka, and Inghram studied the equilibrium

$$V \text{ (gas)} + VO_2 \text{ (gas)} \rightleftharpoons 2VO \text{ (gas)}.$$

For the standard free energy of this reaction at 1945°K they obtained the value −19.2 kcal/mole. The heat effect of this reaction, calculated from the third law of thermodynamics, equaled 800 cal/mole for 298°K.

Below we give the values of the reduced thermodynamic potential $-(F^0 - H_{298})/T$ for VO(gas) and VO (solid) calculated by Brewer and Rosenblatt.

T, °K	$\Phi(VO_{gas})$	$\Phi(VO_{solid})$
1500	65.0	−19.6 cal/deg · mole
2000	67.3	−23.0 cal/deg · mole

Taking the given value (6.4 eV) for the dissociation energy of VO and using the data of Edwards et al. (Edwards, Johnston, Blackburn, 1951) for the vapor pressure of vanadium, we obtain for the free energy of formation of gaseous VO in the range of 1600-2000°K:

$$\Delta F^0_{\text{form}} = 28{,}400 - 18.56\,T \quad \text{cal/mole.}$$

In calculations of the free energy, the molecular constants for VO were taken from the work of Lagerqvist and Selin (1957).

Brewer and Rosenblatt give the following values of the reduced thermodynamic potential $-(F^0 - H_{298})/T$ for gaseous VO_2:

Temperature, °K	$-(F^0 - H_{298})/T$, cal/mole·deg
298	60.0
1000	66.8
1500	71.1
2000	74.6
2500	77.4
3000	79.7

The energy of the atomization of the VO_2 molecule, i.e.,

$$VO_2\ (\text{gas}) = V\ (\text{gas}) + 2O\ (\text{gas}),$$

equals 298 kcal/mole.

Niobium Oxide. The first investigation of the evaporation of niobium oxides was made by Kolchin, Sumarokova, and Chuveleva (1957). These authors state that the lowest oxides of niobium have an appreciable volatility at 1700°C and very high volatility at 1850°C, but there is no quantitative information on vapor pressure in their article.

Golubtsov, Lapitskii, and Shiryaev (1960) measured the saturated vapor pressure of niobium pentoxide and dioxide in the temperature range of 1489-1905°K. In this temperature region niobium oxides have a low vapor pressure and to increase the accuracy of the measurements the authors used a radioactive isotope of niobium Nb^{95}. The vapor pressure was determined with a Knudsen effusion cell and also by a flow method in a stream of air. The radioactive niobium was collected on a collector. A Knudsen cell of tungsten or sintered oxides was used and placed in a vacuum chamber, which gave a pressure in the system of the order of $1 \cdot 10^{-5}$-$5 \cdot 10^{-6}$ mm Hg when the samples were heated (high-frequency heating) to 1500-1600°C.

Special experiments established that when niobium dioxide was heated in vacuum at 1300°C in contact with molybdenum, no changes occurred (the x-ray diffraction pattern showed only lines characteristic of NbO_2). The vapor pressure of niobium dioxide at 1900°K reaches values slightly greater than 10^{-3} mm Hg.

Niobium pentoxide does not evaporate at all in a stream of air at 1350°C. Experiments carried out by the flow method showed that even with a high specific activity of the starting preparation (15 mCi/g), no traces of radioactive deposit were detected on the collector. Niobium pentoxide decomposes in vacuum. X-ray investigation of the product remaining in the effusion chamber after niobium pentoxide had been heated in vacuum showed that there was a considerable amount of niobium dioxide in it. It may be assumed that in vacuum at high temperature (above 1150°C) there is thermal dissociation:

$$Nb_2O_5 = 2NbO_2 + \tfrac{1}{2}O_2.$$

In cases where niobium pentoxide was placed in the Knudsen cell, in actual fact the vapor pressure of niobium dioxide was measured for various degrees of saturation of the vapor in the chamber. At high temperatures, the vapor pressure over Nb_2O_5 was practically the same as over niobium dioxide.

Table 48. Composition of Vapor Above Vanadium and Niobium Oxides

Compound	Type of ion	Ratio of intensities of ion currents	$\Delta H^0_{subl.}$, kcal/mole
VO_2	VO_2^+	1.0	104 ± 2
	VO^+	0.38	105 ± 3
	V^+	0.15	—
V_2O_3	VO^+	1.0	—
	V^+	0.61	—
	VO_2^+	0.03	—
NbO_2	NbO_2^+	1.0	142 ± 3
	NbO^+	0.21	140 ± 5
	Nb^+	0.1	—

Shchukarev, Semenov, and Frantseva (1959) studied the composition of the vapor formed over VO_2, V_2O_3, and NbO_2, deposited on a platinum strip heated to 1500-1880°K. The authors used mass spectrometry with an ionization potential of 50 V. The method was described in the article of Shchukarev and Semenov (1958). The predominant components of the vapor over the three given oxides were VO_2^+, VO^+, and NbO_2^+ (Table 48). The sublimation energies were found from the slopes of lines in the coordinates $\log(I^+T) = f(1/T)$.

The vapor pressure of niobium dioxide was recently determined by Shchukarev, Semenov, and Frantseva (1962a) by the effusion method in the differential variant. A cylindrical effusion chamber of forged molybdenum was heated by electron bombardment. A vacuum of the order of $1 \cdot 10^{-5}$ mm Hg was maintained at temperatures up to 2100°K. The sublimate was condensed on targets cooled with liquid nitrogen.

A preparation corresponding to the composition $NbO_{2.008}$ was obtained by reduction of niobium pentoxide labeled with the isotope Nb^{95} with hydrogen at 1000°C. The vapor pressure of niobium dioxide in the temperature region of 1940-2120°K is described well by the equation

$$\log P \ (\text{mm Hg}) = -30,300 \, T^{-1} + 12.42.$$

The results obtained by Shchukarev, Semenov, and Frantsova differ considerably from those given by Golubtsov and his co-workers (see p. 184). The heat of sublimation of NbO_2 calculated from the data of the latter authors (37.5 kcal/mole) cannot be considered as accurate. Shchukarev and his co-workers give for the heat of sublimation of niobium dioxide the value 138 kcal/mole (in the temperature range studied) and 141 kcal/mole for 0°K. The dissociation energy of the gaseous molecule was found to equal 14.9 eV.

Tantalum Oxides.   Tantalum is often used as the material for containers in the study of refractory systems. At 2000°C tantalum becomes quite a strong reducing agent and the formation of volatile tantalum oxides may distort strongly the results of measuring the vapor pressures of such oxides as Al_2O_3, ZrO_2, ThO_2, etc. The quantitative prediction of the reaction between tantalum and a given oxide requires a detailed study of the evaporation of oxygen compounds of tantalum.

The system Ta—O has not been studied in adequate detail. Schönberg (1954) and Wasilewskii (1953) consider that six suboxide phases exist, but none of them has been isolated in a pure form. The solution of oxygen in tantalum has been demonstrated as this process is accompanied by an increase in the size of the elementary cell of tantalum.

Only two gaseous oxides of tantalum, namely TaO and TaO_2, have been observed in the vapor. The valence state of +5, which is very stable for condensed phases, is not observed in the gas phase. The absence of complex molecules from the saturated vapors over tantalum oxides indicates a marked difference in the chemical bonds in gaseous oxides of tantalum and vanadium, for which the formation of polymeric oxide molecules is characteristic. A higher temperature is evidently required to obtain comparable saturated vapor pressures. Moreover, it is necessary to take into account the difference in free energy of the condensed phases.

According to Ackermann and Thorn (1956), gaseous TaO molecules may be obtained as a product of the reaction of Al_2O_3 and tantalum. By direct observations with a mass spectrometer, Inghram, Chupka, and Berkowitz (1957) detected TaO and TaO_2 molecules in the vapor.

The high reactivity of tantalum at high temperatures has made it possible to study a series of equilibria of the type

$$\text{Oxide(solid)} + \text{Ta(solid)} = \text{Oxide(gas)} + \text{Tantalum oxide(gas).}$$

According to the observations of Wasilewski, tantalum pentoxide reacts very vigorously with tantalum above 1700°C. Even such a stable metal as tungsten reacts with tantalum pentoxide to form volatile tungsten compounds.

To determine the dissociation energy of gaseous TaO and TaO_2, Inghram and his co-workers studied the equilibria

$$\tfrac{1}{5}Ta \text{ (solid)} + \tfrac{2}{5}Ta_2O_5 \text{ (liq)} = TaO_2 \text{ (gas),} \tag{1}$$

$$\tfrac{3}{5}Ta \text{ (solid)} + \tfrac{1}{5}Ta_2O_5 \text{ (liq)} = TaO \text{ (gas).} \tag{2}$$

Chupka, Berkowitz, and Inghram (1957) studied the equilibrium

$$ZrO_2 \text{ (solid)} + Ta \text{ (solid)} = ZrO \text{ (gas)} + TaO \text{ (gas).}$$

Goldstein and his co-workers investigated the equilibrium

$$Nd_2O_3 \text{ (solid.)} + Ta \text{ (solid)} = 2NdO \text{ (gas)} + TaO \text{ (gas).}$$

The values 8.4, 9.34, and 9.7 eV were obtained for the dissociation energy of TaO(gas), depending on the equilibrium studied. The value 15.0 eV was obtained for the dissociation energy of TaO_2.

The uncertainty of the results from investigation of the thermodynamic properties of gaseous TaO and TaO_2 (the data of different authors differ by 10 kcal for the dissociation energy), make it impossible to give reliable equations for the standard free energies of formation of these substances.

For the heat of reaction (2), Inghram gives ΔH_{298} = 146 kcal/mole and $\Delta H_{2000-2300}$ = 131.6 kcal/mole. The heat effect of reaction (1) was calculated by Brewer and Rosenblatt from the data of Inghram, Chupka, and Berkowitz. For ΔH_{298} of this reaction they obtained the value 150.3 kcal/mole. The atomization energy of the molecule TaO_2 equals 351 kcal/mole according to Brewer and Rosenblatt. For the reduced thermodynamic potential of gaseous TaO_2 they give the following values:

Temperature, °K	$-(F^0 - H_{298})/T$, cal/mole·deg
298	63.4
1000	70.0
1500	74.3
2000	77.7
2500	80.6
3000	83.0

3. Evaporation of Chromium, Molybdenum, Tungsten, and Uranium Oxides

Chromium Oxide. It has been observed that crystals of Cr_2O_3 are deposited in the cold parts of the apparatus in the high-temperature oxidation of chromium alloys (see, for example Bandel, 1941; Warshaw, Keith, 1954; Wilms, Rea, 1959), indicating the transfer of chromium through the gas phase. Chromium has a high vapor pressure and it might have been supposed that it was the metal which evaporated from the surface of the object as chromium oxide is involatile up to high temperatures.

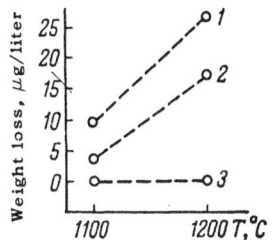

Fig. 99. Loss in weight of Cr$_2$O$_3$ in a stream of wet (1) and dry (2) oxygen and argon (3) at 1100°C and 1200°C.

However, Caplan and Cohen (1961) showed that chromium oxide volatilizes in some conditions even at 1000°C. These authors followed the loss in weight of pressed Cr$_2$O$_3$ tablets which were treated in various ways in addition to being heated. These treatments consisted of passing dry or wet oxygen and dry or wet argon through the furnace. It was found that the volatility of chromium oxide, as observed from the change in weight of the samples, depends substantially on the surrounding atmosphere.

Figure 99 shows the loss in weight of chromium oxide in dry and wet oxygen and wet (or dry) argon.* As Fig. 99 shows, volatilization did not occur in the absence of oxygen. This leads to the view that it is not Cr$_2$O$_3$ which evaporates, but some higher volatile oxide of chromium. Chromium trioxide CrO$_3$ is known to be a more volatile compound, but it normally exists only at low temperatures and at the temperatures and oxygen pressures used by Caplan and Cohen it will be thermodynamically unstable. However, the reaction leading to the formation of gaseous CrO$_3$,

$$Cr_2O_3 \text{ (solid)} + {}^3/_2 O_2 \text{ (gas)} = 2CrO_3 \text{ (gas)},$$

involves an increase in the number of gas molecules, and this means an increase in entropy. Moreover, in a dynamic system with a fast gas stream, considerable transfer of material may occur even with a very low equilibrium partial pressure.

To confirm the existence of CrO$_3$ molecules in the vapor, we can quote the work of Grimley, Burns, and Inghram (1961), who showed by direct observations with a mass spectrometer that Cr$_2$O$_3$ evaporates under oxidizing conditions with the formation of CrO$_3$.

The effect of moisture in the experiments of Caplan and Cohen which were described has not been elucidated. The evaporation of many involatile oxides is known to be caused by the formation of hydrate forms, which evaporate. However, the moisture present in the argon should facilitate the evaporation of Cr$_2$O$_3$, but this does not occur. It may be surmised that the moisture reacts with the CrO$_3$ formed, whose hydrate has a high volatility, but this hypothesis requires further proof. Finally, it may be surmised that the moisture acts catalytically, promoting the surface oxidation reaction or facilitating the removal of the oxidation products from the surface. The solution of these problems requires further investigations.

Molybdenum Oxides. The results of the investigation of evaporation of molybdenum oxides, like other transition elements of group VI, demonstrated the great complexity of the composition of the vapor, which reflects the complexity of the composition of the condensed oxide phases.

The well-characterized oxygen compounds of molybdenum in condensed phases are the dioxide MoO$_2$ and the trioxide MoO$_3$. Between these compounds on the Mo−O phase diagram there are several more compounds and homogeneous phases of variable composition. There are reports of the existence of such oxides as Mo$_5$O$_{12}$, Mo$_2$O$_5$, Mo$_3$O$_8$, and Mo$_4$O$_{11}$. Hägg and Magneli (Hägg, Magneli, 1945; Magneli, Anderson, Blomberg, Kihlborg, 1952) observed phases of variable composition close to the compositions Mo$_2$O$_{2.75}$ and MoO$_{2.85}$.

Working with a molybdenum Knudsen cell into which was placed molybdenum trioxide, Berkowitz, Chupka, and Inghram (1957c) found only the polymeric molecules (MoO$_3$)$_3$, (MoO$_3$)$_4$, and (MoO$_3$)$_5$ in the vapor with a mass spectrometer. The amounts of these molecules fell in the sequence given.

The evaporation of molybdenum dioxide was first studied by Blackburn, Hoch, and Johnston (1958) by Knudsen's effusion method. These authors came to the conclusion that the evaporation is accompanied by disproportionation of molybdenum dioxide to MoO$_3$(gas) and Mo(solid); MoO$_2$ simultaneously evaporates as such

$$MoO_2 \text{ (solid)} \rightarrow MoO_2 \text{ (gas)}.$$

---

* The loss in weight of chromium oxide was the same in wet and dry argon. — Ed.

Table 49. Enthalpies, Entropies, and Free Energies of Reactions Occurring in the Evaporation of MoO_2 for 1600°K

Reaction	ΔH_T^0, kcal/mole	ΔS_T^0, en. units	ΔF_T^0, kcal/mole
$\frac{3}{2}MoO_2(\text{solid}) = MoO_3(\text{gas}) + \frac{1}{2}Mo(\text{solid})$ ..	121.8	46.4	47.6
$3MoO_2(\text{solid}) = (MoO_3)_2(\text{gas}) + Mo(\text{solid})$ ...	133.4	50.6	52.4
$MoO_2(\text{solid}) = MoO_2(\text{gas})$	134.4	48.4	56.9
$\frac{9}{2}MoO_2(\text{solid}) = (MoO_3)_3(\text{gas}) + \frac{3}{2}Mo(\text{solid})$.	142.6	50.6	61.6
$2MoO_3(\text{gas}) = (MoO_3)_2(\text{gas})$	−110.2	−42.1	−42.8
$3MoO_3(\text{gas}) = (MoO_3)_3(\text{gas})$	−222.8	−88.6	−81.1

The same authors studied the evaporation of Mo_4O_{11} and a mixture of Mo with MoO_2. In the first case, the main process in the evaporation was

$$\frac{1}{3}x\,Mo_4O_{11} = (MoO_3)_x\ (\text{gas}) + \frac{1}{3}x\,MoO_2\ (\text{solid}),$$

i.e., different polymeric molecules were present in the vapors. In the evaporation of a mixture of Mo with MoO_2, molybdenum monoxide MoO was detected in the vapors together with other compounds.

Burns, De Maria, Drowart, and Grimley (1960) used the mass spectrometric method to investigate the sublimation of molybdenum dioxide. The outer casing of the Knudsen cell of these authors was molybdenum and the inside was lined with aluminum oxide. In the temperature region of 1500-1780°K they detected (in the sequence given), the gaseous oxides MoO_3, $(MoO_3)_2$, MoO_2, and $(MoO_3)_3$. In the ionized gas they observed weak "peaks" corresponding to MoO_5^+, $Mo_2O_4^+$, and $Mo_3O_8^+$. No gaseous compounds of aluminum and oxygen were found in the vapors. There was no reaction between $MoO_2(\text{solid})$ or $Mo(\text{solid})$ and aluminum oxide.

Burns and his co-workers confirmed the disproportionation of MoO_2 on evaporation observed by Blackburn. Although MoO_2 is present in the vapor, the amount of it is considerably less than MoO_3 and $(MoO_3)_2$. For example, the following partial pressures of the components of the vapor were obtained at 1716°K:

$$
\begin{aligned}
MoO_3 &. \ldots \ldots \ldots \ldots \ldots 4.1 \cdot 10^{-6}\ \text{atm} \\
(MoO_3)_2 &\ \ldots \ldots \ldots \ldots \ldots 1.3 \cdot 10^{-6}\ \text{atm} \\
MoO_2 &\ \ldots \ldots \ldots \ldots \ldots 2.7 \cdot 10^{-7}\ \text{atm} \\
(MoO_3)_3 &\ \ldots \ldots \ldots \ldots 7.1 \cdot 10^{-8}\ \text{atm}
\end{aligned}
$$

Thus, the main reaction in the evaporation of MoO_2 is

$$\frac{3}{2}x\,MoO_2\,(\text{solid}) = (MoO_3)_x\ (\text{gas}) + \frac{1}{2}\,x\,Mo\ (\text{solid}),$$

where x = 1, 2, or 3, and there is the secondary reaction

$$MoO_2\ (\text{solid}) = MoO_2\ (\text{gas}).$$

The temperature dependence of the partial pressures of gaseous MoO_3, $(MoO_3)_2$, $(MoO_3)_3$, and MoO_2 is expressed by the following equations:

$$\log P_{MoO_3} = 10.130 - \frac{2.661 \cdot 10^4}{T},$$

$$\log P_{(MoO_3)_2} = 11.067 - \frac{2.915 \cdot 10^4}{T},$$

$$\log P_{(MoO_3)_3} = 11.067 - \frac{3.117 \cdot 10^4}{T},$$

$$\log P_{MoO_2} = 10.574 - \frac{2.936 \cdot 10^4}{T}.$$

Burns, De Maria, Drowart, and Grimley gave the thermodynamic characteristics in Table 49 for the reactions they studied.

Table 50. Reduced Thermodynamic Potential $-(F^0 - H_0^0)/T$ in cal/mole·deg

Oxides	Temperature, °K					
	2000	2100	2200	2300	2400	2500
MoO	67.36	67.84	68.29	(68.73)	69.15	69.56
MoO$_2$	78.80	79.46	80.09	80.70	81.29	81.56
MoO$_3$	83.01	83.87	84.70	85.50	86.25	86.98

These authors give the following values (for 1600°K) for the entropy of gaseous molybdenum oxides:

MoO_3 . . . . 96.6 en. units
$(MoO_3)_2$ . . . . 151.0 en. units
$(MoO_3)_3$ . . . . 201.2 en. units
MoO_2. . . . . 85.5 en. units

De Maria, Burns, Drowart, and Inghram (1960) observed the monomeric gaseous molecules MoO, MoO$_2$, and MoO$_3$ in studying the system Mo$-$Al$_2$O$_3$. Together with these gaseous oxygen compounds of molybdenum, the molecules AlO, Al$_2$O, and Al$_2$O$_3$ were present in the vapor of this system. The amounts of the monomeric molecules fell in the sequence: MoO$_2$, MoO, MoO$_3$. The investigation made it possible for the authors to determine the atomization energies of these compounds:

$$MoO \text{ (gas)} = Mo \text{ (gas)} + O \text{ (gas)}, \quad \Delta H_0^0 = 116 \pm 15 \text{ kcal/mole,}$$

$$MoO_2 \text{ (gas)} = Mo \text{ (gas)} + 2O \text{ (gas)}, \quad \Delta H_0^0 = 262 \pm 10 \text{ kcal/mole,}$$

$$MoO_3 \text{ (gas)} = Mo \text{ (gas)} + 3O \text{ (gas)}, \quad \Delta H_0^0 = 411 \pm 7 \text{ kcal/mole.}$$

De Maria, Burns, Drowart, and Inghram calculated the reduced thermodynamic potential of these gaseous oxides MoO, MoO$_2$, and MoO$_3$, making a series of assumptions on the structure and molecular constants. Table 50 gives the corresponding values of $-(F^0 - H_0^0)/T$.

In 1960, Ackermann, Thorn, Tetenbaum, and Alexander (1960) summarized the results obtained up to that time on the evaporation of molybdenum oxides, and carried out some experiments of their own. These authors give free energies of sublimation for molybdenum trioxide with the formation of gaseous trimer (MoO$_3$)$_3$, tetramer (MoO$_3$)$_4$, and pentamer (MoO$_3$)$_5$.

$$\Delta F_{subl} \text{ (MoO}_3)_3 = 81,000 - 65.0\,T \quad \text{cal/mole,}$$

$$\Delta F_{subl} \text{ (MoO}_3)_4 = 92,700 - 76.3\,T \quad \text{cal/mole,}$$

$$\Delta F_{subl} \text{ (MoO}_3)_5 = 107,000 - 85.7\,T \quad \text{cal/mole.}$$

The evaporation of molybdic anhydride MoO$_3$ was studied by Spitsyn and Zimakov (1961) in connection with the general problem of the effect of radioactive radiation on the evaporation of substances. These authors studied the rate of evaporation of samples of molybdic anhydride containing various amounts (from 1.0 to 28.0 mCi/g) of the isotope Mo^{99} ($T_{0.5}$ = 68.3 h). Zimakov and Spitsyn (1961) subsequently investigated the effect of the addition of radioactive isotopes of some other elements on the rate of evaporation of MoO$_3$.

The rate of evaporation at 700°C was determined from the decrease in weight of the preparation. The radioactivity had a substantial and complex effect on the evaporation. Figure 100 shows the loss in weight of a sample of MoO$_3$ in 2 h in relation to the specific activity. The effect of the radioactivity began to appear at a specific radioactivity of 2.5-3 mCi/g. A further increase in the radioactivity (from 2.5 to 4 mCi/g) led to some fall in the evaporation rate. The authors explained this fall by the production of electric charges on the surface of the radioactive preparations because of the emission of β-particles. The production of the charges

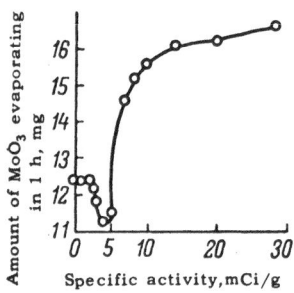

Fig. 100. Relation of the degree of evaporation of Mo*O_3 to the specific radioactivity of the preparations.

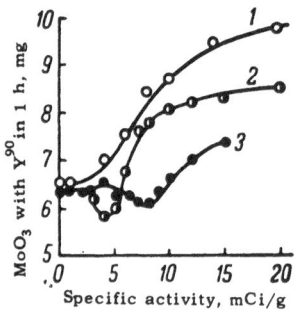

Fig. 101. Relation of the rate of evaporation of MoO_3 with various β -emitters added to the specific radioactivity of the preparations. 1) MoO_3 with Y^{90} added; 2) MoO_3 with Mo^{99} added; 3) MoO_3 with W^{185} added.

leads to an increased interaction between the surface of the solid and molecules of the vapor close to the interphase. When the charge on the surface was earthed, this fall in evaporation was not observed.

The increase in the rate of evaporation of Mo*O_3 with an increase in the specific radioactivity above 4-5 mCi/g depends to a considerable extent on radiation phenomena in the gas phase. A powerful beam of β - particles ionizes the MoO_3 molecules evaporating and produces, in particular, MoO_3^+ ions, promoting their repulsion from the surface. In addition to this, the radioactive radiation strongly changes the structure of the surface of the crystals, promoting its development.

Electron microscope photographs taken by Spitsyn et al. (1961) showed a great difference in the form of the surface of radioactive and nonradioactive preparations of MoO_3. The latter have a smooth flat surface. In the case of radioactive samples, beginning precisely with a specific activity of 4 mCi/g and above, on the surface of MoO_3 there appear irregularities, projections, and cavities, whose number increases with an increase in the specific radioactivity.

Thus, on the example of Mo*O_3 preparations, Spitsyn and Zimakov demonstrated for the first time that the kinetics of evaporation of radioactive solids depend on the level of their radioactivity. A substantial part is played by several factors, namely, the charge of the surface, produced as a result of the emission of charged particles, the change in the character of the crystal surface under the action of radiation, and radiation phenomena in the gas phase.

In their second study, Zimakov and Spitsyn used quartz spring balances to study the rate of evaporation of molybdic anhydride preparations labeled with the isotopes W^{185} (E_{max} = 0.43 MeV) and Y^{90} (E_{max} = 2.18 MeV). It was observed that radiation of different energies has different effects on the evaporation rate. Thus, the evaporation rate of solid molybdic anhydride depends on both the level of the radioactivity of the preparations and on the energy of the radiation of the radioactive material added. Radiation of higher energy produces a stronger change in the rate of evaporation of MoO_3 at the same level of radioactivity.

Figure 101 shows the relation of the rate of evaporation of MoO_3 with various β -emitters added to the specific radioactivity of the preparations. The lack of a minimum in the presence of Y^{90} is apparently connected with the fact that the β -particles of high energy emitted by this isotope have a greater effect on the charge of the crystal surface and ionize more strongly the MoO_3 molecules evaporating, repelling them from the positively charged surface, than when radioactive isotopes with radiation of lower energy were introduced.

Tungsten Oxides. The solid phases in the system W−O are similar to the corresponding phases in the system Mo−O. The existence of two oxides, namely tungsten dioxide and trioxide, is indisputable. On the phase diagram between these oxides it is believed that a whole series of compounds lies: W_2O_5, W_3O_8, W_4O_{11}, and $W_{18}O_{49}$. Hägg and Magneli (1945) observed regions of homogeneity close to $WO_{2.90}$ and $WO_{2.72}$.

A considerable number of molecules of different complexities have been observed in the study of the evaporation of tungsten trioxide and dioxide. Thus, De Maria, Drowart, and Inghram (1959) established the existence of a gaseous monoxide WO; Chupka, Berkowitz, and Giese (1959) found gaseous WO_2 and WO_3; Berkowitz, Chupka, and Inghram (1957a) placed WO_3 in the effusion cell and found in the vapors the polymeric molecule $(WO_3)_3$, $(WO_3)_4$, and $(WO_3)_5$. Blackburn, Hoch, and Johnston (1958) consider that in the evaporation of tungsten dioxide there occurs the disproportionation

$$^9/_2 WO_2 \text{ (solid)} = W_3 O_9 \text{ (gas)} + {}^3/_2 W \text{ (solid)}.$$

The thermodynamic characteristics of monomeric gaseous oxide molecules of tungsten were obtained by De Maria, Burns, Drowart, and Inghram (1960), who studied the evaporation of the system $W-Al_2O_3$. For the atomization energy of gaseous oxide molecules the following values were obtained:

$$WO \text{ (gas)} = W \text{ (gas)} + O \text{ (gas)}, \quad \Delta H_0^0 = 154 \pm 10 \text{ kcal/mole},$$

$$WO_2 \text{ (gas)} = W \text{ (gas)} + 2O \text{ (gas)}, \quad \Delta H_0^0 = 296 \pm 7 \text{ kcal/mole},$$

$$WO_3 \text{ (gas)} = W \text{ (gas)} + 3O \text{ (gas)}, \quad \Delta H_0^0 = 443 \pm 7 \text{ kcal/mole}.$$

The same authors calculated the reduced thermodynamic potential for gaseous tungsten oxides at high temperatures (Table 51).

Thermal data for the sublimation of tungsten trioxide and dioxide were obtained by Blackburn and his co-workers, and also by Berkowitz, Chupka, and Inghram. There are differences between their data. Ackermann and Thorn (1958) consider that the following expressions for the free energy of sublimation of tungsten trioxide are more reliable: 1) sublimation of WO_3 with the formation of gaseous trimer $(WO_3)_3$

$$\Delta F^0_{\text{subl}} = 130,000 - 69.5\,T \text{ cal/mole},$$

2) sublimation of WO_3 with the formation of the gaseous tetramer $(WO_3)_4$

$$\Delta F^0_{\text{subl}} = 151,000 - 81.2\,T \text{ cal/mole}.$$

The free energy of formation of WO_3(gas) was determined by Ackermann and Thorn (1956) by studying the reaction

$$MgO \text{ (solid)} + {}^1/_3 W \text{ (solid)} = Mg \text{ (gas)} + {}^1/_3 WO_3 \text{ (gas)}.$$

For the temperature range of 1920-2020°K they gave the equation

$$\Delta F^0_{\text{form}} = -75,700 + 15.54\,T \text{ cal/mole}.$$

For the free energy of formation of WO_2(gas) the same authors gave the equation

$$\Delta F^0_{\text{form}} = 7360 - 5.80\,T \text{ cal/mole}.$$

In his last paper, Blackburn (1961) measured the vapor pressure over WO_3 in the temperature range of 1040-1500°K, using the methods of Knudsen and Langmuir. The two methods gave practically identical results, and over the given temperature range the vapor pressure varied from $2 \cdot 10^{-12}$ to $2 \cdot 10^{-4}$ atm. If it is assumed that the gas consists completely of W_3O_9 molecules, then for the heat and entropy of evaporation we obtain the values $\Delta H^0_{1300} = 123.6$ kcal/mole and $\Delta S = 65.5$ en. units.

Blackburn studied the vapor pressure over tungsten oxides with different oxygen concentrations in the solid phase in the region of $WO_{2.95} - WO_{1.7}$. As a result of these experiments, Blackburn believes that there exist the following regions of homogeneity in the system $W-O$: $WO_{2.94-2.88}$, $WO_{2.74-2.70}$, and $WO_{2.02-1.98}$.

On treating WO_3 (900-1100°C) and MoO_3 (600-690°C) with water vapor, Glemser and Haeseler (1961) observed the formation of the gaseous hydroxides $W_xO_{3x-1}(OH)_2$ and $Mo_xO_{3x-1}(OH)_2$, whose exact composition they were unable to establish. These compounds exist only in the gas phase and on cooling they decompose to the starting components.

Ackermann and Thorn (1958) made a detailed examination of the reaction of aluminum oxide, magnesium oxide, uranium dioxide, thorium dioxide, beryllium oxide, and zirconium dioxide with tungsten and tantalum, of which effusion cells are usually made.

Table 51. Reduced Thermodynamic Potential $-(F^0 - H_0^0)/T$ for Monomeric Gaseous
Tungsten Oxides, in cal/mole · deg

Oxides	Temperature, °K					
	2000	2100	2200	2300	2400	2500
WO	67.54	68.00	68.44	68.88	69.29	69.69
WO$_2$	81.31	81.78	82.63	83.25	83.85	84.44
WO$_3$	84.99	85.85	86.78	87.47	88.22	88.97

Table 52. Gaseous Products Obtained as a Result of the Reaction of Oxides with Tungsten
and Tantalum

Oxides	Tungsten			Tantalum		
	$T°$, K	Ratio of weight of oxide to weight of metal	Gaseous products	T, °K	Ratio of weight of oxide to weight of metal	Gaseous products
Al$_2$O$_3$ {	2544 2582 2635	1.23±0.01 1.16±0.01 1.13±0.01	Al$_2$O + Al + + O + WO$_x$	2523	0.19±0.01	Al + TaO and small amount of Al$_2$O and TaO$_2$
MgO	2045	0.66±0.01	Mg + WO$_3$	—	—	—
UO$_2$	1600—2500	≥ 50	UO$_2$	2100—2500	≥ 50	UO$_2$
ThO$_2$	2765	≥ 50	ThO$_2$, ThO, O	2753	3.06	ThO + TaO + O
BeO	2462	0.92	Be + WO$_x$ + O (possibly BeO)	—	—	—
ZrO$_2$	2600	≥ 50	ZrO$_2$ + ZrO + O	2560	0.89±0.02	ZrO + TaO + O

If we denote the material of the effusion cell by C and the oxide reacting with it by MeO$_a$, then on condition that no other solid phases are present we may write

$$\text{MeO}_a \text{ (solid)} + \frac{a-b}{c} \text{ C (solid)} = \text{MeO}_b \text{ (gas)} + \frac{a-b}{c} \text{ CO}_c \text{ (gas)}.$$

If we denote the ratio of the weights of MeO$_a$ (solid) and C(solid) reacting by W$_r$ and the ratio of the molecular weights by m$_r$, then

$$\frac{a-b}{c} = \frac{m_r}{W_r} = S.$$

The coefficient S cannot be determined unequivocally solely on the basis of the measurement of W$_r$, as a and b are unknown. It is necessary to know the composition of the vapor to find the stoichiometry of the reaction examined.

Table 52 gives the results of the investigation of the reaction of a series of oxides with tungsten and tantalum. It gives the ratio of the weights of the oxides and metal lost from the effusion cell, and also lists the molecules (or atoms) present in the vapors. Table 52 shows that the ratio of the weights of the aluminum oxide and effusion cell metal evaporating falls with a rise in temperature. Magnesium oxide reacts with tungsten according to the equation

$$\text{MgO (solid)} + \tfrac{1}{3}\text{W (solid)} = \text{Mg (gas)} + \tfrac{1}{3}\text{WO}_3 \text{ (gas)}$$

with a ratio of oxide : metal = 0.658.

Table 53. Reduced Thermodynamic Potential for Gaseous UO, UO_2, and UO_3, cal/mole·deg

Oxides	Temperature, °K					
	2000	2100	2200	2300	2400	2500
UO	68.89	69.31	69.72	70.11	70.49	70.86
UO_2	84.26	84.94	85.60	86.23	86.84	87.44
UO_3	87.25	88.12	88.94	89.75	90.51	91.25

The data in Table 52 seem convincing. Tungsten reacts with aluminum oxide, magnesium oxide, and beryllium oxide and reduces them to lower oxides or even the metals. At the same time, tungsten does not react with UO_2, ThO_2, and ZrO_2, and no gaseous oxides of tungsten were found in the vapor in the investigation of these oxides.

Uranium Oxides. The chemistry of the system uranium—oxygen is very complex because of the formation of oxides of variable composition and also nonstoichiometric oxides, i.e., those which do not satisfy a simple stoichiometric ratio of uranium and oxygen, for example, U_3O_7, U_3O_8, U_4O_9, U_5O_{15}, etc. Makarov (1961) lists the following established higher oxides of uranium: UO_3, U_3O_8, U_2O_5, U_5O_{12}, U_3O_7, U_7O_{16}, and U_4O_9.

The behavior of uranium oxides on evaporation is generally similar to that observed for tungsten oxides, though with uranium there is as yet no indication of the existence of polymeric molecules in the gas phase.

On heating higher uranium oxides UO_{2+x}, Chupka (1957) observed UO, UO_2, and UO_3 in the vapor. Ackermann, Thorn, Alexander, and Tetenbaum (1960) reported that uranium monoxide is found in a small amount and even on evaporation of a mixture of U + UO_2 the bulk of the vapor (98%) consists of the dioxide, whose pressure in the range of 1600-2000°K may be expressed by the equation

$$\log P \ (\text{atm}) = 22.805 - \frac{33,115}{T} - 4.026 \log T.$$

On heating in vacuum, the higher oxides of uranium are converted into the dioxide $UO_{2.00}$, whose vapor pressure in the range of 1600-2000°K is represented by the equation

$$\log P \ (\text{atm}) = 10.417 - \frac{3.7195}{T} \cdot 10^4 + \frac{3.5162}{T^2} \cdot 10^6 + \frac{2.6178}{T^3} \cdot 10^9.$$

Under oxidizing conditions, the higher oxides of uranium evaporate to give the gaseous trioxide

$$\tfrac{1}{3}U_3O_8 \ (\text{solid}) + \tfrac{1}{6}O_2 \ (\text{gas}) = UO_3 \ (\text{gas}).$$

This was demonstrated by Ackermann, Gilles, and Thorn (1956b) by the flow method when oxygen is passed over U_3O_8. The authors consider that uranium trioxide is in the vapor in the form of the monomer, though the possible presence of polymeric molecules cannot be excluded. The vapor pressure of the gaseous trioxide may be expressed by the equation

$$\log P \ (\text{atm}) = \frac{31.3}{R'} - \frac{83.3 \cdot 10^3}{R'T}.$$

Uranium monoxide was obtained as the predominant component of the vapor by De Maria, Burns, Drowart, and Inghram (1960) by heating metallic uranium with aluminum oxide. Atomic uranium and UO_2 were also observed in the vapor. These authors calculated the reduced thermodynamic potential $-(F^0 - H_0^0)/T$ for gaseous uranium oxides (Table 53).

De Maria and his co-workers give the following values for the atomization energy of gaseous uranium oxides:

$$UO \ (\text{gas}) = U \ (\text{gas}) + O \ (\text{gas}), \quad \Delta H_0^0 = 179 \pm 7 \ \text{kcal/mole},$$

$$UO_2 \text{ (gas)} = U \text{ (gas)} + 2O \text{ (gas)}, \quad \Delta H_0^0 = 340 \pm 7 \text{ kcal/mole,}$$

$$UO_3 \text{ (gas)} = U \text{ (gas)} + 3O \text{ (gas)}, \quad \Delta H_0^0 = 493 \pm 7 \text{ kcal/mole.}$$

Uranium dioxide is the only oxide which is hardly reduced by tantalum at high temperatures. Even with complete evaporation of uranium dioxide from a tantalum cell, less than 2% of the UO_2 reacts with the tantalum. This is explained primarily by the low dissociation energy of uranium monoxide. Although the free energy of formation of ThO_2 is more negative than for UO_2, ThO_2 reacts with tantalum as thorium monoxide is more stable than uranium monoxide. Ackermann and Thorn estimated the dissociation energy of uranium monoxide as 7.5-7.9 eV.

Ackermann and Thorn (1961) give the following equations for the temperature dependence of the free energy of formation of uranium dioxide and trioxide:

$$\Delta F^0_{\text{form}} (UO_2 \text{ gas}) = -121,500 + 4.24\,T \text{ cal/mole;}$$

$$\Delta F^0_{\text{form}} (UO_3 \text{ gas}) = -198,000 + 19.0\,T \text{ cal/mole.}$$

The sublimation of uranium dioxide is characterized by the following thermal data: heat of sublimation of UO_2, 137.1 kcal/mole, and entropy of sublimation 36.4 en. units.

Ackermann and Thorn, who observed the anomalous evaporation of UO_2 at temperatures above 1900°C, first ascribed this to the appearance of polymeric gas molecules, but in a communication at the XVI Congress of Pure and Applied Chemistry (Paris, 1957), they no longer mentioned the possible polymerization of UO_2 in the vapor.

Iron Oxide. In his investigations of the melting-point diagrams of systems consisting of zirconium dioxide, aluminum oxide, and magnesium oxide with iron, chromium, and manganese sesquioxides, in cases where the temperature rose above 1800°C, Wartenberg (1928, 1930, 1931, 1932, 1937) observed sublimates, green in the presence of Cr_2O_3 and brown in the presence of Fe_2O_3 and Mn_2O_3. Analogous sublimates had long been observed in open-hearth furnaces and converters and even interfered with production in connection with the use of oxygen in blast furnaces. This problem has been discussed in a long article by Guthmann (1958).

On heating pressed cylinders of a mixture of Fe_3O_4, Mn_3O_4, and Cr_3O_4 in the eutectic ratio in a furnace, Wartenberg (1959) observed that only iron oxides volatilized from these mixtures above 1800°C. Without direct determinations of the composition of the vapor, Wartenberg was only able to put forward certain hypotheses. Considering the existence of the seven-atom molecule Fe_3O_4 improbable, he suggested that FeO molecules were present in the vapor, i.e., the evaporation proceeds in accordance with the reaction

$$2Fe_3O_4 \text{ (liq)} = 6FeO \text{ (gas)} + O_2 \text{ (gas).}$$

He attempted to substantiate this hypothesis by thermodynamic calculations, but nonetheless it is not possible to draw conclusions on the nature of gaseous iron oxides from Wartenberg's work. Subsequent work of Wartenberg's co-workers Glemser and Weizenkorn showed that polynuclear iron and manganese oxides may be present in the vapors, but they did not detect FeO molecules.

In a brief note, Glemser and Weizenkorn (1961) described experiments on the evaporation of Fe_3O_4. The authors used Nernst's method. A sample of Fe_3O_4 was placed in an iridium retort with a capillary side tube, to which was fitted a graduated glass capillary. Inside the capillary was a drop of mercury, which moved with a change in the pressure inside the retort. For prevention of the evaporation of iridum, the inner and outer surfaces of the iridium retort were coated with zirconium dioxide.

In all the experiments at temperatures between 1875 and 2150°C, the volume of gas corresponded to the evaporation of magnetic iron oxide in the form of Fe_3O_4 molecules. Had evaporation occurred in the form of FeO, as the following reaction shows,

$$Fe_3O_4 = 3FeO + \tfrac{1}{2}O_2,$$

the volume of gas would have been 3.5 times as great as that observed.

Glemser and Weizenkorn reported that Fe_3O_4 is the only iron oxide stable above 1800°C, which also exists in a melt. These facts make it probable that a gaseous compound corresponding to the formula Fe_3O_4 exists in the vapor.

Caplan and Cohen (1961) showed that the iron oxide Fe_2O_3 does not evaporate at all, even over a long period at 1200°C. The experiments were carried out with a stream of dry or wet oxygen flowing over Fe_2O_3 tablets, and in neither case was there any reduction in weight in 65 h at the given temperature.

4. Some Generalizations of the Results of Investigating the Evaporation of Oxides

The molecules detected in the vapors of oxides of some elements are shown in Table 54. Here we give the alkali and alkaline earth element groups, the group of scandium, yttrium, and lanthanum, the group of titanium, zirconium, hafnium, and thorium, the group of vanadium, niobium, and tantalum, and the group of chromium, molybdenum, tungsten, and uranium. These groups are sometimes called the A groups.

Table 54 shows that gas molecules, like other parts of the vapor, become increasingly numerous with an increase in the number of the group and, correspondingly, atoms in the gases become less important.

In group IA, free atoms are the main components of the vapor at all temperatures. Gas molecules are observed only in the case of lithium, but even here these molecules form only a tenth of the total amount of vapor. In the group of alkaline earths, the mole fraction of monoxides in the saturated vapors increases with an increase in atomic number (beryllium is an exception); for calcium (and magnesium), the monoxide content of the vapor is 1%, and for barium it is at least 99%. In group IIIA the predominant form of the particles is the monoxides, then atomic oxygen and, finally, the gaseous metal. In this group the stability of the gaseous monoxides largely determines the character of the evaporation process.

In the sublimates of oxides of group IVA elements, both the dioxides and monoxides are present in the vapors, with the latter of less importance than the former. The sublimation processes of oxides of group VA and VIA elements are more complex. Here there is normally more than one condensed phase present and the gaseous molecules have a more complex composition. The compositions of the vapors of vanadium and tantalum pentoxides differ markedly, though these two elements are analogs. Thus, on evaporating, vanadium pentoxide forms predominantly complex molecules: V_4O_{10}, V_4O_8, V_6O_{14}, V_6O_{12}, and V_2O_4. At the same time, tantalum pentoxide gives the gaseous dioxide and monoxide under reducing conditions and the dioxide and oxygen under neutral conditions.

Table 54. Molecules Detected in the Gas Phase of Oxides of A Group Elements

Group	Solid phase	Gas phase	Some examples
I-A	Me_2O	$Me + O_2$, Me_2O	$Li_2O \rightarrow \begin{cases} Li_2O \\ Li + O. \end{cases}$ Others M + O$_2$
II-A	MeO	$Me + O_2$, MeO	$BaO \rightarrow BaO$ (gas)
III-A	Me_2O_3	$MeO + O$, Me	—
IV-A	MeO_2	MeO_2, $MeO + O$	—
V-A	Me_2O_5	$(Me_2O_5)_n$ $MeO_2 + MeO + O$	$V_2O_5 \rightarrow (V_2O_5)_2$ (gas)
VI-A $\Big\{$	MeO_3	$(MeO_3)_n$ $(n = 3, 4, 5)$	$MoO_2 \rightarrow \begin{cases} MoO_3 \text{ (gas)} + Mo \text{ (solid)} \\ MoO_2 \text{ (gas)} \end{cases}$
	MeO_2	$(MeO_3)_n$, MeO_2	$WO_2 \rightarrow (WO_3)_n$ (gas) $+ W$ (solid) $UO_2 \rightarrow UO$ (gas) $+ UO_3$ (gas) (small amounts)

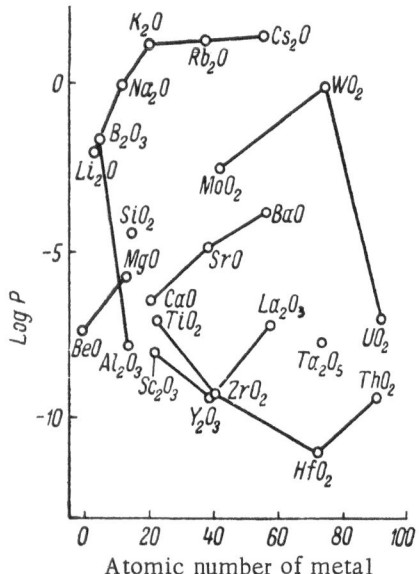

Fig. 102. Vapor pressure of oxides under neutral conditions at 2000°K.

A great difference in the composition of the vapor is also observed in the evaporation of oxides of elements of the molybdenum group. The vapors of molybdenum and tungsten trioxides contain trimers, tetramers, and pentamers of the trioxide, while uranium trioxide evaporates to give the monomer UO_3. Tungsten dioxide sublimes with the formation of polymers of the trioxide and metallic tungsten, while molybdenum dioxide gives monomeric dioxide and trioxide.

This difference is apparently explained by two opposing processes, though the chemical bonds in the gas molecules may remain similar. Thus, the absence of complex molecules from the saturated vapor over tantalum oxide does not indicate that the chemical bonds in gaseous tantalum oxides differ from those in vanadium oxides; the absence of uranium trioxide polymers in no way indicates the lack of chemical similarity of uranium and tungsten. It is most likely that these differences indicate that tantalum pentoxide and uranium trioxide require a higher temperature to give a saturated vapor, and this is connected with the free energies of formation of the condensed phases. It may be said that here there are two opposing tendencies, namely, the tendency to form polymers and the tendency to form a stable condensed phase. In some cases the tendency to form a polymer is so great that the largest polymer, a condensed phase, is formed. Then the evaporation of single molecules is more favored.

To give some idea of the degree of volatility of the different oxides, Fig. 102 shows the total vapor pressures of a series of oxides. As Fig. 102 shows, the least volatile oxides are Li_2O, CaO, Y_2O_3, HfO_2, BeO, and Al_2O_3. Of all the oxides studied up to now, hafnium dioxide is the least volatile. The reasons for the difference in volatility lies in the energy and atomic properties of the corresponding oxygen compounds.

The energetics of the process may be explained best by considering a modified Born-Haber cycle, as is shown in Fig. 103. In this scheme, the processes occurring during evaporation are represented by the diagonal. This cycle shows that if the vapor contains gaseous molecules, the volatility will be greater than calculated for evaporation with decomposition to free atoms. Thus, the volatility of oxides is determined largely by the relative stability of the gaseous molecules.

To predict how an evaporation will proceed (with decomposition to atoms or with the formation of gaseous molecules), strictly speaking, it is necessary to use the free energies of the corresponding processes, but for comparative purposes it is quite possible to use the thermal effects, assuming that the entropy change is approximately the same for all processes. For quantitative characterization of the heat of sublimation with decomposition to the elements $\Delta H^0_{s.e.}$, it is necessary to know $\Delta H^0(Me)$, the heat of sublimation of the metals, $\Delta H^0(MeO_x)$, the heat of formation of the condensed oxide per gram-atom of metal, and the dissociation energy of oxygen $D_0(O_2)$

$$\Delta H^0_{s.e.} = \frac{x}{1+x}\left[\Delta H^0(Me) - \Delta H^0(MeO_x) + \frac{x}{2}D_0(O_2)\right].$$

As Fig. 104 shows, there is complete correspondence between the values of $\Delta H^0_{s.e.}$ and the volatility of the oxides. This correspondence may be followed through all the groups of the periodic table (groups IA-VIA).

In group IA, the volatility increases, while the heat of sublimation with the formation of gaseous elements decreases with an increase in atomic number. For the alkali metals, the gaseous molecules present are so insignificant that they have little effect on the volatility of the oxides. In general, oxides of alkali elements are similar in behavior on evaporation to elements of subgroup B.

Table 55 gives the values of the heat effects characterizing evaporation with the formation of gaseous elements.

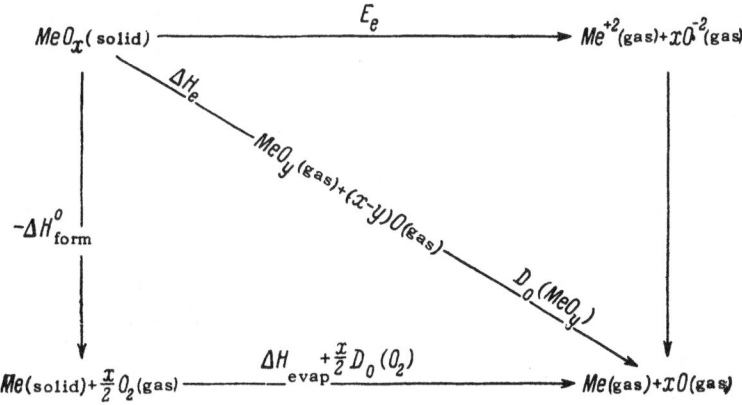

Fig. 103. Thermochemical cycle describing the evaporation of an oxide.

Table 55. Thermal Effects Characterizing Sublimation with the Formation of Gaseous Elements

Oxides	$-\Delta H^0_{298} (MeO_x)$, kcal/g-atom of Me (data from Coughlin)	$\Delta H^0_{298} (Me)$, kcal/g-atom of Me (data from Stahl and Zinke)	$\frac{1}{1+x} \left[-\Delta H^0 (Me) - \Delta H^\circ (MeO_x) + \frac{x}{2} D_0 (O_2) \right]$, kcal	Oxides	$-\Delta H^0_{298} (MeO_x)$, kcal/g-atom of Me (data from Coughlin)	$\Delta H^0_{298} (Me)$, kcal/g-atom of Me (data from Stahl and Zinke)	$\frac{1}{1+x} \left[\Delta H^\circ (Me) - \Delta H^\circ (MeO_x) + \frac{x}{2} D_0 (O_2) \right]$, kcal
Li_2O . . .	71	38	92	Y_2O_3 . . .	228	(102)	167
Na_2O . . .	50	26	70	La_2O_3 . . .	214	100	161
K_2O	43	22	63	SiO_2 . . .	210	105	144
Rb_2O . . .	40	20	59	TiO_2 . . .	226	113	152
Cs_2O . . .	38	19	57	ZrO_2 . . .	262	146	175
BeO	143	78	140	HfO_2 . . .	266	168	184
MgO . . .	144	36	120	ThO_2 . . .	293	134	183
CaO . . .	151	42	126	V_2O_5 . . .	186	123	131
SrO	141	39	120	Nb_2O_5 . . .	228	178	158
BaO	134	42	118	Ta_2O_5 . . .	244	187	165
B_2O_3 . . .	153	128	—	MoO_2 . . .	141	158	139
Al_2O_3 . . .	200	77	146	WO_2 . . .	141	200	153
Sc_2O_3 . . .	(205)	(82)	150	UO_2	259	117	165

Note. Hypothetical values are given in brackets.

In group IIA the volatility increases, while the heat of sublimation $\Delta H^0_{s.e.}$ falls with an increase in atomic number. The dissociation energy of the monoxide increases with an increase in the atomic number, i.e., the effect is opposite to that which we observe in the alkali metal group and in subgroup B. The alkaline earth elements (calcium, strontium, and barium) form a link between the alkali metals and the transition elements: some of their properties, such as volatility, are similar to the alkali elements, while the dissociation energy of the monoxides varies from element to element as in the transition groups.

In group IIIA (Sc, Y, La), yttrium oxide has the minimum volatility and the maximum heat of sublimation. The dissociation energy of the monoxides increases with an increase in the atomic number. The heat of formation of the solid oxides rather than the dissociation energy of the monoxides has a substantial effect on the volatility of the oxides and the value of $\Delta H^0_{s.e.}$ in group IIIA.

In group IVA (Ti, Zr, Hf, Th), hafnium dioxide has the minimum volatility and the maximum heat of sublimation $\Delta H^0_{s.e.}$. The dissociation energy of the monoxides (and probably the dioxides) increases with an increase in atomic number. As a rule, the volatility is determined by the stability of the solid oxide, but since the dissociation energy for gaseous thorium oxides is higher than for gaseous hafnium oxides, thorium oxide is more volatile than hafnium oxide.

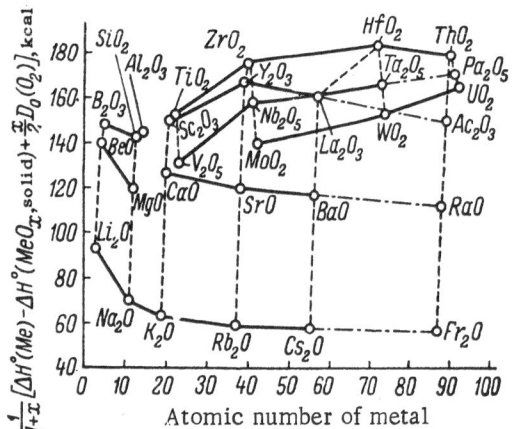

Fig. 104. Heat of sublimation of oxides with the formation of gaseous elements.

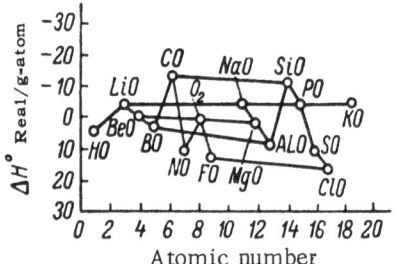

Fig. 105. Heats of formation of gaseous monoxides.

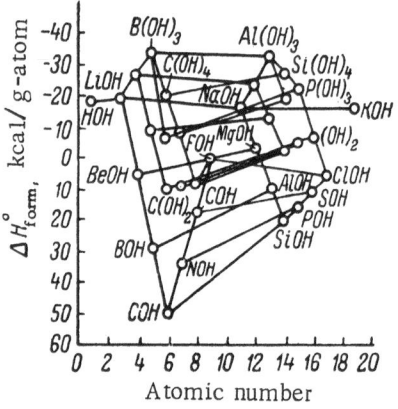

Fig. 106. Heats of formation of gaseous hydroxides.

In group VI (V, Nb, Ta, Pa), the volatility is minimal either for tantalum oxide or for protactinium oxide. Correspondingly, the heat of sublimation is maximal for one or other of these two oxides. The dissociation energy of the monoxides and probably the dioxides increases with the atomic number and consequently may determine the relative volatility of tantalum and protactinium oxides. However, here the heat of formation of the solid oxides is of greater importance. The presence of gaseous polymers in the case of vanadium oxide strengthens the factors responsible for the stability of the condensed oxides.

Apart from chromium oxide, in group VIA the volatility is maximal with tungsten dioxide, but the heat of sublimation increases with an increase in atomic number. Here the volatility is determined by the relative stability of the gaseous polymers of tungsten and molybdenum trioxides, while the heat of sublimation $\Delta H^0_{s.e.}$ is determined largely by the heat of sublimation of the metals.

Up to now we have examined the volatility of the oxides in the separate groups of the periodic table. Now it is necessary to show how the volatility changes in the rows of the periodic table .

In row K, the volatility falls from lithium oxide to beryllium oxide, and then increases from beryllium oxide to boron oxide. The heat of sublimation with simultaneous formation of free atoms $\Delta H^0_{s.e.}$ and the dissociation energy of the gaseous oxides increase in this series with an increase in the atomic number of the element. The dissociation energy of boron oxide is so high that its volatility is higher than that of beryllium oxide.

In row L the volatility falls between sodium and aluminum oxides and then increases. The heat of sublimation $\Delta H^0_{s.e.}$ and the stability of the gaseous oxides increases along the row from left to right. The increase in the dissociation energy (stability) results in the fact that silica is more volatile than alumina.

In row M the volatility falls from potassium oxide to scandium oxide and increases from scandium oxide to vanadium oxide. The heat of sublimation $\Delta H^0_{s.e.}$ is maximal for scandium oxide. The dissociation energy of the monoxides apparently reaches a maximum at titanium oxide. The other gaseous oxides, beginning with titanium, become more stable since the volatility is even greater than would be expected on the assumption that there is decomposition to atoms of the elements or to gaseous monoxides and oxygen.

In row N the volatility falls from rubidium oxide to yttrium oxide, and increases from yttrium oxide to molybdenum oxide. The heat of sublimation $\Delta H^0_{s.e.}$ increases from rubidium oxide to zirconium oxide and falls from zirconium oxide to molybdenum oxide. The dissociation energy of the monoxides is apparently maximal for zirconium oxide. Other gaseous dioxides and trioxides become stable, beginning with zirconium,

Table 56. Heats of Formation of Some Gaseous Hydroxides
at 298.15°K, in kcal/g-atom

Compound	ΔH°_{form}	Author
LiOH	−20.7	Bŭchler, 1959.
LiOH	−21	O'Brien, Perrin, Perrine, 1960.
BeOH	+5	The same
Be(OH)$_2$ . . .	−27	»
BOH	+29	»
B(OH)$_2$	−9	»
B(OH)$_3$	−34	»
COH	+50	»
C(OH)$_2$	+10	»
C(OH)$_3$	−7	»
C(OH)$_4$	−21	»
NaOH	−17	Bŭchler, 1959.
NaOH	−17	O'Brien, Perrin, Perrine, 1960.
MgOH	−3.3	Bulewicz, Sugden, 1959.
MgOH	−4	O'Brien, Perrin, Perrine, 1960.
Mg(OH)$_2$ . . .	−25	The same
AlOH	+10	»
Al(OH)$_2$ . . .	−14	»
Al(OH)$_3$ . . .	−33	»
SiOH	+20	»
Si(OH)$_2$ . . .	−3	»
Si(OH)$_3$ . . .	−20	»
Si(OH)$_4$ . . .	−28	»
POH	+15	»
P(OH)$_2$	−6	»
P(OH)$_3$	−23	»
KOH	−17	»

and the volatility of the oxides is even greater than that calculated on the assumption that on evaporation there is decomposition to the gaseous elements or gaseous monoxides and oxygen. For these reasons, zirconium dioxide is somewhat more volatile than yttrium oxide.

In row O the volatility falls from cesium to hafnium, and increases from hafnium to titanium. The dissociation energy probably increases, reaching a maximum at hafnium monoxide. The change in these characteristics of oxides in row P is evidently the same as in row O. In reviewing the evaporation of the oxides of the elements, we used the work of Ackermann, Thorn, and Winslow (1961b).

The heat of formation of gaseous oxides obeys definite rules if we arrange these compounds in the groups and rows of the periodic table.

O'Brien, Perrin, and Perrine (1960) give the heats of formation of the following gaseous molecules: HO, LiO, BeO, BO, CO, NO, O$_2$, FO, NaO, MgO, AlO, SiO, PO, SO, ClO, and KO (Fig. 105). As Fig. 105 shows, the relation of the heat of formation of the monoxides to the atomic numbers of the elements is represented by a broken line with minima at BO and AlO. The authors even consider it possible to change literature data if the corresponding values do not fall on the curve for the heat of formation.

Of the oxides listed, for HO, BO, CO, NO, MgO, AlO, SiO, PO, SO, and ClO, the heats of formation are taken from available collections of thermal constants, largely from Report No. 6297 of the National Bureau of Standards, USA (1959) and Bureau of Mines Bulletin, USA (1954).

For LiO, BeO, NaO, and FO, O'Brien and his co-workers give data which differ considerably from available literature data. In their opinion, LiO and NaO should have a more negative heat of formation than the OH group, for which the heat of formation is taken as 5 kcal/g-atom (298.15°K). For LiO(gas) these authors take ΔH^0_{form} = −4 kcal/g-atom (according to the Bureau of Standards summary, +7 kcal/g-atom); for NaO, ΔH^0_{form} = −4 kcal/g-atom [according to the data of Brewer and Margrave (1955), +1 kcal/g-atom].

Table 57. Heat of Formation of Some Gaseous Oxides at 298.15°K, in kcal/g-atom

Compound	ΔH^0_{form}	Authors or reviews
LiO	+ 7	National Bureau of Standards (N. B. S.), Report No. 6297
LiO	− 4	O'Brien et al.
NaO	+ 1	N. B. S., Report No. 6297
NaO	− 4	O'Brien et al.
KO	− 4	The same
K_2O	−12	The same
BeO	+15.2	N. B. S., Report No. 6297
BeO	0	O'Brien et al.
MgO	+ 2.1	N. B. S., Report No. 6297
MgO	+ 2	O'Brien et al.
BO	+ 2.9	N. B. S., Report No. 6252 (1958)
BO	+ 3	O'Brien et al.
B_2O_3	−41.5	N. B. S., Report No. 6252
B_2O_3	−42	O'Brien et al.
AlO	+ 8.5	N. B. S., Report No. 6297
AlO	+ 9	O'Brien et al.
Al_2O	−13.1	N. B. S., Report No. 6297
Al_2O	−13	O'Brien et al.
Al_2O_3	−39	The same
SiO	−10.9	Bureau of Mines Bulletin No. 542 (1954)
SiO	−11	O'Brien et al.
PO	− 4.8	Bureau of Mines Bulletin No. 542
PO	− 5	O'Brien et al.

BeO(gas) is given a more negative value on the assumption that this molecule has a strong double bond. As Fig. 105 shows, the heat of formation of BeO should equal zero for it to fall on a straight line connecting the corresponding values for LiO and BO. Spectral investigations give completely different values for the heats of dissociation of BeO(gas). Thus, according to Gaydon (1953), the heat of dissociation of BeO(gas) equals 124 kcal/mole (heat of formation 14 kcal/mole). For gaseous potassium oxide KO, Fig. 105 gives the same heat of formation as for LiO and NaO, i.e., −4 kcal/g-atom.

On the basis of the rule for the change in the heat of formation of gaseous molecules with an increase in the atomic number, O'Brien, Perrin, and Perrine (1960) give heats of formation of the different hydroxides (Fig. 106). In the construction of this graph, the authors had very few literature data and drew the curves on the basis of purely speculative conclusions. Many of the gaseous hydroxides whose heats of formation are given by O'Brien et al. have not even been shown to exist experimentally. Tables 56 and 57 give the heats of formation of a large number of oxide and hydroxide molecules taken from the article of O'Brien, Perrin, and Perrine.

Summarized Data on the Free Energy of Formation of Gaseous Metal Oxides. As a result of many investigations of the vapor pressure of oxides, a large amount of thermodynamic data has now been obtained, so that it is possible to draw graphs of the free energy of formation of oxide molecules known in the gaseous state.

Quite a full summary of the free energies in relation to temperature is given graphically by Gleiser (1961). The author gives the relation $\Delta F = f(T)$ per gram-mole of oxygen (Fig. 107) and per gram-atom of metal (Fig. 108). Some data given by Gleiser differ from those given above in our book.

For elements giving polymeric molecules in the vapor, the free energies of their formation are known. Thus, the free energies of formation of BeO, $(BeO)_2$, $(BeO)_3$, $(BeO)_4$, $(BeO)_5$, and $(BeO)_6$ are given for the beryllium oxides. As is shown by Fig. 107, the free energy of formation becomes more negative with a change from $2BeO(gas)$ to $\frac{1}{3}(BeO)_6(gas)$.

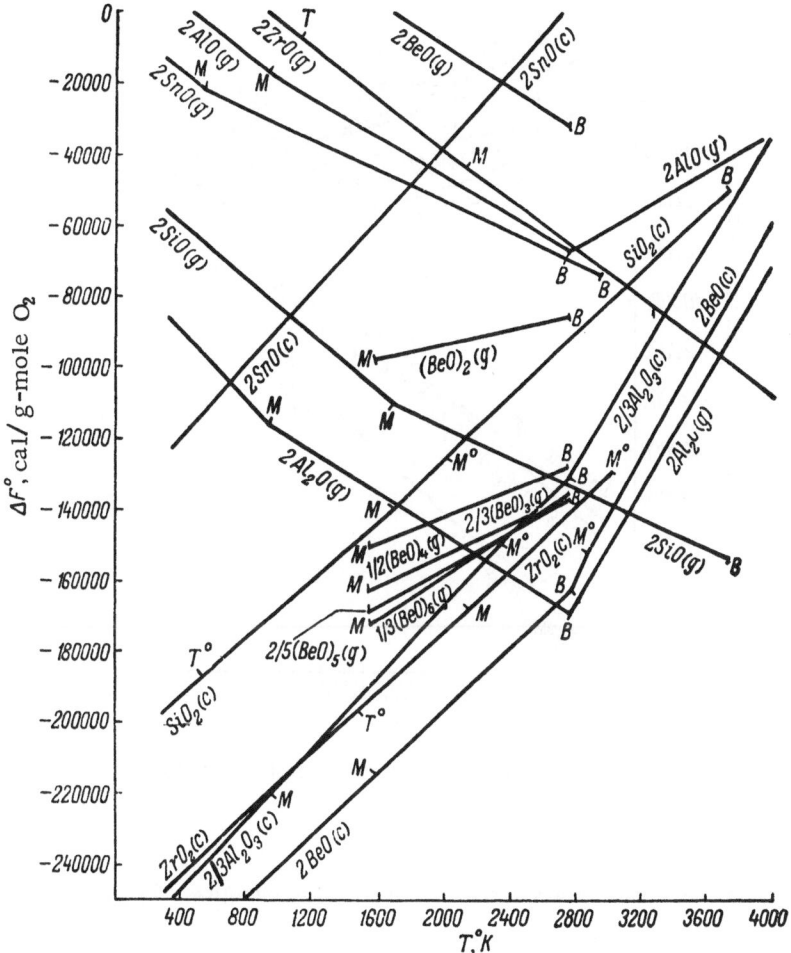

Fig. 107. Free energies of formation of gaseous oxides per mole of oxygen. c) Condensed phase (liquid or solid); g) gas; B) boiling point of metal; M, M⁰) melting points of metal and oxide; T, T⁰) transition points of metal and oxide.

In accordance with the data of Chupka, Berkowitz, and Giese (1959), the evaporation of beryllium oxide gives a mixture of several oxides (see p. 142). In this case, the boiling point of the oxide cannot be obtained from two lines giving temperature dependences of ΔF [e.g., of BeO(solid) and (BeO)$_2$(gas)]. The intersection point of the two lines (one for the solid oxide and the other for the gaseous oxide) gives the boiling point only if the gas phase contains only one oxide of the same molecular formula as the condensed phase.

For example, for SnO, the gas phase contains only one type of molecule, SnO, i.e., of the same composition as the solid phase. From the intersection of the lines for the condensed phase 2SnO(solid) and gaseous phase 2SnO(gas), we obtain the normal boiling point of tin monoxide at 1 atm pressure, 1760°K (Fig. 107). Here it is assumed that the gas and condensed phase have the same empirical formula and that the monoxide is stable and does not undergo decomposition with the formation of higher oxides.

The graphs given make it possible to find the vapor pressure of tin monoxide at other temperatures. For this purpose we find for the required temperature the free energies of the reactions

$$2Sn\,(solid) + O_2\,(gas) = 2SnO\,(gas),$$
$$2Sn\,(solid) + O_2\,(gas) = 2SnO\,(solid),$$

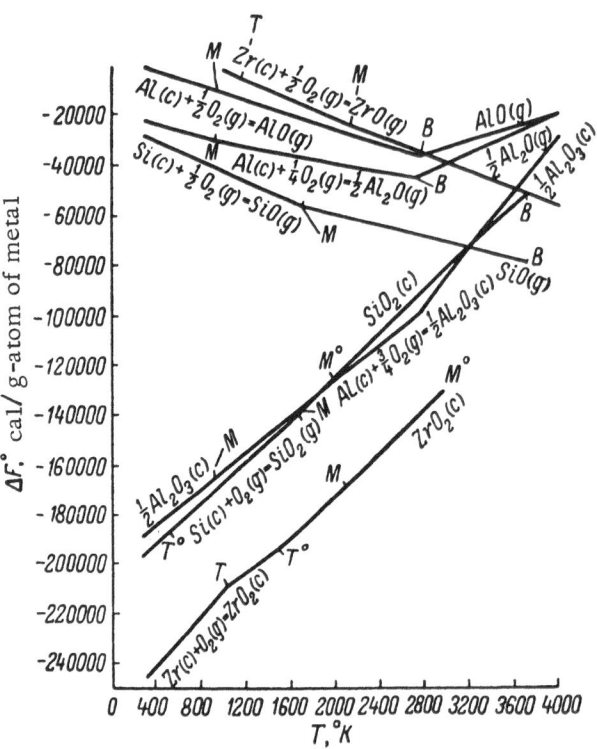

Fig. 108. Free energies of formation of gaseous oxides
per 1 g-atom of metal. The symbols are as for Fig. 107.

which equal ΔF_1^0 and ΔF_2^0, respectively. The difference in these free energies, $\Delta F_1^0 - \Delta F_2^0$ will be the free energy of the reaction

$$2 SnO\,(\text{solid}) = 2 SnO\,(\text{gas}).$$

The free energy $\Delta F_1^0 - \Delta F_2^0$ is related to the vapor pressure of the monoxide P_{SnO} by the equation

$$\Delta F_1^0 - \Delta F_2^0 = -4.575\,T \log P_{SnO}^2.$$

In addition to tin monoxide, oxides of lead, titanium, and some others are assumed to give single forms
of gaseous molecules.

For beryllium oxide, which gives several forms of gaseous molecules, from the graphs (Figs. 107 and 108)
it is possible to obtain the partial pressure of each polymer. Taking for the given temperature the free energies of formation of, for example, the polymer $(BeO)_2$,

$$2 Be\,(\text{solid}) + O_2\,(\text{gas}) = (BeO)_2\,(\text{gas})$$

and solid beryllium oxide

$$2 Be\,(\text{solid}) + O_2\,(\text{gas}) = 2 BeO\,(\text{solid}),$$

equal to ΔF_1^0 and ΔF_2^0, respectively, we may obtain an expression for the difference in free energies, from which
we may determine the partial pressure of the dimer $(BeO)_2$.

$$\Delta F_1^0 - \Delta F_2^0 = -4.575\,T \log P_{(BeO)_2}^2.$$

The total vapor pressure of beryllium oxide will be represented by the sum of the partial pressures calculated for each form of molecule, i.e., BeO, $(BeO)_2$, $(BeO)_4$, $(BeO)_5$, and $(BeO)_6$. The quantitative ratio of these

molecules may be obtained by assuming that the polymer with the lowest free energy of formation at the given temperature will be in the predominant amount, and that with the highest free energy of formation will be in the smallest amount.

Some gaseous oxides are stable only in a definite temperature region and undergo disproportionation, for example, at other temperatures. The graphs given (Figs. 107 and 108) make it possible to find the regions of stability and instability. Thus, for silicon oxides, the solid phase corresponds to the composition SiO_2 and the gaseous phase to SiO. To determine the stability of SiO(gas), i.e., that there is no disproportionation according to the equation

$$2SiO \, (gas) = Si \, (solid) + SiO_2 \, (solid),$$

it is necessary to consider the two reactions

$$2Si \, (solid) + O_2 \, (gas) = 2SiO \, (gas), \tag{1}$$

$$Si \, (solid) + O_2 \, (gas) = SiO_2 \, (solid), \tag{2}$$

whose free energies are given in Fig. 107. For the first equation the free energy will be ΔF_1^0 and for the second, ΔF_2^0. By subtracting (2) from (1), we have

$$Si \, (solid) + SiO_2 \, (solid) = 2SiO \, (gas).$$

The free energy of this reaction will be $\Delta F_T^0 = \Delta F_1^0 - \Delta F_2^0$. Below the point of intersection of the lines for SiO_2 (solid) and SiO(gas), which corresponds to a temperature of 2100°K, the difference $\Delta F_1^0 - \Delta F_2^0 > 0$. Hence, it follows that at a pressure of 1 atm and below 2100°K, SiO(gas) will be unstable and decompose to silicon and silica. Above 2100°K, the difference $\Delta F_1^0 - \Delta F_2^0 < 0$ and silicon monoxide is stable (see also Chapter VII).

Figures 107 and 108 give the temperature dependences of ΔF on condition that the gases are at a pressure equal to 1 atm. In many actual practical systems we are dealing with a low pressure. To find the free energy for a low pressure it is possible to use the following arguments.

We assume that for the reaction

$$y \, Me \, (solid) + O_2 \, (gas, \, 1 \, atm \,) = \frac{y}{x} \, Me_x O_{2\frac{x}{y}} \, (solid)$$

the free energy equals ΔF_1^0. It is assumed that the activities of the gases may be taken as equal to their pressures. The free energy of the process

$$O_2 \, (gas, \, 1 \, atm \,) = O_2 (\, gas \, at \, P \, atm),$$

i.e., the change of oxygen from a state at a pressure of 1 atm to a state at a pressure of P atm, will be

$$\Delta F = 4.575T \log P - 4.575T \, \log 1 = 4.575T \log P.$$

In the equation given above for the formation of the oxide from the metal and oxygen, instead of O_2 (gas, 1 atm), it is possible to take O_2 (gas, P atm) and then the free energy of the reaction

$$y \, Me \, (solid) + O_2 \, (gas, \, P \, atm \,) = \frac{y}{x} \, Me_x O_{2\frac{x}{y}} \, (solid)$$

$\Delta F_T^0 = \Delta F_1^0 - 4.575 \log P$. If $P < 1$, then $RT \ln P < 0$ and the line for ΔF will be higher on the diagram, i.e., $\Delta F_T^0 > \Delta F_1^0$.

These arguments were applied in the case where oxygen is in the gas phase. Similarly, it is possible to calculate the free energy for a low gas pressure when oxide molecules are present in the gas phase. Here, also, a fall in pressure reduces the value of ΔF_T^0. As an example of the magnitude of the correction, Table 58 gives the calculated values of $4.575T \log P$ for values of P of 10^{-3} and 10^{-6} atm.

Table 58. Calculated Values of ΔF_T^0 = 4.575 T log P for Pressures P of 10^{-3} and 10^{-6} atm, in cal

$T°K$	$P = 10^{-3}$ atm	$P = 10^{-6}$ atm
298.15	4093	8186
1000	13728	27456
2000	27456	54912
3000	41184	82368
4000	54912	109826

Figure 107 makes it possible to obtain directly an answer to the questions: What metal will reduce other oxides under given conditions? At what temperatures and pressures will a gaseous oxide dissociate (more correctly, disproportionate) to a condensed oxide and the metal? At what temperature and pressure will a condensed oxide dissociate to the metal and oxygen?

For refractory oxides it is important to know the temperature at which a solid oxide decomposes to form oxygen and a gaseous oxide. This question may be answered by means of Fig. 108, which shows the temperature dependence of the free energy of formation of some oxides per g-atom of metal. As with Fig. 107, the standard free energy of formation of the oxides is given here. The standard state of the gas is the gas at a pressure of 1 atm and the standard states of condensed metals and oxides are pure phases.

The point of intersection of the line for the condensed oxide and the line of the same oxide in the gaseous state gives the temperature above which the condensed oxide evaporates, giving oxygen and the gaseous oxide at a pressure of 1 atm. For example, at an oxygen pressure of 1 atm, SiO_2(solid) is stable up to 3200°K. Above this temperature, SiO_2 decomposes in accordance with the reaction

$$SiO_2 \text{ (solid)} = SiO \text{ (gas)} + \frac{1}{2} O_2 \text{ (gas)}.$$

In this case, the correction for a pressure differing from 1 atm is made as described above (see p. 203).

From Fig. 108 it is readily seen that a fall in pressure leads to a fall in the temperature at which a refractory begins to decompose with the formation of a gaseous oxide.

BIBLIOGRAPHY

Ackermann, R. J., P. W. Gilles, and R. J. Thorn. J. Am. Chem. Soc. 78: 1767 (1956).

Ackermann, R. J., P. W. Gilles, and R. J. Thorn. J. Chem. Phys. 25: 1089 (1956).

Ackermann, R. J., and R. J. Thorn. J. Am. Chem. Soc. 78: 4169 (1956).

Ackermann, R. J., and R. J. Thorn. XVI Congrès Internat. Chim. Pure et Appl., 1957. Mémoires Presentés a la Section Chimie Minéral, Paris (1958), pp. 667-684.

Ackermann, R. J., and R. J. Thorn. "Vaporization of oxides," Progr. Ceram. Sci. 1: 85 (1961).

Ackermann, R. J., R. J. Thorn, M. Tetenbaum, and C. Alexander. J. Phys. Chem. 64: 350 (1960).

Ackermann, R. J., R. J. Thorn, and G. H. Winslow. Abstracts of Scientific Papers Presented at the Eighteenth International Congress on Pure and Applied Chemistry (1961), p. 125.

Allen, N. P., O. Kubaschewskii, and von Goldbeek. J. Electrochem. Soc. 98: 417 (1951).

Ariya, S.M., et al. Zh. Neorgan. Khim. 2: 18 (1957); Vestn. Leningr. Univ., Ser. Fiz. i Khim., No. 22: 96 (1958).

Bandel, G. Arch. Eisenhüttenw. 15: 271 (1941).

Beckett, R.W., et al. Preliminary Report on the Thermodynamic Properties of Lithium, Beryllium, Magnesium, Aluminum, and Their Compounds with Oxygen, Hydrogen, Fluorine, and Chlorine, Natl. Bur. Std. (U. S.) Rept. 6297 (January 1, 1959).

Berkowitz, J., W. Chupka, and M. G. Inghram. J. Chem. Phys. 27: 85 (1957).

Berkowitz, J., W. A. Chupka, and M. G. Inghram. J. Phys. Chem. 61: 1569 (1957).

Berkowitz, J., W. A. Chupka, and M. G. Inghram. J. Chem. Phys. 26: 842 (1957).

Blackburn, P. E. Abstracts of Scientific Papers Presented at the Eighteenth International Congress on Pure and Applied Chemistry (1961), pp. 105-106.

Blackburn, P. E., M. Hoch, and H. L. Johnston. J. Phys. Chem. 62: 769 (1958).

Brewer, L., and J. Margrave. J. Phys. Chem. 59: 421 (1955).

Brewer, L., and G. M. Rosenblatt. Chem. Rev. 61(3): 257-263 (1961).

Büchler, A. American Institute of Chemical Engineers Preprint 10 for Symposium on Thermodynamics of Jet and Rocket Propulsion, Thermodynamic Properties of Some Gaseous Metal Compounds (1959).

Bulewicz, E. M., and T. M. Sugden. Trans. Faraday Soc. 55: 720 (1959).

Burns, R. P., G. De Maria, J. Drowart, and R. T. Grimley. J. Chem. Phys. 32: 1363 (1960).

Caplan, D., and M. Cohen. J. Electrochem. Soc. 108: 438-442 (1961).

Chupka, W. A. Argonne National Laboratory Reports ANL-5753 and ANL-5786 (1957).

Chupka, W. A., J. Berkowitz, and C. F. Giese. J. Chem. Phys. 30: 827 (1959).

Chupka, W. A., J. Berkowitz, and M. G. Inghram. J. Chem. Phys. 26: 1207 (1957).

Coughlin, J. N. Contributions to the Data on Theoretical Metallurgy XII, Heats and Free Energies of Formation of Inorganic Oxides, U. S. Bureau of Mines Bulletin, No. 542 (1954).

Darnell, A. J., and W. A. McCollum. High-Temperature Reactions of Thorium and Thoria and the Vapor Pressure of Thoria, U. S. Atomic Energy Comm. NAA-SR-6498; see Chem. Abstr. 56(1): 32 (1962).

Darnell, A. J., W. A. McCollum, and T. A. Milne. J. Phys. Chem. 64: 341-346 (1960).

De Maria, G., R. P. Burns, J. Drowart, and M. G. Inghram. J. Chem. Phys. 32(5): 1373-1377 (1960).

Devries, R. C., and R. Roy. Bull. Am. Ceram. Soc. 33: 370 (1954).

Domagala, R. F., and D. J. McPherson. J. Metals, AIME Trans. 200: 238 (1954).

Drowart, J., G. De Maria, R. P. Burns, and M. G. Inghram. J. Chem. Phys. 32(5): 1366-1372 (1960).

Edwards, J. W., H. L. Johnston, and P. E. Blackburn. J. Am. Chem. Soc. 73: 4727 (1951).

Edwards, J. W., H. L. Johnston, and W. E. Ditmars. J. Am. Chem. Soc. 75: 2467 (1953).

Evans, W. H., D. D. Wagman, and E. J. Rosen. Thermochemistry and Thermodynamic Functions of Some Boron Compounds, Natl. Bur. Std. (U. S.) Rept. 6252 (1958).

Gaydon, A. G. Dissociation Energy and Spectra of Diatomic Molecules, London (1953).

Gilles, P. W. "High-temperature chemistry," Ann. Rev. Phys. Chem., Vol. 12 (1961).

Gleiser, M. Trans. Met. Soc. AIME 221(2): 300 (1961).

Glemser, O., and R. von Haeseler. Abstracts Scientific Papers Presented at the Eighteenth International Congress on Pure and Applied Chemistry (1961), pp. 113-114.

Glemser, O., and H.-H. Weizenkorn. Naturwissenschaften 48(23): 715-716 (1961).

Golubtsov, I. V., A. V. Lapitskii, and V. K. Shiryaev. Izv. Vysshikh Uchebn. Zavedenii, Khim. i Khim. Tekhnol. 3(4): 571-574 (1960).

Grimley, R. T., R. P. Burns, and M. G. Inghram. J. Chem. Phys. 34: 664 (1961).

Groves, W. A., M. Hoch, and H. L. Johnston. J. Phys. Chem. 59: 127 (1955).

Guthmann, K. Radex Rundschau (1958), pp. 3-30, 253-276, 323-347.

Hägg, G., and A. Magneli. Arkiv Kemi, Mineral., Geol. 19: 1 (1945).

Hampson, G. C., and A. J. Stosick. J. Am. Chem. Soc. 60: 1814 (1938).

Hoch, M., and H. L. Johnston. J. Am. Chem. Soc. 76: 4833 (1954).

Hoch, M., M. Nakata, and H. L. Johnston. J. Am. Chem. Soc. 76: 2651 (1954).

Huber, E. J., E. Holley, and E. H. Meierkord. J. Am. Chem. Soc. 74: 3406 (1952).

Inghram, M. G., W. A. Chupka, and J. Berkowitz. Proceedings of the 1956 International Conference on Astrophysics (1956).

Inghram, M. G., W. A. Chupka, and J. Berkowitz. J. Chem. Phys. 27: 569 (1957).

Inghram, M. G., and J. Drowart. Mass Spectrometry Applied to High Temperature Chemistry, Proceedings of an International Symposium of High-Temperature Technology. [Russian translation]: Investigations at High Temperature, IL (1962).

Katzin, L. I. J. Am. Chem. Soc. 80: 5908 (1958).

Kelley, K. K. Contributions to the Data of Theoretical Metallurgy. XIII, U. S. Bureau of Mines Bulletin, No. 584 (1960).

Kolchin, O. P., V. N. Sumarokova, and N. P. Chuveleva. At. Energ. (USSR) 3: 575 (1957).

Krishnamurty, S. G. Proc. Phys. Soc. (London) 64A: 852 (1951).

Kubaschewskii, O. Trans. Brit. Ceram. Soc. 60(1): 71 (1961).

Kubaschewskii, O., and W. A. Dench. J. Inst. Metals 82: 87 (1953/'54).

Lagerqvist, A., and L. E. Selin. Arkiv Fysik 12: 553 (1957).

Magneli, A., G. Anderson, B. Blomberg, and L. Kihlborg. Anal. Chem. 24: 1998 (1952).

Makarov, E. S. Dokl. Akad. Nauk SSSR 139(3): 612-615 (1961).

Margrave, J. In: Physicochemical Measurements at High Temperatures, J. O'M. Bockris, J. L. White, and J. D. Mackenzie (eds.) (1959).

O'Brien, C. J., J. R. Perrin, and J. Perrine. Kinetics, Equilibria, and Performance of High-Temperature
    Systems, G. S. Bahn and E. E. Zukoski (eds.) (1960), pp. 5-17.

Osborne, D. W., and E. W. Westrum. J. Chem. Phys. 21:1884 (1953).

Phillips, J. G. Astrophys. J. 115:567 (1952).

Polyakov, A. Ya. Zh. Fiz. Khim. 20:1021 (1946).

Rostoker, W. The Metallurgy of Vanadium [Russian translation] (1959).

Schönberg, N. Acta Chem. Scand. 8:240 (1954).

Shapiro, E. J. Am. Chem. Soc. 74:5233 (1952).

Shchukarev, S. A., and G. A. Semenov. Dokl. Akad. Nauk SSSR 120(5):1059-1061 (1958).

Shchukarev, S. A., G. A. Semenov, and K. E. Frantseva. Zh. Neorgan. Khim. 4(11):2638 (1959).

Shchukarev, S. A., G. A. Semenov, and K. E. Frantseva. Dokl. Akad. Nauk SSSR 145(1):119 (1962).

Shchukarev, S. A., G. A. Semenov, and K. E. Frantseva. Izv. Vysshikh Uchebn. Zavedenii, Khim. i Khim.
    Tekhnol. 5:2 (1962).

Skinner, G. B., and H. L. Johnston. J. Am. Chem. Soc. 73:4549 (1951).

Spitsyn, V. I., L. I. Zemlyakova, I. E. Mikhailenko, V. V. Gromov, and I. E. Zimakov. Dokl. Akad. Nauk SSSR
    139(5) (1961).

Spitsyn, V. I., and I. E. Zimakov. Dokl. Akad. Nauk SSSR 139(3):654-657 (1961).

Starodubtsev, S. V., and Yu. I. Timokhina. Zh. Tekhn. Fiz. 19:606 (1949).

U. S. Bureau of Mines Bulletin, No. 542 (1954).

Warshaw, I., and M. L. Keith. J. Am. Chem. Soc. 37:161 (1954).

Wartenberg, H. Z. Anorg. Allgem. Chem. 176:349-362 (1928); 178:183 (1930); 196:375 (1931); 208:375
    (1932); 230:261 (1937); 232:183 (1937).

Wartenberg, H. Arch. Eisenhüttenw. 30(10):585-587 (1959).

Wasilewskii,R. J. J. Am. Chem. Soc. 75:1001 (1953).

Weinreich and Danforth. Phys. Rev. 88:953 (1952).

Wilms, G. R., and T. W. Rea. J. Less-Common Metals 1:411 (1959).

Zimakov, I. E., and V. I. Spitsyn. Dokl. Akad. Nauk SSSR 141(6):1400-1402 (1961).

CHAPTER XI

FIBROUS CRYSTALS OF HIGHLY REFRACTORY OXIDES

Fibrous crystals of highly refractory oxides are attracting increasing attention, mainly in connection with the promising product of specially strong nonbrittle materials with them.

White (1961) reported that a material made from fibrous crystals cemented with an appropriate filler should be similar in properties to glass-reinforced plastics, i.e., be strong and simultaneously highly refractory.

Brenner (1962) reported that it is possible to prepare very strong solids by embedding Al_2O_3 fibers in metallic matrices. The strength of such materials at high temperature will be determined largely by the temperature dependence of the strength of the oxide fibers (Jech, MacDaniels, Weeton, 1959). According to Sutton (1962), after reinforcement with α-Al_2O_3. "whiskers," silver has five times its previous strength.

In Russian technical literature there is no established name for peculiar crystals of fibrous form. In English literature they are called "whiskers." Sometimes the word "whiskers" (USA) is also used in Russian scientific articles to define the crystals examined. We will use the term "whiskers."

Fibrous crystals have now been prepared from very different materials, namely metals and alloys, oxides, halides, etc. Metallic whiskers have been studied in most detail. A review of work on crystallographic and physical properties of fibrous crystals has been given in articles by Nadgornyi et al. (Nadgornyi, Osipyan, Perkass, and Rozenberg, 1959; Nadgornyi, 1962).

Fibrous crystals are obtained by various methods, namely by crystallization from the gas phase, from molds, and from solutions, as a result of the chemical decomposition of some compounds and the oxidation of metals, in electrolysis, and even directly from massive crystals, for example by splitting them along cleavage planes. In an article of Nabarro and Jackson (1958b) there is an exhaustive review (up to 1958) of methods of obtaining crystalline whiskers. Work has been carried out in many countries in recent years to find new methods of producing fibrous crystals.

For the practical application of whiskers it is extremely important to look for methods of producing long fine crystals. As yet, oxide whiskers do not exceed 2.5-3 cm in length. Fibrous crystals of rock salt up to 20 cm long were obtained by Esenski and Khartmann (1962) on porous materials impregnated with saturated NaCl solution. Esenski and Khartmann showed that the elastic torsion of fibrous NaCl crystals is ten times that of normal crystals and the maximum stress reaches 8100 g/mm².

The most important property of fibrous crystals is their high mechanical strength. For some samples of whiskers the experimental strength was found to be close to the theoretical. This high strength is explained by the idealness of the structure and the absence of defects and dislocations. However, it has been established that whiskers may contain dislocations, usually in small numbers. Amelinckx (1961) observed dislocations in transparent fibrous NaCl crystals; Webb (1958) used an x-ray diffraction microscope to establish the existence of screw dislocations in aluminum oxide whiskers.

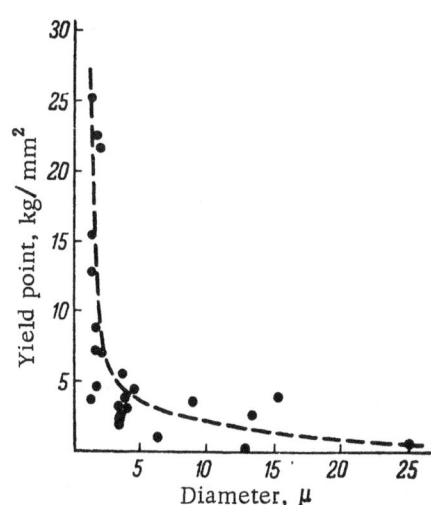

Fig. 109. Yield point of fibrous NaCl crystals grown in the presence of cane sugar.

Table 59. Elastic Properties of Fibrous Crystals (Tensile Tests)

Substance	Maximum elastic deformation, %	Preparation method	Authors
SiO_2	5.2	Condensation of vapor	Brenner, 1958
Al_2O_3	3.0	The same	Brenner, 1958
MoO_3	1.0	The same	Brenner, 1958
ZnO (bending)	1.5	Chemical reaction	Pearson, Read, Feldman, 1957
Graphite	2.0	Condensation of vapor	Bacon, 1958
LiF	0.6	Precipitation	Brenner, 1958
NaCl	2.6	The same	Gyulai, 1954
Si	2.0	Condensation of vapor	Eisner, 1955
Fe	4.9	Reduction of halide	Brenner, 1956

In a report on a conference held in Cambridge in 1958 on the mechanical properties of whiskers and thin films, Gordon and Menter (1958) pointed out that the participants in the discussion came to the conclusion that there is the possibility of the high mechanical strength of whiskers, even when there are dislocations in them. The specific strength of fibrous crystals depends strongly on their thickness. A sharp increase in strength is observed at "diameters" less than $2-3 \mu$.

Great attention is being paid to the elastic-plastic properties of fibrous crystals. The yield point of fibrous crystals is very high. For crystals with one or several screw dislocations lying along the axis of the crystal, immediately after the beginning of plastic deformation the flow stress falls to values characteristic of normal defect crystals because of a sharp rise in the dislocation density (Klassen-Neklyudova and Rozhanskii, 1962).

The magnitude of the yield point of fibrous crystals is usually inversely proportional to their diameter. Figure 109 shows the effect of the size of NaCl crystals on their strength according to the data of Gordon (1958). It may be considered that at fiber diameters of 10^{-4} cm the yield point approaches the theoretical limit.

The maximum elastic deformation of many fibrous crystals has been determined. In a review article, Brenner (1958) gives data for some metals, salts, and oxides (Table 59).

Fibrous oxide crystals have been investigated little as yet, but new papers are appearing in the literature which show that it is possible to obtain whiskers of not only pure oxides, but also complex compounds and, in particular, silicates. We will not consider here artificial fibrous crystals of hydrosilicates such as asbestos, the production of which has already been undertaken on industrial scales.

In a special symposium of the American Ceramic Society (May 1962), problems in the production of highly refractory fibrous crystals were discussed. It was reported here that in addition to aluminum oxide whiskers, which were already known, it is possible to obtain crystalline fibers of titanium dioxide and modified and unmodified zirconium dioxide. According to Scheffler (1962), the last three forms of whisker were obtained from a melt containing B_2O_3. Kirchner and Knoll (1962) described the production of silicon carbide fibers by pyrolysis of methyltrichlorosilane in hydrogen. The strength of these fibers reached 1140 kg/mm^2.

Fibrous Crystals of Aluminum Oxide. The first detailed study of the formation of fibrous crystals of aluminum oxide from the gas phase was made by Webb and Forgeng (1957). These authors heated aluminum or the intermetallic compound $TiAl_3$ to 1300-1450°C and treated them with a stream of wet ($1.2 \cdot 10^{-4}$ atm H_2O) hydrogen. Fluffy crystals were deposited on the walls of the crucible at some distance (about 2 cm) from the charge.

More voluminous deposits of crystals were observed in cases where the compound $TiAl_3$ was used instead of metallic aluminum. The crystals were acicular and lamellar forms and the acicular hexagonal crystals had a length of 1 to 30 mm and were from 3 to 50 μ across. The thickness of the lamellar crystals was 0.5-10 μ with a length up to 10 mm.

The possibility of the chemical action of water vapor at the given pressure on aluminum was demonstrated by Webb and Forgeng by thermodynamic calculations. From the thermodynamic data presented in the review of Coughlin (1954) for the reaction

$$2Al\,(liq) + 3H_2O\,(gas) = Al_2O_3\,(solid) + 3H_2\,(gas),$$

it is possible to obtain the following temperature dependence of the free energy

$$\Delta F^0 = -225\,750 + 38.47T \text{ cal.}$$

This equation makes it possible to find the equilibrium ratio $P_{H_2O} : P_{H_2}$ at various temperatures. Thus, for 1500°K the ratio $P_{H_2O} : P_{H_2}$ is about 10^{-9} and at 1800°K, of the order of 10^{-7}. In the experiments described, the ratio $P_{H_2O} : P_{H_2}$ was estimated at 10^{-3}-10^{-4}, i.e., the oxidation of aluminum by water vapor is thermodynamically possible.

In order to determine which of the compounds of aluminum with oxygen is present in the gas phase under the conditions of the experiments described, it is necessary to examine the free energies of the reactions

$$2Al\,(liq) + H_2O\,(gas) = Al_2O\,(gas) + H_2\,(gas),$$
$$Al\,(liq) + H_2O\,(gas) = AlO\,(gas) + H_2\,(gas).$$

The value given by Webb and Forgeng for the change in free energy for the reaction with the formation of AlO was shown by Hargreaves to be inaccurate and, therefore, the conclusion of these authors that the AlO molecule is responsible for the transport of aluminum oxide through the gas phase is incorrect.

Using the data of Brewer and Searcy (1951), Hargreaves carried out more accurate thermodynamic calculations of the possible reactions. Below we give the reactions for forming volatile oxides and the subsequent reactions for forming aluminum oxide crystals from the gas phase:

$$2Al\,(liq) + 3H_2O\,(gas) = Al_2O_3\,(solid) + 3H_2\,(gas), \tag{1}$$

$$2Al\,(liq) + H_2O\,(gas) = Al_2O\,(gas) + H_2\,(gas), \tag{2}$$

$$Al\,(liq) + H_2O\,(gas) = AlO\,(gas) + H_2\,(gas), \tag{3}$$

$$Al_2O\,(gas) + 2H_2O\,(gas) = Al_2O_3\,(solid) + 2H_2\,(gas), \tag{4}$$

$$2Al_2O\,(gas) + H_2O\,(gas) = Al_2O_3\,(solid) + H_2\,(gas) + 2Al\,(liq), \tag{5}$$

$$3Al_2O\,(gas) = Al_2O_3\,(solid) + 4Al\,(liq), \tag{6}$$

$$2AlO\,(gas) + H_2O\,(gas) = Al_2O_3\,(solid) + H_2\,(gas), \tag{7}$$

$$3AlO\,(gas) = Al_2O_3\,(solid) + Al\,(liq). \tag{8}$$

Figure 110 gives the equilibrium partial pressures of Al_2O and AlO in accordance with reactions (2)-(8), and also the equilibrium vapor pressure of aluminum at 1500 and 1800°K. Figure 110 shows that the vapor of Al_2O molecules has the greater relative pressure and, therefore, it is through this compound that fibrous and lamellar crystals of aluminum oxide (sapphire) will be formed.

Sears and De Vries (1960) also consider that the formation of fibrous crystals of aluminum oxide is based on the primary reaction

$$Al_2O_3\,(solid) + 2H_2\,(gas) = Al_2O\,(gas) + 2H_2O\,(gas)$$

with subsequent decomposition of Al_2O.

Ackermann and Thorn (1961) calculated the partial pressure of Al_2O formed by the reaction

$$Al_2O_3\,(solid) + 2H_2\,(1 \text{ atm}) = Al_2O\,(gas) + 2H_2O\,(gas).$$

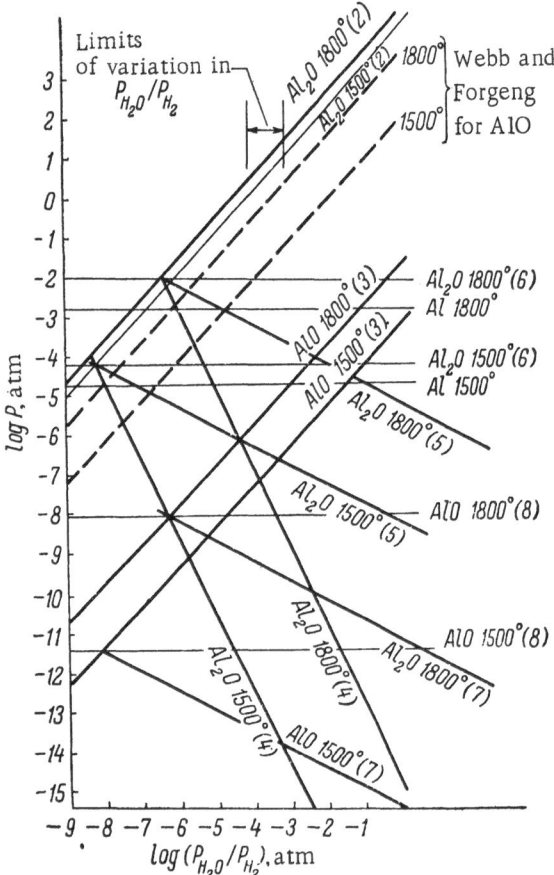

Fig. 110. Equilibrium partial pressures of Al_2O and AlO in accordance with reactions (2)-(8) and equilibrium vapor pressure of aluminum at 1500 and 1800°K.

These calculations were based on the value obtained by these authors for the free energy of formation of gaseous Al_2O

$$\Delta F^0 = -47,200 - 9.27T \quad \text{cal/mole.}$$

It was found that at 2273°K the partial pressure of gaseous Al_2O is 0.3 mm Hg, i.e., a value great enough to guarantee the transfer of a considerable amount of material through the gas phase. The amount of Al_2O transferred may be regulated by the moisture content of the hydrogen fed.

Diamond, Efimenko, Hampson, and Walker (1961) reported that fibrous Al_2O_3 crystals are formed on sintered aluminum oxide if the latter is heated at 1700°C in a furnace with a graphite tube in an argon atmosphere.

May (1959) obtained fine single crystals of α-Al_2O_3 from the gas phase on a molybdenum foil in a tubular furnace with a molybdenum heater. Through a porous tube of aluminum oxide was passed a stream of hydrogen to protect the molybdenum winding against oxidation. Crystals of α-Al_2O_3 grew at a temperature above 1500°C through the gas phase, which could have contained Al_2O vapor, formed as a result of the action of hydrogen on the aluminum oxide tube:

$$Al_2O_3 \text{ (solid)} + 2H_2 \rightleftarrows Al_2O \text{ (gas)} + 2H_2O \text{ (gas).}$$

By special experiments in which a sapphire rod was heated in a stream of dry or wet hydrogen, it was shown that the loss in weight of the rod was greater in dry hydrogen, meaning that it is precisely hydrogen which acts on alumina, for example by the reaction above. Even after an hour, May obtained single crystals of sapphire in the form of plates approximately 100 μ long with a thickness from 0.1 to 10 μ. On the surface of the thicker plates there was spiral growth of crystals, indicating the deposition of material by crystallization in screw dislocations.

Sears, De Vries, and Huffine (1961) proposed an apparatus for producing larger aluminum oxide whiskers. In this apparatus it was possible to obtain whiskers up to 20 mm long. The source of Al_2O_3 was an alumina rod, which was set in a spiral of tungsten wire. The fibrous crystals were deposited on the inner walls of a cylinder, which contained the Al_2O_3 rod. Hydrogen was circulated in the cylinder. Sears et al. obtained aluminum oxide crystals which, as could be seen under a binocular microscope (magnification of 40) consisted of fine ribbons twisted at the top. The authors associated the formation of twisted crystals with the presence of screw dislocations.

Edwards and Happel (1962a) obtained fibrous crystals of aluminum oxide on a single crystal of Al_2O_3 by heating to 1400°C metallic aluminum lying close to the crystal in a stream of moist hydrogen. The crystals of aluminum oxide grew crystallographically coherently with the backing in the direction of screw dislocations.

Brenner (1962) determined the strength of aluminum oxide whiskers over a wide temperature range (25-2030°C). The Al_2O_3 whiskers were grown by the method proposed by Webb and Forgeng (1957). A porcelain boat filled with aluminum was heated at 1250°C in a stream of moist hydrogen (dew point approximately -55°C). Bundles of intertwined crystals, small polycrystalline particles, and well-developed whiskers from a few microns to about 100 μ in diameter were obtained. The whiskers were at some distance from the aluminum charge, i.e., were formed from the gas phase. They had a hexagonal section with a central hole along the crystal.

Brenner tested the alumina whiskers on an instrument he described previously (Brenner, 1957). The mechanical strength was determined under both dynamic and static conditions. At room temperature the tensile strength depended on the size and reached 1000 kg/mm². With a rise in temperature the strength fell and became less dependent on the size. Brenner considered that the breaking mechanism changes at 1000-1600°C. Above 630°C the tensile strength becomes time dependent (slow breaking). This dependence is observed in an atmosphere of hydrogen and oxygen and therefore cannot be explained by corrosion stresses.

Fibrous Crystals of Beryllium Oxide. Fibrous crystals of BeO several microns long were obtained by Scott (1959) by heating a single crystal of beryllium in air for one hour at 800°C. Grossweiner and Seifert (1952) described the preparation of BeO whiskers from the gas phase when beryllium oxide was treated with water vapor.

Interesting results on the crystallization of beryllium oxide from the gas phase were obtained by Budnikov and Shishkov (1961). The growth of fibrous crystals from the gas phase begins at temperatures considerably below the melting point of the oxide and was observed even at 1600°C. In the experiments of Budnikov and Shishkov, polycrystalline beryllium oxide was placed at the bottom of a graphite beaker, while the crystals grew on the graphite lid, the temperature of which was 10-50° lower than the temperature of the sample. Budnikov and Shishkov obtained crystals of very different forms with a predominance of plates and rods. Of particular interest was the formation of dendritic branching and long fibrous or ribbon crystals, which these authors called whiskers.

Budnikov and Shishkov demonstrated that the formation of crystals from the gas phase proceeds by condensation of material at the tip of the growing crystal. Single crystals of beryllium oxide in the form of whiskers are very strong and may undergo large elastic deformations (for example, bending through 180°) without breaking. Budnikov and Shishkov emphasized that the method of growing oxide crystals from the gas phase is promising, but consider that it is still necessary to search for methods of increasing the crystal growth rate and developing methods of directing the condensation process at high temperatures.

According to the observations of Ryschkewitsch (1960), when beryllium oxide is heated in the presence of a directed stream of water vapor, there is volatilization of the product, probably in the form of the hydroxide Be(OH)$_2$, which condenses in the cold parts of the furnace and gives fibrous crystals of beryllium oxide. Finer crystals are obtained with sharp cooling (quenching) of the oxide. Long flat twinned crystals were also obtained. Ryschkewitsch estimated the bending strength of the fibrous crystals at 150,000 kg/cm².

Edwards and Happel (1962b) obtained fibrous and lamellar crystals of beryllium oxide by heating a silica boat with metallic beryllium in a furnace with a silica tube. Whiskers and plates of beryllium oxide were deposited on the outside of the boat and the walls of the tube when beryllium flakes were heated in a hydrogen atmosphere at 1500°C for 16 h. The authors described a new form of whisker, namely a thin fibrous crystal, hollow inside (from a fraction of a micron to 20 μ in diameter, and from several microns to 1 mm long) at the top of which is a ball of metallic luster (diameter from several microns to 100 μ). These whiskers, which the authors called "flagpole whiskers," grow at places where there was previously metallic beryllium.

Normal whiskers often had a zigzag or branched form. They did not have internal pores and reached up to 1 cm in length and from 1 to 10 μ in diameter. The rate of growth of the whiskers was estimated at 0.2 μ per sec. The lamellar crystals had a thickness up to 10 μ and a width up to 200 μ. "Microplates" of beryllium oxide with a thickness of the order of 1000 Å or less were also found.

Edwards and Happel considered that the "flagpole" and normal whiskers were formed in different ways. The former were evidently formed without transfer of beryllium through the gas phase. The beryllium oxide on which the "flagpole whiskers" grew was obtained as a result of the reaction of liquid beryllium and silica (the material of the boat)

$$\text{Be (liq)} + \frac{1}{2}\text{SiO}_2\text{(solid)} = \text{BeO (solid)} + \frac{1}{2}\text{Si (liq)}.$$

The standard free energy of this process of -33 kcal/mole (Coughlin, 1954), makes this reaction possible.

Normal BeO whiskers grow at some distance from the position of the beryllium and their formation must involve the transfer of the material through the gas phase. It is possible to visualize three processes leading to the transfer of beryllium oxide.

1. The evaporation of beryllium oxide (formed in the middle of the boat by the given reaction of beryllium with silica) and its subsequent condensation. This process is improbable, as the vapor pressure of BeO at 1500°C is only $6.6 \cdot 10^{-11}$ atm (Erway, Seifert, 1951).

2. The action of water vapor on beryllium oxide with the formation of a gaseous hydroxide

$$\text{BeO (solid)} + \text{H}_2\text{O (gas)} = \text{Be(OH)}_2\text{(gas)},$$

which on reaching the site of growth of the whiskers gives beryllium oxide by the reverse reaction. However, the authors point out that the equilibrium pressure of gaseous Be(OH)_2 is too great ($1.3 \cdot 10^{-4}$ atm) for the following reaction to occur:

$$\text{Be(OH)}_2\text{(gas)} = \text{BeO (solid)} + \text{H}_2\text{O (gas)}.$$

3. The reaction

$$\text{Be (gas)} + \text{H}_2\text{O (gas)} = \text{BeO (solid)} + \text{H}_2\text{(gas)}.$$

The equilibrium vapor pressures of beryllium and water under the conditions of the experiments examined are 0.6 and 0.1 mm Hg, respectively, and the change in the standard free energy of the reaction given is -65 kcal per mole. All this indicates that this reaction is possible.

The following mechanism is proposed for the formation of "flagpole whiskers": water vapor reacts with a sphere of liquid beryllium and the fibrous crystal formed "lifts up" the sphere of beryllium, which also serves as the material for growth of the whisker by the reaction

$$\text{Be (liq)} + \text{H}_2\text{O (gas)} = \text{BeO (solid)} + \text{H}_2\text{(gas)}.$$

Fibrous Crystals of Magnesium Oxide. Fibrous crystals of magnesium oxide were obtained by Speros and Schupp (1960) by heating a cylindrical single crystal of magnesium oxide to 1400-2000°K in an atmosphere of hydrogen, carbon dioxide, or a mixture of these gases. The surface of the magnesium oxide crystals was gradually covered with a layer of fine whiskers of MgO, which finally enveloped the starting crystal like a "cocoon." The authors considered the following possible reversible reactions, leading to the transfer of magnesium oxide through the gas phase:

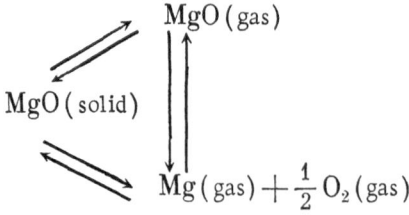

Hulse (1961) describes fibrous crystals of magnesium oxide obtained by deformative compression of MgO single crystals, when there is splitting along the cleavage planes. Such crystals may be called "fracture whiskers" in contrast to "growth whiskers." The first fracture whiskers of LiF were obtained by Gilman and Johnston, whose work is reported in an article by Nabarro and Jackson (1958). Venables (1960) reported that when the compounds InSb, HgSe, Mg$_2$Ge, and MgO are split along the cleavage planes there are formed fibrous fragments adjacent to the cleavage steps.

Dikin and Shpunt (1962) observed analogous phenomena in a series of ionic crystals, NaCl, NaI, KI, LiF, and $NaNO_3$ and two metals, Bi and Sb. These authors determined the strength of fracture whiskers of LiF and came to the conclusion that fibrous crystals obtained directly from large crystals have the same unusual mechanical properties as growth whiskers.

Whiskers obtained by mechanical splitting out from large crystals often have cross-sectional profiles which are irregular and nonuniform down the length. However, by selection of the conditions for fracture of the crystal, it is possible to obtain linear fragments with cross sections which are approximately regular and uniform down the length. Strelkov and Shpunt (1962) obtained whiskers from optical lithium fluoride. Their length varied from 0.1 to 0.7 mm, and the cross section from 0.5 to 20 μ. The elastic deformation limit of the LiF whiskers studied by these authors increased substantially in the range with a diameter less than 2-3 μ.

The results of the investigation of whiskers of lithium fluoride, which is similar in structure and mechanical properties to periclase, is of great interest for understanding the peculiarities of fracture whiskers of magnesium oxide.

Hulse deformed right up to fracture a single crystal of magnesium oxide in the [001] direction. The whiskers were formed parallel to the <100> crystallographic direction. The fibrous crystals had a diameter in the range of 1-3 μ, bent readily, and their maximum observable strength according to Hulse was 245,000 kg per cm^2.

Hulse considered that the magnesium oxide whiskers were formed as a result of plastic deformation of single crystals acted on by compressive forces.

Fibrous Crystals of Silica. Some time ago, Weiss and Weiss (1954) obtained fibrous silica by oxidation of gaseous silicon monoxide at 1200-1400°C.

An interesting method of preparing fibrous silica was described by Haller (1961). A heated mixture of $SiCl_4$ and H_2O vapor, diluted with nitrogen, was directed at a hot (1100°C) platinum surface, where fibrous silica was formed. These fibers had a round section with a diameter from 1 to 50 μ and a length from 10 μ to several millimeters. The refractive index of the fibers was close to 1.46, i.e., it practically coincided with that of vitreous silica. X-ray powder patterns showed bands characteristic of amorphous bodies and weak, but quite clear diffraction lines, indicating a certain degree of crystallinity. Silica fibers obtained by the method described had different forms, namely branched, with bulges, bends, etc.

Jaccodine and Kline (1961) obtained fibrous crystals of silica by heating fused quartz in a tubular furnace through which a stream of dry nitrogen was passed (1 liter/min). Close to the site of formation of the whiskers was placed material containing silicon. At a temperature of about 1425°C, in 24 h there grew thin crystals, from a few microns to a few millimeters in length, and also aggregations of dendrites. The authors consider that the presence of a layer of devitrified quartz is of great importance for the formation of crystal nuclei. This layer is formed as a result of prolonged heat treatment at a high temperature.

Abrahams and Stockbridge (1962) consider that fibrous crystals obtained from fused quartz are single crystals of high-temperature cristobalite, grown along the [111] axis. The authors made a detailed x-ray investigation of crystals 17 and 35 μ in diameter and obtained the constant of a cubic lattice $a = 6.99 \pm 0.02$ A [Lukesh (1952) gives $a = 7.0459$ A].

Fibrous crystals of silica are known to be obtained by hydrothermal processes (Buchler, Walker, 1949, 1950). Abrahams and Stockbridge studied a single-crystal whisker obtained by the hydrothermal method and found that it was α-quartz.

Fibrous Crystals of Niobium Pentoxide were studied by Markali (1961). Whiskers of Nb_2O_5 were prepared by oxidation of niobium in an oxygen atmosphere (pressure 10^{-2} mm Hg) at 800-900°C. The authors consider that the whiskers are formed as a result of the plastic flow of this compound. During the oxidation of metallic niobium, as a result of the large difference in the volumes of the metal and the oxide, in the oxide layer there arise high stresses so that in some sections plastic deformation of the oxide occurs. Plastic flow, and this means growth of the crystal, proceeds at a high rate. Thus, in the case examined, the growth of the crystal occurs at its base and not at its top.

A very high degree of elastic deformation (about 10%) was obtained for Nb_2O_5 whiskers. This deformation cannot be explained by extension and compression of chemical bonds in the crystal.

A detailed investigation of the formation of fibrous crystals of copper oxide by oxidation of copper was carried out by Gulbransen, Copan, and Andrew (1961).

Borchardt, Phillips, and Gambino (1961) stated that the fibrous crystals of molybdenum trioxide they obtained were the most flexible of all the oxide crystals obtained up to that time. MoO_3 whiskers were obtained from vapor of this compound, formed by heating MoO_3 powder in a tubular furnace at 800°C in an air atmosphere. After several hours, lamellar and acicular crystals had been deposited in the cold parts of the tube. The acicular crystals had a length of 5-10 mm and a rectangular cross section of the order of tenths of a millimeter.

The crystals were readily bent at room temperature in air to a spindle-like form. They could be bent only along the a axis. Although the fibrous MoO_3 crystals did not break even at the sharpest part of the bend, they had a tendency to split along their length. This splitting occurred readily with coarser whiskers, but even fine whiskers could also be split after several sharp bends. The authors associate the unusual flexibility of fibrous crystals of molybdenum trioxide with the layer structure of this oxide.

Sears (1962) described artificial fibrous crystals of forsterite Mg_2SiO_4, which were obtained as a tangled ball like cotton. The crystals were very fine (0.2 μ) and in a strongly bent state. The elastic deformation of these crystals was even higher than for niobium pentoxide. Sears reported that this deformation, as was shown by bending experiments, reached at least 11%.

Kubo (1961) obtained fibrous crystals of zinc oxide by heating ZnF_2 powder in a platinum crucible in air. The lower part of the crucible was heated to 1050°C with a Bunsen burner. Clear, yellowish, hexagonal, acicular crystals of zinc oxide were obtained at the top of the crucible, where the temperature was about 950°C.

Timofeeva and Yamgin (1956) prepared crystals of various compounds from the gas phase by using a charge containing fluoride salts. When a charge containing potassium fluoride, magnesium fluoride, aluminum oxide, and silica was heated to 1200-1300°C, on the walls of the crucible above the level of the melt were obtained crystals of lamellar and octahedral forms, which were found to be corundum and magnesium spinel. This method can probably be used to obtain fibrous crystals.

The authors considered that from the charge there evaporated fluorides, which reacted with oxygen to give aluminum oxide or the spinel

$$2AlF_3 + 3O_2 = Al_2O_3 + 3F_2O,$$
$$MgF_2 + 2AlF_3 + 4O_2 = MgO \cdot Al_2O_3 + 4F_2O.$$

The mechanism was probably the same in the experiments of Kubo described above. The action of water vapor is considered improbable, as the charge was first dried carefully and the furnace was heated for a long time at high temperature. This reaction scheme is confirmed by the formation of corundum or spinel crystals on heating of amorphous aluminum fluoride or a melt of magnesium and aluminum fluorides, respectively. Single crystals of zinc oxide and zinc spinel $ZnO \cdot Al_2O_3$ were also obtained by the method proposed.

Timofeeva and Zalesskii (1959) obtained crystals of cobalt and manganese ferrites up to 15-20 mm long from a gas phase by using a solution of these compounds in sodium borate (borax). The ferrites passed into the gas phase together with the solvent and crystals were deposited above the level of the melt along the edges and walls of the crystallizer.

BIBLIOGRAPHY

Abrahams, S. C., and C. D. Stockbridge. Nature 193(4816): 670 (1962).

Ackermann, R. J., and R. J. Thorn. In: Progress in Ceramic Science, J.E. Burke (ed.), Pergamon Press, Inc., New York (1961), p. 56.

Amelinckx, S. Mechanical Properties of Engineering Ceramics, 30th Ed., W. W. Kriegel and H. Palmour (eds.), Interscience Publishers, Inc., New York (1961).

Bacon, R. Growth and Perfection of Crystals, Proceedings of International Conference on Crystal Growth held at Cooperstown, August 27-29, 1958, John Wiley and Sons, Inc., New York (1958), pp. 197-203.

Borchardt, H. J., W. L. Phillips, and J. R. Gambino. J. Am. Ceram. Soc. 44(4): 198 (1961).

Brenner, S. S. J. Appl. Phys. 27: 1484 (1956).

Brenner, S. S. J. Appl. Phys. 28: 1023 (1957).

Brenner, S. S. Growth and Perfection of Crystals, Proceedings of International Conference on Crystal Growth held at Cooperstown, August 27-29, 1958, John Wiley and Sons, Inc., New York (1958), pp. 157-188.

Brenner, S. S. J. Appl. Phys. 33(1): 33 (1962).

Buchler, E., and A. C. Walker. Sci. Monthly 69: 148 (1949).

Buchler, E., and A. C. Walker. Industr. Eng. Chem. 42: 1369 (1950).

Budnikov, P. P., and N. V. Shishkov. Dokl. Akad. Nauk SSSR 138(5): 1093 (1961).

Coughlin, J. P. U. S. Bureau of Mines Bulletin, No. 542 (1954).

Diamond, J. J., J. Efimenko, R. F. Hampson, and R. W. Walker. Reactivity of Solids, Proceedings of the Fourth International Symposium (1961), p. 275.

Dikin, L. S., and A. S. Shpunt. Fiz. Tverd. Tela 4(2): 556 (1962).

Eisner, R. S. Acta Metal. 3: 419 (1955).

Edwards, P. L., and R. J. Happel. J. Appl. Phys. 33(3): 826-827 (1962).

Edwards, P. L., and R. J. Happel. J. Appl. Phys. 33(3): 943-948 (1962).

Erway, N. D., and R. L. Seifert. J. Electrochem. Soc. 98: 83 (1951).

Esenski, B., and É. Khartmann. Kristallografiya 7(3): 433-436 (1962).

Gordon, J. E., and J. W. Menter. Nature 182(4631): 296-299 (1958).

Gordon, J. E. Growth and Perfection of Crystals, Proceedings of International Conference on Crystal Growth held at Cooperstown, August 27-29, 1958, John Wiley and Sons, Inc., New York (1958), p. 219.

Grossweiner, L. T., and R. L. Seifert. J. Am. Chem. Soc. 74: 2701 (1952).

Gulbransen, E. A., T. P. Copan, and K. F. Andrew. J. Electrochem. Soc. 108(2): 119 (1961).

Gyulai, Z. Z. Physik 138: 317 (1954).

Haller, W. Nature 191(4789): 662 (1961).

Hargreaves, C. M. J. Appl. Phys. 32(5): 936 (1961).

Hulse, C. O. J. Am. Ceram. Soc. 44(11): 572 (1961).

Jaccodine, and R. K. Kline. Nature 189(4761): 298 (1961).

Jech, R. W., D. L. MacDaniels, and J. W. Weeton. Composite Materials and Composite Structures, Sagamore Ordnance Materials Research Conference, 1959 (1959), p. 116.

Kirchner, H. P., and P. Knoll. Bull. Am. Ceram. Soc. 4: 292 (1962).

Klassen-Neklyudova, M. V., and V. N. Rozhanskii. Kristallografiya 7(4): 499 (1962).

Kubo, J. J. Phys. Soc. Japan 16: 2358 (1961).

Lukesh, J. Am. Mineralogist 27: 226 (1942).

Markali, J. Mechanical Properties of Engineering Ceramics, 30th Ed., W. W. Kriegel and H. Palmour (eds.), Interscience Publishers, Inc., New York (1961), pp. 93-102.

May, J. E. J. Am. Ceram. Soc. 42(8): 391 (1959).

Nabarro, F. R. N., and P. J. Jackson. Growth and Perfection of Crystals, Proceedings of International Conference on Crystal Growth held at Cooperstown, August 27-29, 1958, John Wiley and Sons, Inc., New York (1958), p. 84.

Nabarro, F. R. N., and P. J. Jackson. Growth and Perfection of Crystals, Proceedings of International Conference on Crystal Growth held at Cooperstown, August 27-29, 1958, John Wiley and Sons, Inc., New York (1958), pp. 13-101.

Nadgornyi, É. M., Yu. A. Osipyan, M. D. Perkass, and V. M. Rozenberg, Usp. Fiz. Nauk 67: 625 (1959).

Nadgornyi, É. M. Usp. Fiz. Nauk 77: 2 (1962).

Pearson, G. L., N. T. Read, and W. L. Feldman. Acta Met. 5: 181 (1957).

Ryschkewitsch, E. Trans. Brit. Ceram. Soc. 59: 303 (1960).

Scheffler, L. F. Bull. Am. Ceram. Soc. 4: 291 (1962).

Scott, V. D. Acta Cryst. 12 : 136 (1959).

Sears, G. W. J. Chem. Phys. 36(3) : 862 (1962).

Sears, G. W., and R. C. De Vries. J. Chem. Phys. 32 : 93 (1960).

Sears, G. W., R. C. De Vries, and C. Huffine. J. Chem. Phys. 34(6) : 2142 (1961).

Speros, D. M., and L. J. Schupp. J. Phys. Chem. Solids 15(1-2) : 157-166 (1960).

Strelkov, P. G., and A. A. Shpunt. Fiz. Tverd. Tela 4(8) : 2260 (1962).

Sutton, W. H. Bull. Am. Ceram. Soc. 4 : 292 (1962).

Timofeeva, V. A., and I. I. Yamzin. Tr. Inst. Kristallogr. Akad. Nauk SSSR 12 : 67-72 (1956).

Timofeeva, V. A., and A.V. Zalesskii. In collection: Growth of Crystals, Vol. 2 (1959), p. 88.

Venables, J. D. J. Appl. Phys. 31(8) : 1503 (1960).

Webb, W. W., and W. D. Forgeng. J. Appl. Phys. 28(12) : 1449 (1957).

Webb, W. W. Growth and Perfection of Crystals, Proceedings of International Conference on Crystal Growth held at Cooperstown, August 27-29, 1958, John Wiley and Sons, Inc., New York (1958), pp. 230-238.

Weiss, A., and A. Weiss. Naturwissenschaften 41 : 12 (1954); Z. Anorg. Allgem. Chem. 276 : 95 (1954).

White, J. Trans. Brit. Ceram. Soc. 60 : 11 (1961).